To See the Unseen

A History of Planetary Radar Astronomy

The past 50 years have brought forward a unique capability to research and expand scientific knowledge of the Solar System through the use of radar to conduct planetary astronomy. This technology involves the aiming of a carefully controlled radio signal at a planet (or some other Solar System target, such as a planetary satellite, an asteroid, or a ring system), detecting its echo, and analyzing the information that the echo carries.

This capability has contributed to the scientific knowledge of the Solar System in two fundamental ways. Most directly, planetary radars can produce images of target surfaces otherwise hidden from sight and can furnish other kinds of information about target surface features. Radar also can provide highly accurate measurements of a target's rotational and orbital motions. Such measurements are obviously invaluable for the navigation of Solar System exploratory spacecraft, a principal activity of NASA since its inception in 1958.

Andrew J. Butrica has written a comprehensive and illuminating history of this little-understood but surprisingly significant scientific activity. Quite rigorous and systematic in its methodology, *To See the Unseen* explores the development of the radar astronomy specialty in the larger community of scientists.

More than just discussing the development of this field, however, Butrica uses planetary radar astronomy as a vehicle for understanding larger issues relative to the planning and execution of "big science" by the Federal government. His application of the "social construction of science" and Kuhnian paradigms to planetary radar astronomy is a most welcome and sophisticated means of making sense of the field's historical development.

Andrew J. Butrica received his Ph.D. in the history of science and technology at Iowa State University. He is a research historian in Franklin Park, New Jersey, specializing in the history of science. In 1990, Praeger Publishers issued his *Out of Thin Air: A History of Air Products and Chemicals, Inc., 1940–1990*.

About the cover: "Big Dish Antenna" painting by Paul Arlt. Courtesy of the NASA Art Program, no. 74-HC-467.

TO SEE THE UNSEEN

NASA SP-4218

TO SEE THE UNSEEN

A History of Planetary Radar Astronomy

by Andrew J. Butrica

The NASA History Series

National Aeronautics and Space Administration
NASA History Office
Washington, D.C.   1996

Library of Congress Cataloguing-in-Publication Data

To See the Unseen: A History of Planetary Radar Astronomy / Andrew J. Butrica
p. cm.—(The NASA history series) (NASA SP: 4218)

  Includes bibliographical references and indexes.
  1. Planetology—United States. 2. Planets—Exploration. 3. Radar in
Astronomy.   I. Title. II. Series. III. Series: NASA SP: 4218.
QB602.9.B87   1996   95-35890
523.2'028-dc20   CIP

To my dear friends and former colleagues at the Center for Research in History of Science and Technology: Bernadette Bensaude-Vincent, Christine Blondel, Paulo Brenni, Yves Cohen, Jean-Marc Drouin, Irina and Dmitry Gouzevitch, Anna Guagnini, Andreas Kahlow, Stephan Lindner, Michael Osborne, Anne Rasmussen, Mari Williams, Anna Pusztai, and above all Robert Fox.

Contents

From Locksley Hall

For I dipt into the future,
        far as human eye could see,
Saw the Vision of the world,
        and all the wonder that would be;

Saw the heavens fill with commerce,
        argosies of magic sails,
Pilots of the purple twilight,
        dropping down with costly bales;

Heard the heavens fill with shouting,
        and there rained a ghastly dew
From the nations' airy navies
        grappling in the central blue;

Far along the world-wide whisper
        of the south-wind rushing warm,
With the standards of the peoples
        plunging through the thunder-storm;

Till the war-drum throbbed no longer,
        and the battle-flags were furled
In the Parliament of man,
        the Federation of the world.

There the common sense of most
        shall hold a fretful realm in awe,
And the kindly earth shall slumber,
        lapt in universal law.

So I triumphed ere my passion
        sweeping through me left me dry,
Left me with the palsied heart,
        and left me with the jaundiced eye;

Eye, to which all order festers,
        all things here are out of joint:
Science moves, but slowly slowly,
        creeping on from point to point:

Alfred Baron Tennyson
(1842)

Acknowledgments

Let me begin with a confession and some explanations. Before beginning this project, I knew nothing about planetary radar astronomy. I quickly realized that I was not alone. I discovered, too, that most people confuse radar astronomy and radio astronomy. The usual distinction made between the two is that radar astronomy is an "active" and radio astronomy a "passive" form of investigation. The differentiation goes much deeper, however; they represent two disparate forms of scientific research.

Radio astronomy is more akin to the methods of natural history, in which observation and classification constitute the principal methods of acquiring knowledge. Radio astronomers search the cosmos for signals that they then examine, analyze, and classify. Radar astronomy, on the other hand, is more like a laboratory science. Experimental conditions are controlled; the radar astronomer determines the parameters (such as frequency, time, amplitude, phase, and polarization) of the transmitted signals.

The control of experimental parameters was only one of many aspects of planetary radar astronomy that captivated my interest, and I gradually came to find the subject and its practitioners irresistibly fascinating. I hope I have imparted at least a fraction of that fascination. Without the planetary radar astronomers, writing this book would have been a far less enjoyable task. They were affable, stimulating, cooperative, knowledgeable, and insightful.

The traditional planetary radar chronology begins with the earliest successful attempts to bounce radar signals off the Moon, then proceeds to the detection of Venus. I have deviated from tradition by insisting that the field started in the 1940s and 1950s with the determination by radar that meteors are part of the solar system. Meteor, auroral, solar, lunar, and Earth radar research, as well as radar studies of planetary ionospheres and atmospheres and the cislunar and interplanetary media are specializations in themselves, so were not included in this history of planetary radar astronomy in any comprehensive fashion. What has defined radar astronomy as a scientific activity has changed over time, and the nature of that change is part of the story told here.

This history was researched and written entirely under a contract with the California Institute of Technology (Caltech) and the Jet Propulsion Laboratory (JPL), as a subcontract with the National Aeronautics and Space Administration (NASA). This history would not have come into existence without the entrepreneurial energies of JPL's Nicholas A. Renzetti, who promoted the project and found the money to make it happen. It is also to his credit that he found additional support for a research trip to England and for attendance at a conference in Flagstaff, as well as for the transcription of additional interviews. As JPL technical manager, he administered all technical aspects of the contract. I hope this work meets and exceeds his expectations. During my frequent and sometimes extended visits to JPL, Nick provided secretarial, telephone, photocopying and other supplies and services, as well as a professional environment in which to work. I also want to thank the JPL secretarial personnel, especially Dee Worthington, Letty Rivas, and Judy Hoeptner, as well as Penny McDaniel of the JPL Photo Lab, who was so resourceful in finding pictures.

Teresa L. Alfery, JPL contract negotiator, deserves more than a few words of thanks. Working out the contract details could have been an insufferable experience, were it not for her. Moreover, she continued her cordial and capable performance through several contract modifications.

The contract also came under the purview of the NASA History Office, which provided the author office supplies and services during visits there. More importantly, Chief Historian Roger D. Launius offered encouragement and support in a manner that was both professional and congenial. It was a pleasure to work with Roger. This history owes not inconsequential debt to him and the staff of the History Office, especially Lee Saegesser, archivist, who lent his extensive and unique knowledge of the NASA History Office holdings.

I also want to acknowledge certain individuals who helped along the way. Before this project even began, Joseph N. Tatarewicz afforded it a rich documentary source at the NASA History Office by rescuing the papers of William Brunk, which hold a wealth of information on the Arecibo Observatory and other areas relevant to planetary astronomy at NASA. Joe also was a valuable source of facts and wisdom on the history of the space program and an invaluable guide to the planetary geological community.

This history also owes a debt to Craig B. Waff. His extensive collection of photocopied materials greatly facilitated my research, as did his manuscript histories of the Deep Space Network and Project Galileo. Craig generously offered a place to stay during my first visits to California and was my JPL tour guide.

The staff of the JPL Archives deserves an exceptional word of appreciation. They do not know the word "impossible" and helped facilitate my research in a manner that was always affable and competent. In particular, I want to acknowledge the director, Michael Q. Hooks, for assembling a superb team, John F. Bluth, for his command of the JPL oral history collection and our informative talks about JPL history, and Julie M. Reiz, for her help in expediting access to certain collections.

I also wish to thank those librarians, archivists, historians, and others who expedited my research in, or who provided access to, special documentary collections: Helen Samuels and Elizabeth Andrews, MIT Institute Archives and Special Collections; Mary Murphy, Lincoln Laboratory Library Archives; Ruth Liebowitz, Phillips Laboratory; Richard Bingham, Historical Archives, U.S. Army Communications-Electronics Command, Ft. Monmouth, NJ; Richard P. Ingalls and Alan E. E. Rogers, NEROC, Haystack Observatory; George Mazuzan, NSF Historian's File, Office of Legislation and Public Affairs, National Science Foundation; Eugene Bartell, administrative director, National Astronomy and Ionosphere Center, Cornell University; Jane Holmquist, Astrophysics and Astronomy Library, Princeton University; and August Molnar, president of the American Hungarian Foundation.

In addition, I want to acknowledge those individuals who made available materials in their possession: Julia Bay, Bryan J. Butler, Donald B. Campbell, Von R. Eshleman, Thomas Gold, Paul E. Green, Jr., Raymond F. Jurgens, Sir Bernard Lovell, Steven J. Ostro, Gordon H. Pettengill, Nicholas A. Renzetti, Martin A. Slade, and William B. Smith. Credit also goes to those individuals who reviewed part or all of this manuscript: Louis Brown, Ronald E. Doel, George S. Downs, John V. Evans, Robert Ferris, Richard M. Goldstein, Paul E. Green, Jr., Roger D. Launius, Sir Bernard Lovell, Steven J. Ostro, Gordon H. Pettengill,

Robert Price, Alan E. E. Rogers, Irwin I. Shapiro, Richard A. Simpson, Martin A. Slade, and Joseph N. Tatarewicz.

There are numerous people at NASA involved in the mechanics of publishing who helped in myriad ways in the preparation of this history. J.D. Hunley, of the NASA History Office, edited and critiqued the text before he departed to take over the History Program at the Dryden Flight Research Center; and his replacement, Stephen J. Garber, helped in the final proofing of the work. Nadine Andreassen of the NASA History Office performed editorial and proofreading work on the project; and the staffs of the NASA Headquarters Library, the Scientific and Technical Information Program, and the NASA Document Services Center provided assistance in locating and preparing for publication the documentary materials in this work. The NASA Headquarters Printing and Design Office developed the layout and handled printing. Specifically, we wish to acknowledge the work of Jane E. Penn, Patricia Lutkenhouse Talbert, Kimberly Jenkins, Lillian Gipson and James Chi for their design and editorial work. In addition, Michael Crnkovic, Craig A. Larsen, and Larry J. Washington saw the book through the publication process.

Finally, I want to recognize the friendship of fellow cat lover Joel Harris, the cordial and entertaining SETI evening spent at the Griffith Observatory with Mike Klein, Judy Hoeptner, and company (without forgetting the Renaissance Festival!), the stimulating conversations with Adrienne Harris, and the friendly folk dancers of Pasadena, as well as the contra dancers of Highland Park and Franklin Park, and Ghislaine, the most important one of all in many ways.

Introduction

Planetary radar astronomy has not attracted the same level of public attention as, say, the Apollo or shuttle programs. In fact, few individuals outside those scientific communities concerned with planetary studies are aware of its existence as an ongoing scientific endeavor. Yet, planetary radar has contributed fundamentally and significantly to our knowledge of the solar system.

As early as the 1940s, radar revealed that meteors are part of the solar system. After the first detections of Venus in 1961, radar astronomers refined the value of the astronomical unit, the basic yardstick for measuring the solar system, which the International Astronomical Union adopted in 1964, and they discovered the rotational rate and direction of Venus for the first time. Next, radar astronomers determined the correct orbital period of Mercury and calculated an accurate value for the radius of Venus, a measurement that Soviet and American spacecraft had failed to make reliably. Surprisingly, radar studies of Saturn revealed that its rings were not swarms of minute particles, but rather consisted of icy chunks several centimeters or more in diameter. Planetary radar also provided further proof of Albert Einstein's theory of General Relativity, as well as the "dirty snowball" theory of comets. The only images of Venus' surface available to researchers are those made from radar observations. The ability of planetary radar astronomy to characterize the surfaces of distant bodies has advanced our general knowledge of the topography and geology of the terrestrial planets, the Galilean moons of Jupiter, and the asteroids. The Viking project staff utilized radar data to select potential landing sites on Mars. More recently, radar revealed the surprising presence of ice on Mercury and furnished the first three-dimensional images of an asteroid.

Again, these achievements seldom have attracted the attention of the media. The initial American radar detections of the Moon in 1946 and of Venus in 1961 attracted notice in daily newspapers, weekly news magazines, news reels, and cartoons. Only in recent years have the accomplishments of radar astronomy returned to the front-page of the news. The images of Venus sent back by Magellan received full media coverage, and images of the asteroid Toutatis appeared on the front-page of the *New York Times*.

Planetary radar astronomy has shared its anonymity with other applications of radar to space research. The NASA radar-equipped SEASAT satellite provided unprecedented images of Earth's oceans; European, Canadian, and Japanese satellites, as well as a number of space shuttles, have imaged Earth with radar. The radars of NASA's Deep Space Network also have played a major role in tracking space launches and spacecraft on route to planets as distant as Saturn and Neptune. Among the more down-to-Earth, visible and even pervasive applications of radar are those for air traffic control and navigation, the surveillance of automobile traffic speeds, and the imaging of weather patterns reported daily on television and radio.

Planetary radar astronomy is part of the great wave of progress in solid-state and digital electronics that has marked the second half of the twentieth century. For instance, the earliest planetary radar experiment marked the first use of a maser (a solid-state microwave amplifying device) outside the laboratory. Although radio astronomy has long claimed the first maser application for itself, namely in April 1958 by Columbia University and the Naval Research Laboratory, two months earlier, MIT's Lincoln Laboratory used a maser in its first attempt to bounce radar waves off Venus. The same radar experiment also saw one of the first uses of a digital tape recorder, as well as the incorporation of a digital computer and other digital data processing equipment into a civilian radar system.

The origins of this solid-state and digital electronics progress, as well as of planetary radar astronomy, are rooted in electronic research and development that started as early as the 1930s. The first radar astronomy experiments, which were carried out on meteors and the Moon in the 1940s, relied on equipment designed and built for military defense during World War II and were based on research conducted during the 1930s.

Planetary radar astronomy, and so too radar itself, had its origins in Big Science. British war preparations during the 1930s concentrated large amounts of scientific, technological, financial, and human resources into a single effort. Part of that effort was a massive radar research and development program that produced an impressive range of defensive and offensive radars. In a secret mission known only at the highest levels of government, Britain gave the United States one of the key devices born of that large-scale radar effort, the magnetron. In turn, the magnetron formed the technological base for an American radar research and development effort on a scale equal to that of the Manhattan Project, which historians traditionally have recognized as the beginning of Big Science.

The history of planetary radar astronomy in the United States *is* the history of Big Science. Without Big Science, planetary radar astronomy would be impossible and unthinkable. That is one of the main contentions of this book. The radar astronomy experiments of the 1940s and 1950s, as well as much of pre-war radar development, were intimately linked to ionospheric research, which was then undergoing a rapid publication rate typical of Big Science.

Also, the evolutions of planetary radar and radio astronomy converged. The search for research instruments free of military constraints brought planetary radar astronomers closer to radio astronomy during the 1960s, a time when radio astronomy was undergoing a rapid growth that transformed it into Big Science. Planetary radar and radio astronomy shared instruments and a common interest in electronic hardware and techniques, though ironically the instrumentation needs of the two communities ultimately provided little basis for cohabitation.

In the end, military Big Science was far more important than either radio astronomy or ionospheric science. Planetary radar astronomy emerged in the late 1950s thanks to Cold War defense research that furnished the essential instruments of planetary radar experimentation. The vulnerability of the United States to aircraft and ICBM attacks with nuclear explosives necessitated the creation of a network of ever more powerful and sensitive defensive radars. What President Dwight D. Eisenhower called the military-industrial complex, and what historian Stuart Leslie calls the military-industrial-academic complex,[1] provided the radar instrument for the first attempts at Venus. The military-industrial or military-industrial-academic complex served as the social matrix which nurtured military and other Big Science research. Planetary radar astronomy eventually found itself part of a different, though at times interlocking, complex centered on the civilian enterprise to explore space, that is, what one might call the NASA-industrial-academic complex.

---

1. Stuart W. Leslie, *The Cold War and American Science: The Military-Industrial-Academic Complex at MIT and Stanford* (New York: Columbia University Press, 1993).

The emergence of space as Big Science under the financial and institutional aegis of NASA, and the design and construction of a worldwide network of antennas to track launches and communicate with spacecraft, furnished instruments for planetary radar research as early as 1961. Within a decade, NASA became the *de facto* underwriter of all planetary radar astronomy. Data on the nature of planetary surface features and precise reckoning of both the astronomical unit and planetary orbits were highly valuable to an institution whose primary goal was (and whose budgetary bulk paid for) the designing, building, and launching of vessels for the exploration of the solar system. Association with NASA Big Science enhanced the tendency of radar astronomers to emphasize the utility of their research and promoted mission-oriented, as opposed to basic, research.

The history of planetary radar astronomy is intrinsically interesting and forms the frame-work of this book. It also says something about Big Science. Defining Big Science, or even Little Science, is not easy though. After all, how true are the images of the Little Scientist as "the lone, long-haired genius, moldering in an attic or basement workshop, despised by society as a nonconformist, existing in a state of near poverty, motivated by the flame burning within him," and the Big Scientist as "honored in Washington, sought after by all the research corporations of the 'Boston ring road,' part of an elite intellectual brother-hood of co-workers, arbiters of political as well as technological destiny"?[2]

Since the publication in 1963 of Derek J. De Solla Price's ground-breaking *Little Science, Big Science*, historians have attempted to define Big Science.[3] Their considerable efforts have clarified the meaning of the term, though without producing a universally authori-tative definition. If large-scale expensive research instruments are the measure, then one might count the island observatory of Tycho Brahe in the sixteenth century, or the giant electrical machines built in eighteenth-century Holland. If Big Science is a large grouping of investigators from several disciplines working together on a common project, then the gathering of mathematicians, chemists, and physicists at Thomas Edison's West Orange laboratory was Big Science. A long-term research project, such as the quest for an AIDS cure, or one that entails elaborate organization, such as the Manhattan Project, might be termed Big Science too.

Defining Big Science is the intellectual equivalent of trying to nail Jell-O to the wall. For the purposes of this book, we shall call Big Science the large-scale organization of science and scientists, underwritten by an imposing pledge of (usually) public funds and centered around a complex scientific instrument. In his search to understand Big Science, Derek Price decided to "turn the tools of science on itself," charting the historical growth of sci-ence by means of a variety of statistical indicators obtained from the Institute for Scientific Information in Philadelphia. Price concluded that scientific activity (as measured by the amount of literature published) has grown exponentially over the last three hundred years, doubling in size about every fifteen years.[4] We also shall define a rapid growth in scientific literature greater than the Price rate (doubling every fifteen years) as indicating

2. Derek J. DeSolla Price, *Little Science, Big Science... and Beyond* (New York: Columbia University Press, 1986), p. 2.
3. Price, *Little Science, Big Science... and Beyond*, p. 15.
4. Price, *Little Science, Big Science* (New York: Columbia University Press, 1963). This discussion of Big Science draws on Peter Galison and Bruce Hevly, eds., *Big Science: The Growth of Large-Scale Research* (Stanford: Stanford University Press, 1992); James H. Capshaw and Karen A. Rader, "Big Science: Price to the Present," *Osiris*, ser. 2, vol. 7 (1992): 3–25; and Joel Genuth, "Microwave Radar, the Atomic Bomb, and the Background to U.S. Research Priorities in World War II," *Science, Technology, and Human Values* 13 (1988): 276–289.

an emerging Big Science field. Whatever it is, Big Science has become the dominant form of contemporary American science. Moreover, because of its scale and scope, the conduct of Big Science necessarily intrudes into many areas of society, and in turn, society, through political, economic, and other activity, shapes the conduct of Big Science.

The interdependency of institutional factors, funding patterns, science, technology, and techniques found in Big Science has been the subject of extensive study by historians and sociologists of science and technology. Scholars traditionally have concerned themselves with both science and technology and their interactions. Such studies came to be termed "internalist," meaning that they dealt solely with the inner workings of science and technology. In contrast stood the so-called "externalist" approaches, which emphasized the social, economic, political, and other factors neglected by the "internalists."

Starting around 1980, sociologists of science, such as Michel Callon, developed new approaches, which were introduced into the history of technology by Thomas P. Hughes. These new approaches came to be called generically the "social construction of technology." The "technosocial networks" of Callon and the "systems" of Hughes consider the "internalist" and "externalist" aspects of technology as constituting a single continuum or "seamless web". Inventors, scientists, instruments, financing, institutions, politics, laws, and so forth are all equally part of the "technosocial network" or "system".[5]

The chief advantage of replacing the "internalist" and "externalist" dualism with the unitarian approach of the social construction school is the more sophisticated and certainly more complex view of the scientific, technical, economic, political, institutional, legal, and other aspects of Big Science that it offers. Moreover, by stressing that all components of a technosocial network are equal and necessary, the social construction approach dissuades us from emphasizing any one factor, "internal" or "external", over all others.

The social construction approach is useful for creating a taxonomy of the factors that shape Big Science. Nonetheless, although they served as a guiding principle in the writing of this book, social construction case studies do not go far enough; they fail to address the question that is, after chronicling the achievements of radar astronomy, central to this book—namely the conduct of Little Science in the context of Big Science. Furthermore, in all the discussions of Big Science, with few exceptions, the symbiotic relationship between Big Science and Little Science has been overlooked. This relationship is especially relevant to the organization of science within NASA space missions. The scientists who conduct experiments from those spacecraft typify Little Science: they work individually or in small collaborative groups, often with graduate assistants, and have relatively small budgets and limited laboratory equipment. Participation in NASA spacecraft missions induces these Little Scientists to function as part of a Big Science endeavor. The scientists are organized into both working groups around a single scientific instrument and disciplinary groups. They participate in the design of experiments and in

5.   For a discussion of this evolution, see John M. Staudenmaier, "Recent Trends in the History of Technology," *The American Historical Review* 95 (1990): 715–725, as well as Hughes, "The Seamless Web: Technology, Science, Etcetera, Etcetera," *Social Studies of Science* 16 (1986): 281–292. The primary social construction works are Wiebe E. Bijker, Hughes, and Trevor Pinch, eds., *The Social Construction of Technological Systems: New Directions in the Sociology and History of Technology* (Cambridge: MIT Press, 1987), and Bijker and John Law, eds., *Shaping Technology/Building Society: Studies in Sociotechnical Change* (Cambridge: MIT Press, 1992).

the decisions to drop or modify certain experiments, as well as in the design of the instruments themselves. The overall scale of operation and budget is beyond that normally encountered by Little Scientists.

One noteworthy exception to the lack of literature dealing with the relationship between Big Science and Little Science is historian John Krige's study of British nuclear physics research in the period immediately following World War II. The Labor Government of Clement Attlee set out to equip the universities of Birmingham, Glasgow, Liverpool, Cambridge, and Oxford with particle accelerators for conducting high-energy nuclear physics research. The accelerator program involved the kinds of large-scale budgets and instruments that typify Big Science; however, research was conducted in a manner more typical of Little Science. Large multidisciplinary teams, in which physicists and engineers rubbed shoulders, did not form; rather the physicists remained individual academic researchers.[6]

Krige's case of "Big Equipment but not Big Science" finds its parallel in planetary radar astronomy. Big Science was the sine qua non of planetary radar astronomy, but planetary radar astronomy was not Big Science. It was, and remains, Little Science in terms of manpower, instruments, budget, and publications. Planetary radar astronomy took root within the interstices of Big Science, but rather than expand over time, it actually shrank.

The field attained its largest size, in terms of personnel, instruments, and publications, during the 1960s. Although one can count five active instruments between 1961 and 1964, the greatest number to ever carry out planetary radar experiments, only three subsequently sustained active research programs. That number fell to two instruments after 1975. For much of the period between 1978 and 1986, only one instrument, indeed the only instrument to have an established and secure planetary radar astronomy research program, the Arecibo Observatory, was steadily active.

The number of active planetary radar astronomers has declined since the 1960s too. As a group, they tend not to reproduce as easily or as abundantly as other scientists, and many practitioners in the long run find something else to do. Two paths—artifacts of the field's evolution—lead to a career in planetary radar astronomy. Many follow the traditional university path—doctoral research on a planetary radar topic, followed by a research position that permits them to perform planetary radar experiments. Of the current practitioners, the most recent Ph.D. was granted in 1994, the second most recent in 1978. The path more followed: practitioners were hired to conduct planetary radar experiments.

The declining instrument and manpower numbers are reflected in the planetary radar astronomy publication record (see Appendix: Planetary Radar Astronomy Publications). Price has shown that science publications have doubled about every fifteen years over the last three centuries. The planetary radar publication curve differs markedly from that normal growth pattern, suggesting a ceiling condition that has limited growth. The nature of that ceiling condition, as well as the causal factors for the declining size of the planetary radar enterprise, are part of the story of how planetary radar Little Science has been conducted within the framework of American Big Science. The association of planetary radar

---

6.   John Krige, "The Installation of High-Energy Accelerators in Britain after the War: Big Equipment but not 'Big Science,'" in Michelangelo De Maria, Mario Grilli, and Fabio Sebastiani, eds., *The Restructuring of Physical Sciences in Europe and the United States, 1945–1960* (Teaneck, NJ: World Scientific, 1989), pp. 488–501.

Little Science with NASA Big Science ultimately affected the conduct of planetary radar astronomy. Radar astronomers always had argued the utility of their efforts for space research; NASA mission-oriented support of planetary radar astronomy only reinforced that utilitarian inclination. As the story unfolds, other factors that shaped and amplified the utilitarian tendency of radar astronomers will rise to the surface.

Its relationship with NASA Big Science also transformed planetary radar astronomy from an exclusively ground-based scientific activity to one that was conducted in space as well. During the 1960s, planetary radar astronomers distinguished their ground-based research from that conducted from spacecraft, which they characterized as space exploration as opposed to astronomy. Starting in the following decade, when NASA became its sole underwriter, planetary radar astronomy began to engage the planetary geology community largely through its ability to image and otherwise characterize planetary surfaces. NASA funded specific radar imaging projects. At the same time, NASA began planning two missions to Venus, Pioneer Venus and Magellan, in order to capture in radar images the features of that planet's surface. Its opaque atmosphere keeps Venus's surface hidden from sight and bars exploration with optical methods.

Pioneer Venus and Magellan ultimately had a profound impact on the practice of planetary radar astronomy. In addition to enlarging the community of scientists using radar imagery and other data to encompass both geologists and astronomers, those two NASA missions erased the turf boundary between space exploration and ground-based planetary radar astronomy. Although Magellan in particular also gave radar astronomers a taste of Big Science, planetary radar astronomy did not permanently shift from Little to Big Science. Radar imaging from a spacecraft had limited prospects. Ultimately, the greatest consequence of Magellan for planetary radar astronomy was that it effectively ended ground-based radar observations of Venus, the chief object of radar research.

The plan of this book is to relate the history of planetary radar astronomy from its origins in radar to the present day and secondarily to bring to light that history as a case of "Big Equipment but not Big Science". Chapter One sketches the emergence of radar astronomy as an ongoing scientific activity at Jodrell Bank, where radar research revealed that meteors were part of the solar system. The chief Big Science driving early radar astronomy experiments was ionospheric research. Chapter Two links the Cold War and the Space Race to the first radar experiments attempted on planetary targets, while recounting the initial achievements of planetary radar, namely, the refinement of the astronomical unit and the rotational rate and direction of Venus.

Chapter Three discusses early attempts to organize radar astronomy and the efforts at MIT's Lincoln Laboratory, in conjunction with Harvard radio astronomers, to acquire antenna time unfettered by military priorities. Here, the chief Big Science influencing the development of planetary radar astronomy was radio astronomy. Chapter Four spotlights the evolution of planetary radar astronomy at the Jet Propulsion Laboratory, a NASA facility, at Cornell University's Arecibo Observatory, and at Jodrell Bank. A congeries of funding from the military, the National Science Foundation, and finally NASA marked that evolution, which culminated in planetary radar astronomy finding a single Big Science patron, NASA.

Chapter Five analyzes planetary radar astronomy as a science using the theoretical framework provided by philosopher of science Thomas Kuhn. Chapter Six explores the shift in planetary radar astronomy beginning in the 1970s that resulted from its financial and institutional relationship with NASA Big Science. This shift saw the field 1) transform from an exclusively ground-based scientific activity to one conducted in space, as well as on Earth, and 2) capture the interest of planetary scientists from both the astronomy and geology communities. Chapter Seven relates how the Magellan mission was the culmination of this evolution. Chapters Eight and Nine discuss the research carried out at ground-based facilities by this transformed planetary radar astronomy, as well as the upgrading of the Arecibo and Goldstone radars.

The conclusion serves a dual purpose. It responds to the concern for the future of planetary radar astronomy expressed by many of the practitioners interviewed for this book, as well as to the author's wish to provide a slice of applied history that might be of value to both radar astronomers and policy makers. The conclusion also appraises planetary radar as a case of "Big Equipment but not Big Science". It considers the factors that have limited the size of planetary radar, its utilitarian nature, and its dependency on large-scale technological enterprises.

A technical essay appended to this book provides an overview of planetary radar techniques, especially range-Doppler mapping, for the general reader. Furthermore, the text itself explains certain, though not all, technical aspects of radar astronomy. The author assumed that the reader would have a familiarity with general technical and scientific terminology or would have access to a scientific dictionary or encyclopedia. For those readers seeking additional, and especially more technically-oriented, information on planetary radar astronomy, the technical essay includes a list of articles on the topic written by radar practitioners.

Chapter One

A Meteoric Start

During the 1940s, investigators in the United States and Hungary bounced radar waves off the Moon for the first time, while others made the first systematic radar studies of meteors. These experiments constituted the initial exploration of the solar system with radar. In order to understand the beginnings of radar astronomy, we first must examine the origins of radar in radio, the decisive role of ionospheric research, and the rapid development of radar technology triggered by World War II.

As early as 20 June 1922, in an address to a joint meeting of the Institute of Electrical Engineers and the Institute of Radio Engineers in New York, the radio pioneer Guglielmo Marconi suggested using radio waves to detect ships:[1]

> *As was first shown by Hertz, electric waves can be completely reflected by conducting bodies. In some of my tests I have noticed the effects of reflection and deflection of these waves by metallic objects miles away.*
>
> *It seems to me that it should be possible to design apparatus by means of which a ship could radiate or project a divergent beam of these rays in any desired direction, which rays, if coming across a metallic object, such as another steamer or ship, would be reflected back to a receiver screened from the local transmitter on the sending ship, and thereby immediately reveal the presence and bearing of the other ship in fog or thick weather.*
>
> *One further advantage of such an arrangement would be it would have the ability to give warning of the presence and bearing of ships, even should these ships be unprovided with any kind of radio.*

By the time Germany invaded Poland in September 1939 and World War II was underway, radio detection, location, and ranging technologies and techniques were available in Japan, France, Italy, Germany, England, Hungary, Russia, Holland, Canada, and the United States. Radar was not so much an invention, springing from the laboratory bench to the factory floor, but an ongoing adaptation and refinement of radio technology. The apparent emergence of radar in Japan, Europe, and North America more or less at the same time was less a case of simultaneous invention than a consequence of the global nature of radio research.[2]

Although radar is identified overwhelmingly with World War II, historian Sean S. Swords has argued that the rise of high-performance and long-range aircraft in the late 1930s would have promoted the design of advanced radio navigational aids, including radar, even without a war.[3] More decisively, however, ionospheric research propelled radar development in the 1920s and 1930s. As historian Henry Guerlac has pointed out, "Radar was developed by men who were familiar with the ionospheric work. It was a relatively straightforward adaptation for military purposes of a widely-known scientific technique,

---

1. Guglielmo Marconi, "Radio Telegraphy," *Proceedings of the Institute of Radio Engineers* 10 (1922): 237.
2. Charles Süsskind, "Who Invented Radar?" *Endeavour* 9 (1985): 92–96; Henry E. Guerlac, "The Radio Background of Radar," *Journal of the Franklin Institute* 250 (1950): 284–308.
3. Swords, *A Technical History of the Beginnings of Radar* (London: Peter Peregrinus Press, 1986), pp. 270–271.

which explains why this adaptation—the development of radar—took place simultaneously in several different countries."[4]

The prominence of ionospheric research in the history of radar and later of radar astronomy cannot be ignored. Out of ionospheric research came the essential technology for the beginnings of military radar in Britain, as well as its first radar researchers and research institutions. After the war, as we shall see, ionospheric research also drove the emergence of radar astronomy.

Chain Home

Despite its scientific origins, radar made its mark and was baptized during World War II as an integral and necessary instrument of offensive and defensive warfare. Located on land, at sea, and in the air, radars detected enemy targets and determined their position and range for artillery and aircraft in direct enemy encounters on the battlefield. Other radars identified aircraft to ground bases as friend or foe, while others provided navigational assistance and coastal defense. World War II was the first electronic war, and radar was its prime agent.[5]

In 1940, nowhere did radar research achieve the same advanced state as in Britain. The British lead initially resulted from a decision to design and build a radar system for coastal defense, while subsequent research led to the invention of the cavity magnetron, which placed Britain in the forefront of microwave radar. The impetus to achieve that lead in radar came from a realization that the island nation was no longer safe from enemy invasion.

For centuries, Britain's insularity and navy protected it from invasion. The advent of long-range airplanes that routinely outperformed their wooden predecessors spelled the end of that protection. Existing aircraft warning methods were ineffectual. That Britain was virtually defenseless against an air assault became clear during the summer air exercises of 1934. In simulated night attacks on London and Coventry, both the Air Ministry and the Houses of Parliament were successfully "destroyed," while few "enemy" bombers were intercepted.[6]

International politics also had reached a critical point. The Geneva Disarmament Conference had collapsed, and Germany was rearming in defiance of the Treaty of Versailles. Under attack from Winston Churchill and the Tory opposition, the British government abandoned its disarmament policy and initiated a five-year expansion of the Royal Air Force. Simultaneously, the Air Ministry Director of Scientific Research, Henry Egerton Wimperis, created a committee to study air defense methods.

Just before the Committee for the Scientific Survey of Air Defence first met on 28 January 1935, Wimperis contacted fellow Radio Research Board member Robert (later Sir) Watson-Watt. Watson-Watt, who oversaw the Radio Research Station at Slough, was a scientist with twenty years of experience as a government researcher. Ionospheric research had been a principal component of Radio Research Station studies, and Watson-Watt fostered the development there of a pulse-height technique.[7]

---

4.   Guerlac, "Radio Background," p. 304.

5.   Alfred Price, *Instruments of Darkness: The History of Electronic Warfare*, 2d. ed. (London: MacDonald and Jane's, 1977); Tony Devereux, *Messenger Gods of Battle, Radio, Radar, Sonar: The Story of Electronics in War* (Washington: Brassey's, 1991); David E. Fisher, *A Race on the Edge of Time: Radar—the Decisive Weapon of World War II* (New York: McGraw-Hill, 1988).

6.   H. Montgomery Hyde, *British Air Policy Between the Wars, 1918–1939* (London: Heinemann, 1976), p. 322. See also Malcolm Smith, *British Air Strategy Between the Wars* (Oxford, Clarendon Press, 1984).

7.   Swords, p. 84; Edward G. Bowen, *Radar Days* (Bristol: Adam Hilger, 1987), pp. 4–5, 7 and 10; Robert Watson-Watt, *The Pulse of Radar: The Autobiography of Sir Robert Watson-Watt* (New York: Dial Press, 1959), pp. 29–38, 51, 69, 101, 109–110, 113; A.P. Rowe, *One Story of Radar* (Cambridge: Cambridge University Press, 1948), pp. 6–7; Reg Batt, *The Radar Army: Winning the War of the Airwaves* (London: Robert Hale, 1991), pp. 21–22. The Radio Research Board was under the Department of Scientific and Industrial Research, created in 1916.

The pulse-height technique was to send short pulses of radio energy toward the ionosphere and to measure the time taken for them to return to Earth. The elapsed travel time of the radio waves gave the apparent height of the ionosphere. Merle A. Tuve, then of Johns Hopkins University, and Gregory Breit of the Carnegie Institution's Department of Terrestrial Magnetism in Washington, first developed the technique in the 1920s and undertook ionospheric research in collaboration with the Naval Research Laboratory and the Radio Corporation of America.[8]

In response to the wartime situation, Wimperis asked Watson-Watt to determine the practicality of using radio waves as a "death ray." Rather than address the proposed "death ray," Watson-Watt's memorandum reply drew upon his experience in ionospheric research. Years later, Watson-Watt contended, "I regard this Memorandum on the 'Detection and Location of Aircraft by Radio Methods' as marking the birth of radar and as being in fact the invention of radar."[9] Biographer Ronald William Clark has termed the memorandum "the political birth of radar." Nonetheless, Watson-Watt's memorandum was really less an invention than a proposal for a new radar application.

The memorandum outlined how a radar system could be put together and made to detect and locate enemy aircraft. The model for that radar system was the same pulse-height technique Watson-Watt had used at Slough. Prior to the memorandum in its final form going before the Committee, Wimperis had arranged for a test of Watson-Watt's idea that airplanes could reflect significant amounts of radio energy, using a BBC transmitter at Daventry. "Thus was the constricting 'red tape' of official niceties slashed by Harry Wimperis, before the Committee for the Scientific Survey of Air Defence had so much as met," Watson-Watt later recounted. The success of the Daventry test shortly led to the authorization of funding (£12,300 for the first year) and the creation of a small research and development project at Orford Ness and Bawdsey Manor that drew upon the expertise of the Slough Radio Research Station.

From then onwards, guided largely by Robert Watson-Watt, the foundation of the British radar effort, the early warning Chain Home, materialized. The Chain Home began in December 1935, with Treasury approval for a set of five stations to patrol the air approaches to the Thames estuary. Before the end of 1936, and long before the first test of the Thames stations in the autumn of 1937, plans were made to expand it into a network of nineteen stations along the entire east coast; later, an additional six stations were built to cover the south coast.

Born Robert Alexander Watson Watt in 1892, he changed his surname to "Watson-Watt" when knighted in 1942. See the popularly-written biography of Watson-Watt, John Rowland, *The Radar Man: The Story of Sir Robert Watson-Watt* (London: Lutterworth Press, 1963), or Watson-Watt, *Three Steps to Victory* (London: Odhams Press Ltd., 1957). An account of Watson-Watt's research at Slough is given in Watson-Watt, John F. Herd, and L.H. Bainbridge-Bell, *The Cathode Ray Tube in Radio Research* (London: His Majesty's Stationery Office, 1933).

8. By "apparent height of the ionosphere," I mean what ionosphericists call virtual height. Since the ionosphere slows radio waves before being refracted back to Earth, the delay is not a true measure of height. The Tuve-Breit method preceded that of Watson-Watt and was a true send-receive technique, while that of Watson-Watt was a receive-only technique.

Tuve "Early Days of Pulse Radio at the Carnegie Institution," *Journal of Atmospheric and Terrestrial Physics* 36 (1974): 2079–2083; Oswald G. Villard, Jr., "The Ionospheric Sounder and its Place in the History of Radio Science," *Radio Science* 11 (1976): 847–860; Guerlac, "Radio Background," pp. 284–308; David H. DeVorkin, *Science With a Vengeance: How the Military Created the U.S. Space Sciences after World War II* (New York: Springer-Verlag, 1992), pp. 12, 301 and 316; C. Stewart Gillmor, "Threshold to Space: Early Studies of the Ionosphere," in Paul A. Hanle and Von Del Chamberlin, eds., *Space Science Comes of Age: Perspectives in the History of the Space Sciences* (Washington: National Air and Space Museum, Smithsonian Institution, 1981), pp. 102–104; J.A. Ratcliffe, "Experimental Methods of Ionospheric Investigation, 1925–1955," *Journal of Atmospheric and Terrestrial Physics* 36 (1974): 2095–2103; Tuve and Breit, "Note on a Radio Method of Estimating the Height of the Conducting Layer," *Terrestrial Magnetism and Atmospheric Electricity* 30 (1925): 15–16; Breit and Tuve, "A Radio Method of Estimating the Height of the Conducting Layer," *Nature* 116 (1925): 357; and Breit and Tuve, "A Test of the Existence of the Conducting Layer," *Physical Review* 2d ser., vol. 28 (1926): 554–575; special issue of *Journal of Atmospheric and Terrestrial Physics* 36 (1974): 2069–2319, is devoted to the history of ionospheric research.

9. Watson-Watt, *Three Steps*, p. 83; Ronald William Clark, *Tizard* (London: Methuen, 1965), pp. 105–127.

The Chain Home played a crucial role in the Battle of Britain, which began in July 1940. The final turning point was on 15 September, when the Luftwaffe suffered a record number of planes lost in a single day. Never again did Germany attempt a massive daylight raid over Britain. However, if radar won the day, it lost the night. Nighttime air raids showed a desperate need for radar improvements.

The Magnetron

In order to wage combat at night, fighters needed the equivalent of night vision—their own on-board radar, but the prevailing technology was inadequate. Radars operating at low wavelengths, around 1.5 meters (200 MHz), cast a beam that radiated both straight ahead and downwards. The radio energy reflected from the Earth was so much greater than that of the enemy aircraft echoes that the echoes were lost at distances greater than the altitude of the aircraft. At low altitudes, such as those used in bombing raids or in air-to-air combat, the lack of radar vision was grave. Microwave radars, operating at wavelengths of a few centimeters, could cast a narrower beam and provide enough resolution to locate enemy aircraft.[10]

Although several countries had been ahead of Britain in microwave radar technology before the war began, Britain leaped ahead in February 1940, with the invention of the cavity magnetron by Henry A. H. Boot and John T. Randall at the University of Birmingham.[11] Klystrons were large vacuum tubes used to generate microwave power, but they did not operate adequately at microwave frequencies. The time required for electrons to flow through a klystron was too long to keep up with the frequency of the external oscillating circuit. The cavity magnetron resolved that problem and made possible the microwave radars of World War II. As Sean Swords asserted, "The emergence of the resonant-cavity magnetron was a turning point in radar history."[12] The cavity magnetron launched a line of microwave research and development that has persisted to this day.

The cavity magnetron had no technological equivalent in the United States, when the Tizard Mission arrived in late 1940 with one of the first ten magnetrons constructed. The Tizard Mission, known formally as the British Technical and Scientific Mission, had been arranged at the highest levels of government to exchange technical information between Britain and the United States. Its head and organizer, Henry Tizard, was a prominent physics professor and a former member of the committee that had approved Watson-Watt's radar project. As James P. Baxter wrote just after the war's end with a heavy handful of hyperbole, though not without some truth: "When the members of the Tizard Mission brought one [magnetron] to America in 1940, they carried the most valuable cargo ever brought to our shores. It sparked the whole development of microwave radar and constituted the most important item in reverse Lease-Lend."[13]

---

10.   Swords, pp. 84–85; Bowen, pp. 6, 21, 26 and 28; Batt, pp. 10, 21–22, 69 and 77; Rowe, pp. 8 and 76; R. Hanbury Brown, *Boffin: A Personal Story of the Early Days of Radar, Radio Astronomy, and Quantum Optics* (Bristol: Adam Hilger, 1991), pp. 7–8; P.S. Hall and R.G. Lee, "Introduction to Radar," in P.S. Hall, T.K. Garland-Collins, R.S. Picton, and R.G. Lee, eds., *Radar* (London: Brassey's, 1991), pp. 6–7; Watson-Watt, *Pulse*, pp. 55–59, 64–65, 75, 113–115 and 427–434; Watson-Watt, *Three Steps*, pp. 83 and 470–474; Bowen, "The Development of Airborne Radar in Great Britain, 1935–1945," in Russel W. Burns, ed., *Radar Development to 1945* (London: Peter Peregrinus Press, 1988), pp. 177–188. For a description of the technology, see B.T. Neale, "CH—the First Operational Radar," in Burns, pp. 132–150.

11.   Boot and Randall, "Historical Notes on the Cavity Magnetron," *IEEE Transactions on Electron Devices* ED-23 (1976): 724–729; R.W. Burns, "The Background to the Development of the Cavity Magnetron," in Burns, pp. 259–283.

12.   Swords, p. xi.

13.   Baxter, *Scientists Against Time* (Boston: Little, Brown and Company, 1946), p. 142; Swords, pp. 120, 259, and 266; Clark, especially pp. 248–271.

In late September 1940, Dr. Edward G. Bowen, the radar scientist on the Tizard Mission, showed a magnetron to members of the National Defense Research Committee (NDRC), which President Roosevelt had just created on 27 June 1940. One of the first acts of the NDRC, which later became the Office of Scientific Research and Development, was to establish a Microwave Committee, whose stated purpose was "to organize and consolidate research, invention, and development as to obtain the most effective military application of microwaves in the minimum time."[14]

A few weeks after the magnetron demonstration, the NDRC decided to create the Radiation Laboratory at MIT. While the MIT Radiation Laboratory accounted for nearly 80 percent of the NDRC Microwave Division's contracts, an additional 136 contracts for radar research, development, and prototype work were let out to sixteen colleges and universities, two private research institutions, and the major radio industrial concerns, with Western Electric taking the largest share. The MIT Radiation Laboratory personnel skyrocketed from thirty physicists, three guards, two stock clerks, and a secretary for the first year to a peak employment level of 3,897 (1,189 of whom were staff) on 1 August 1945. The most far-reaching early achievement, accomplished in the spring of 1941, was the creation of a new generation of radar equipment based on a magnetron operating at 3 cm. Experimental work in the one cm range led to numerous improvements in radars at 10 and 3 cm.[15]

Meanwhile, research and development of radars of longer wavelengths were carried out by the Navy and the Army Signal Corps, both of which had had active ongoing radar programs since the 1930s. The Navy started its research program at the Naval Research Laboratory (NRL) before that of the Signal Corps, but radar experimenters after the war used Signal Corps equipment, especially the SCR-270, mainly because of its wide availability. A mobile SCR-270, placed on Oahu as part of the Army's Aircraft Warning System, spotted incoming Japanese airplanes nearly 50 minutes before they bombed United States installations at Pearl Harbor on 7 December 1941. The warning was ignored, because an officer mistook the radar echoes for an expected flight of B-17s.[16]

Historians view the large-scale collection of technical and financial resources and manpower at the MIT Radiation Laboratory engaged in a concerted effort to research and develop new radar components and systems, along with the Manhattan Project, as

---

14.   Guerlac, *Radar in World War II*, The History of Modern Physics, 1800–1950, vol. 8 (New York: Tomash Publishers for the American Institute of Physics, 1987), vol. 1, p. 249; Swords, pp. 90 and 119; Batt, pp. 79–80; Bowen, pp. 159–162: Watson Watt, *Pulse*, pp. 228–229 and 257; Watson-Watt, *Three Steps*, 293.

In addition to Tizard and Bowen, the Mission team consisted of Prof. J.D. Cockcroft, Col. F.C. Wallace, Army, Capt. H.W. Faulkner, Navy, Capt. F.L. Pearce, Royal Air Force, W.E. Woodward Nutt, Ministry of Aircraft Production, Mission Secretary, Prof. R.H. Fowler, liaison officer for Canada and the United States of the Department of Scientific and Industrial Research, and Col. H.F.G. Letson, Canadian military attache in Washington.

15.   Guerlac, *Radar in World War II*, 1:258–259, 261, 266 and 507–508, and 2:648 and 668. See also the personal reminiscences of Ernest C. Pollard, *Radiation: One Story of the MIT Radiation Laboratory* (Durham: The Woodburn Press, 1982). Interviews (though not all are transcribed) of some Radiation Laboratory participants are available at the IEEE Center for the History of Electrical Engineering (CHEE), Rutgers University. CHEE, *Sources in Electrical History 2: Oral History Collections in U.S. Repositories* (New York: IEEE, 1992), pp. 6–7. The British also developed magnetrons and radar equipment operating at microwave frequencies concurrently with the MIT Radiation Laboratory effort.

16.   Guerlac, *Radar in World War II*, 1:247–248 and 117–119. For the Navy, see L.A. Hyland, "A Personal Reminiscence: The Beginnings of Radar, 1930–1934," in Burns, pp. 29–33; Robert Morris Page, *The Origin of Radar* (Garden City, NY: Anchor Books, Doubleday & Company, 1962); Page, "Early History of Radar in the U.S. Navy," in Burns, pp. 35–44; David Kite Allison, *New Eye for the Navy: The Origin of Radar at the Naval Research Laboratory* (Washington: Naval Research Laboratory, 1981); Guerlac, *Radar in World War II*, 1:59–92; Albert Hoyt Taylor, *The First Twenty-five Years of the Naval Research Laboratory* (Washington: Navy Department, 1948). On the Signal Corps, see Guerlac, *Radar in World War II*, 1:93–121; Harry M. Davis, *History of the Signal Corps Development of U.S. Army Radar Equipment* (Washington: Historical Section Field Office, Office of the Chief Signal Officer, 1945); Arthur L. Vieweger, "Radar in the Signal Corps," *IRE Transactions on Military Electronics* MIL-4 (1960): 555–561.

signalling the emergence of Big Science. Ultimately, from out of the concentration of personnel, expertise, materiel, and financial resources at the successor of the Radiation Laboratory, Lincoln Laboratory, arose the first attempts to detect the planet Venus with radar. The Radiation Laboratory Big Science venture, however, did not contribute immediately to the rise of radar astronomy.

The radar and digital technology used in those attempts on Venus was not available at the end of World War II, when the first lunar and meteor radar experiments were conducted. Moreover, the microwave radars issued from Radiation Laboratory research were far too weak for planetary or lunar work and operated at frequencies too high to be useful in meteor studies. Outside the Radiation Laboratory, though, U.S. Army Signal Corps and Navy researchers had created radars, like the SCR-270, that were more powerful and operated at lower frequencies, in research and development programs that were less concentrated and conducted on a smaller scale than the Radiation Laboratory effort.

Wartime production created an incredible excess of such radar equipment. The end of fighting turned it into war surplus to be auctioned off, given away, or buried as waste. World War II also begot a large pool of scientists and engineers with radar expertise who sought peacetime scientific and technical careers at war's end. That pool of expertise, when combined with the cornucopia of high-power, low-frequency radar equipment and a pinch of curiosity, gave rise to radar astronomy.

A catalyst crucial to that rise was ionospheric research. In the decade and a half following World War II, ionospheric research underwent the kind of swift growth that is typical of Big Science. The ionospheric journal literature doubled every 2.9 years from 1926 to 1938, before stagnating during the war; but between 1947 and 1960, the literature doubled every 5.8 years, a rate several times faster than the growth rate of scientific literature as a whole.[17] Interest in ionospheric phenomena, as expressed in the rapidly growing research literature, motivated many of the first radar astronomy experiments undertaken on targets beyond the Earth's atmosphere.

Project Diana

Typical was the first successful radar experiment aimed at the Moon. That experiment was performed with Signal Corps equipment at the Corps' Evans Signal Laboratory, near Belmar, New Jersey, under the direction of John H. DeWitt, Jr., Laboratory Director. DeWitt was born in Nashville and attended Vanderbilt University Engineering School for two years. Vanderbilt did not offer a program in electrical engineering, so DeWitt dropped out in order to satisfy his interest in broadcasting and amateur radio. In 1929, after building Nashville's first broadcasting station, DeWitt joined the Bell Telephone Laboratories technical staff in New York City, where he designed radio broadcasting transmitters. He returned to Nashville in 1932 to become Chief Engineer of radio station WSM. Intrigued by Karl Jansky's discovery of "cosmic noise," DeWitt built a radio telescope and searched for radio signals from the Milky Way.

In 1940, DeWitt attempted to bounce radio signals off the Moon in order to study the Earth's atmosphere. He wrote in his notebook: "It occurred to me that it might be possible to reflect ultrashort waves from the moon. If this could be done it would open up wide possibilities for the study of the upper atmosphere. So far as I know no one has ever

17.   Gillmor, "Geospace and its Uses: The Restructuring of Ionospheric Physics Following World War II," in DeMaria, Grilli, and Sebastiani, pp. 75–84, especially pp. 78–79.

18.   DeWitt notebook, 21 May 1940, and DeWitt biographical sketch, HL Diana 46 (04), HAUSACEC. There is a rich literature on Jansky's discovery. A good place to start is Woodruff T. Sullivan III, "Karl Jansky and the Discovery of Extraterrestrial Radio Waves," in Sullivan, ed., *The Early Years of Radio Astronomy: Reflections Fifty Years after Jansky's Discovery* (New York: Cambridge University Press, 1984), pp. 3–42.

sent waves off the earth and measured their return through the entire atmosphere of the earth."[18]

On the night of 20 May 1940, using the receiver and 80-watt transmitter configured for radio station WSM, DeWitt tried to reflect 138-MHz (2-meter) radio waves off the Moon, but he failed because of insufficient receiver sensitivity. After joining the staff of Bell Telephone Laboratories in Whippany, New Jersey, in 1942, where he worked exclusively on the design of a radar antenna for the Navy, DeWitt was commissioned in the Signal Corps and was assigned to serve as Executive Officer, later as Director, of Evans Signal Laboratory.

On 10 August 1945, the day after the United States unleashed a second atomic bomb on Japan, military hostilities between the two countries ceased. DeWitt was not demobilized immediately, and he began to plan his pet project, the reflection of radio waves off the Moon. He dubbed the scheme Project Diana after the Roman mythological goddess of the Moon, partly because "the Greek [sic] mythology books said that she had never been cracked."

In September 1945, DeWitt assembled his team: Dr. Harold D. Webb, Herbert P. Kauffman, E. King Stodola, and Jack Mofenson. Dr. Walter S. McAfee, in the Laboratory's Theoretical Studies Group, calculated the reflectivity coefficient of the Moon. Members of the Antenna and Mechanical Design Group, Research Section, and other Laboratory groups contributed too.

No attempt was made to design major components specifically for the experiment. The selection of the receiver, transmitter, and antenna was made from equipment already on hand, including a special crystal-controlled receiver and transmitter designed for the Signal Corps by radio pioneer Edwin H. Armstrong. Crystal control provided frequency stability, and the apparatus provided the power and bandwidth needed. The relative velocities of the Earth and the Moon caused the return signal to differ from the transmitted signal by as much as 300 Hz, a phenomenon known as Doppler shift. The narrow-band receiver permitted tuning to the exact radio frequency of the returning echo. As DeWitt later recalled: "We realized that the moon echoes would be very weak so we had to use a very narrow receiver bandwidth to reduce thermal noise to tolerable levels....We had to tune the receiver each time for a slightly different frequency from that sent out because of the Doppler shift due to the earth's rotation and the radial velocity of the moon at the time."[19]

The echoes were received both visually, on a nine-inch cathode-ray tube, and acoustically, as a 180-Hz beep. The aerial was a pair of "bedspring" antennas from an SCR-271 stationary radar positioned side by side to form a 32-dipole array antenna and mounted on a 30-meter (100-ft) tower. The antenna had only azimuth control; it had not been practical to secure a better mechanism. Hence, experiments were limited to the rising and setting of the Moon.

19.   DeWitt to Trevor Clark, 18 December 1977, HL Diana 46 (04); "Background Information on DeWitt Observatory" and "U.S. Army Electronics Research and Development Laboratory, Fort Monmouth, New Jersey," March 1963, HL Diana 46 (26), HAUSACEC. For published full descriptions of the equipment and experiments, see DeWitt and E. King Stodola, "Detection of Radio Signals Reflected from the Moon," *Proceedings of the Institute of Radio Engineers* 37 (1949): 229–242; Jack Mofenson, "Radar Echoes from the Moon," *Electronics* 19 (1946): 92–98; and Herbert Kauffman, "A DX Record: To the Moon and Back," *QST* 30 (1946): 65–68.

Figure 1

The "bedspring" mast antenna, U.S. Army Signal Corps, Ft. Monmouth, New Jersey, used by Lt. Col. John H. DeWitt, Jr., to bounce radar echoes off the Moon on 10 January 1946. Two antennas from SCR-271 stationary radars were positioned side by side to form a 32-dipole array aerial and were mounted on a 100-ft (30-meter) tower. (Courtesy of the U.S. Army Communications-Electronics Museum, Ft. Monmouth, New Jersey.)

The Signal Corps tried several times, but without success. "The equipment was very haywire," recalled DeWitt. Finally, at moonrise, 11:48 A.M., on 10 January 1946, they aimed the antenna at the horizon and began transmitting. Ironically, DeWitt was not present: "I was over in Belmar having lunch and picking up some items like cigarettes at the drug store (stopped smoking 1952 thank God)."[20] The first signals were detected at 11:58 A.M., and the experiment was concluded at 12:09 P.M., when the Moon moved out of the radar's range. The radio waves had taken about 2.5 seconds to travel from New Jersey to the Moon and back, a distance of over 800,000 km. The experiment was repeated daily over the next three days and on eight more days later that month.

The War Department withheld announcement of the success until the night of 24 January 1946. By then, a press release explained, "the Signal Corps was certain beyond doubt that the experiment was successful and that the results achieved were pain-staking-ly [sic] verified."[21]

As DeWitt recounted years later: "We had trouble with General Van Deusen our head of R&D in Washington. When my C.O. Col. Victor Conrad told him about it over the telephone the General did not want the story released until it was confirmed by outsiders for fear it would embarrass the Sig[nal]. C[orps]." Two outsiders from the Radiation Laboratory, George E. Valley, Jr. and Donald G. Fink, arrived and, with Gen. Van Deusen, observed a moonrise test of the system carried out under the direction of King Stodola. Nothing happened. DeWitt explained: "You can imagine that at this point I was dying. Shortly, a big truck passed by on the road next to the equipment and immediately the echoes popped up. I will always believe that one of the crystals was not oscillating until it was shaken up or there was a loose connection which fixed itself. Everyone cheered except the General who tried to look pleased."[22]

Although he had had other motives for undertaking Project Diana, DeWitt had received a directive from the Chief Signal Officer, the head of the Signal Corps, to develop radars capable of detecting missiles coming from the Soviet Union. No missiles were available for tests, so the Moon experiment stood in their place. Several years later, the Signal Corps erected a new 50-ft (15-meter) Diana antenna and 108-MHz transmitter for ionospheric research. It carried out further lunar echo studies and participated in the tracking of Apollo launches.[23]

The news also hit the popular press. The implications of the Signal Corps experiment were grasped by the War Department, although *Newsweek* cynically cast doubt on the War Department's predictions by calling them worthy of Jules Verne. Among those War Department predictions were the accurate topographical mapping of the Moon and planets, measurement and analysis of the ionosphere, and radio control from Earth of "space ships" and "jet or rocket-controlled missiles, circling the Earth above the stratosphere." *Time* reported that Diana might provide a test of Albert Einstein's Theory of Relativity. In contrast to the typically up-beat mood of *Life*, both news magazines were skeptical, and

---

20.   DeWitt replies to Clark questions, HL Diana 46 (04), HAUSACEC.

21.   HL Radar 46 (07), HAUSACEC; Harold D. Webb, "Project Diana: Army Radar Contacts the Moon," *Sky and Telescope* 5 (1946): 3–6.

22.   DeWitt to Clark, 18 December 1977, HL Diana 46 (04), HAUSACEC; Guerlac, *Radar in World War II*, 1:380 and 382, 2:702.

23.   DeWitt, telephone conversation, 14 June 1993; Materials in folders HL Diana 46 (25), HL Diana 46 (28), and HL Diana 46 (33), *USASEL Research & Development Summary* vol. 5, no. 3 (10 February 1958): 58, in "Signal Corps Engineering Laboratory Journal/R&D Summary," and *Monmouth Message*, 7 November 1963, n.p., in "Biographical Files," "Daniels, Fred Bryan," HAUSACEC; Daniels, "Radar Determination of the Scattering Properties of the Moon," *Nature* 187 (1960): 399; and idem., "A Theory of Radar Reflection from the Moon and Planets," *Journal of Geophysical Research* 66 (1961): 1781–1788.

rightly so; yet all of the predictions made by the War Department, including the relativity test, have come true in the manner of a Jules Verne novel.[24]

Zoltán Bay

Less than a month after DeWitt's initial experiment, a radar in Hungary replicated his results. The Hungarian apparatus differed from that of DeWitt in one key respect; it utilized a procedure, called integration, that was essential to the first attempt to bounce radar waves off Venus and that later became a standard planetary radar technique. The procedure's inventor was Hungarian physicist Zoltán Bay.

Bay graduated with highest honors from Budapest University with a Ph.D. in physics in 1926. Like many Hungarian physicists before him, Bay spent several years in Berlin on scholarships, doing research at both the prestigious Physikalisch-Technische-Reichanstalt and the Physikalisch-Chemisches-Institut of the University of Berlin. The results of his research tour of Berlin earned Bay the Chair of Theoretical Physics at the University of Szeged (Hungary), where he taught and conducted research on high intensity gas discharges.

Bay left the University of Szeged when the United Incandescent Lamps and Electric Company (Tungsram) invited him to head its industrial research laboratory in Budapest. Tungsram was the third largest manufacturer of incandescent lamps, radio tubes, and radio receivers in Europe and supplied a fifth of all radio tubes. As laboratory head, Zoltán Bay oversaw the improvement of high-intensity gas discharge lamps, fluorescent lamps, radio tubes, radio receiver circuitry, and decimeter radio wave techniques.[25]

Although Hungary sought to stay out of the war through diplomatic maneuvering, the threat of a German invasion remained real. In the fall of 1942, the Hungarian Minister of Defense asked Bay to organize an early-warning system. He achieved that goal, though the Germans occupied Hungary anyway. In March 1944, Bay recommended using the radar for scientific experimentation, including the detection of radar waves bounced off the Moon. The scientific interest in the experiment arose from the opportunity to test the theoretical notion that short wavelength radio waves could pass through the ionosphere without considerable absorption or reflection. Bay's calculations, however, showed that the equipment would be incapable of detecting the signals, since they would be significantly below the receiver's noise level.

The critical difference between the American and Hungarian apparatus was frequency stability, which DeWitt achieved through crystal control in both the transmitter and receiver. Without frequency stability, Bay had to find a means of accommodating the frequency drifts of the transmitter and receiver and the resulting inferior signal-to-noise ratio. He chose to boost the signal-to-noise ratio. His solution was both ingenious and far-reaching in its impact.

Bay devised a process he called cumulation, which is known today as integration. His integrating device consisted of ten coulometers, in which electric currents broke down a watery solution and released hydrogen gas. The amount of gas released was directly proportional to the quantity of electric current. The coulometers were connected to the output of the radar receiver through a rotating switch. The radar echoes were expected

24.   "Diana," *Time* Vol. 47, no. 5 (4 February 1946): 84; "Radar Bounces Echo off the Moon to Throw Light on Lunar Riddle," *Newsweek* vol. 27, no. 5 (4 February 1946): 76–77; "Man Reaches Moon with Radar," *Life* vol. 20, no. 5 (4 February 1946): 30.

25.   Zoltán Bay, *Life is Stronger*, trans. Margaret Blakey Hajdu (Budapest: Püski Publisher, 1991), pp. 5 and 17–18; Francis S. Wagner, *Zoltán Bay, Atomic Physicist: A Pioneer of Space Research* (Budapest: Akadémiai Kiadó, 1985), pp. 23–27, 29, 31–32; Wagner, *Fifty Years in the Laboratory: A Survey of the Research Activities of Physicist Zoltán Bay* (Center Square, PA: Alpha Publications, 1977), p. 1.

to return from the Moon in less than three seconds, so the rotating switch made a sweep of the ten coulometers every three seconds. The release of hydrogen gas left a record of both the echo signal and the receiver noise. As the number of signal echoes and sweeps of the coulometers added up, the signal-to-noise ratio improved. By increasing the total number of signal echoes, Bay believed that any signal could be raised above noise level and made observable, regardless of its amplitude and the value of the signal-to-noise ratio.[26] Because the signal echoes have a more-or-less fixed structure, and the noise varies from pulse to pulse, echoes add up faster than noise.

Despite the conceptual breakthrough of the coulometer integrator, the construction and testing of the apparatus remained to be carried out. The menace of air raids drove the Tungsram research laboratory into the countryside in the fall of 1944. The subsequent siege of Budapest twice interrupted the work of Bay and his team until March 1945. The Ministry of Defense furnished Bay with war surplus parts for a 2.5-meter (120-MHz) radar

Figure 2
Antenna built and used by Zoltán Bay to bounce radar echoes off the Moon in February and May 1946. (Courtesy of Mrs. Julia Bay)

manufactured by the Standard Electrical Co., a Hungarian subsidiary of ITT. Work was again interrupted when the laboratory was dismantled and all equipment, including that for the lunar radar experiment, was carried off to the Soviet Union. For a third time, construction of entirely new equipment started in the workshops of the Tungsram Research Laboratory, beginning August 1945 and ending January 1946.

Electrical disturbances in the Tungsram plant were so great that measurements and tuning had to be done in the late afternoon or at night. The experiments were carried out on 6 February and 8 May 1946 at night by a pair of researchers. Without the handicap of operating in a war zone, Bay probably would have beaten the Signal Corps to the Moon, although he could not have been aware of DeWitt's experiment. More importantly, though, he invented the technique of

26.   Bay, "Reflection of Microwaves from the Moon," *Hungarica Acta Physica* 1 (1947): 1–6; Bay, *Life is Stronger*, pp. 20 & 29; Wagner, *Zoltán*, pp. 39–40; Wagner, *Fifty Years*, pp. 1–2.

long-time integration generally used in radar astronomy. As the American radio astronomers Alex G. Smith and Thomas D. Carr wrote some years later: "The additional tremendous increase in sensitivity necessary to obtain radar echoes from Venus has been attained largely through the use of long-time integration techniques for detecting periodic signals that are far below the background noise level. The unique method devised by Bay in his pioneer lunar radar investigations is an example of such a technique."[27]

Both Zoltán Bay and John DeWitt had fired shots heard round the world, but there was no revolution, although others either proposed or attempted lunar radar experiments in the years immediately following World War II. Each man engaged in other projects shortly after completing his experiment. Bay left Hungary for the United States, where he taught at George Washington University and worked for the National Bureau of Standards, while DeWitt re-entered radio broadcasting and pursued his interest in astronomy.[28]

As an ongoing scientific activity, radar astronomy did not begin with the spectacular and singular experiments of DeWitt and Bay, but with an interest in meteors shared by researchers in Britain, Canada, and the United States. Big Science, that is, ionospheric physics and secure military communications, largely motivated that research. Moreover, just as the availability of captured V-2 parts made possible rocket-based ionospheric research after the war,[29] so war-surplus radars facilitated the emergence of radar astronomy. Like the exploration of the ionosphere with rockets, radar astronomy was driven by the availability of technology.

Meteors and Auroras

Radar meteor studies, like much of radar history, grew out of ionospheric research. In the 1930s, ionospheric researchers became interested in meteors when it was hypothesized that the trail of electrons and ions left behind by falling meteors caused fluctuations in the density of the ionosphere.[30] Edward Appleton and others with the Radio Research Board of the British Department of Scientific and Industrial Research, the same organization with which Watson-Watt had been associated, used war-surplus radar furnished by

27.   Smith and Carr, *Radio Exploration of the Planetary System* (New York: D. Van Nostrand, 1964), p. 123; Bay, "Reflection," pp. 2, 7–15 and 18–19; P. Vajda and J.A. White, "Thirtieth Anniversary of Zoltán Bay's Pioneer Lunar Radar Investigations and Modern Radar Astronomy," *Acta Physica Academiae Scientiarum Hungaricae* 40 (1976): 65–70; Wagner, *Zoltán*, pp. 40–41. Bay, *Life is Stronger*, pp. 103–124, describes the looting and dismantling of the Tungsram works by armed agents of the Soviet Union.

28.   DeWitt, telephone conversation, 14 June 1993; DeWitt biographical sketch, HL Diana 46 (04), HAUSACEC; Wagner, *Zoltán*, p. 49; Wagner, *Fifty Years*, p. 2.

Among the others were Thomas Gold, Von Eshleman, and A.C. Bernard Lovell. Gold, retired Cornell University professor of astronomy, claims to have proposed a lunar radar experiment to the British Admiralty during World War II; Eshleman, Stanford University professor of electrical engineering, unsuccessfully attempted a lunar radar experiment aboard the U.S.S. Missouri in 1946, while returning from the war; and Lovell proposed a lunar bounce experiment in a paper of May 1946. Gold 14 December 1993, Eshleman 9 May 1994, and Lovell, "Astronomer by Chance" manuscript, February 1988, Lovell materials, p. 183.

Even earlier, during the 1920s, the Navy unsuccessfully attempted to bounce a 32-KHz, 500-watt radio signal off the Moon. A. Hoyt Taylor, *Radio Reminiscences: A Half Century* (Washington: NRL, 1948), p. 133. I am grateful to Louis Brown for pointing out this reference.

29.   See DeVorkin, passim.

30.   A.M. Skellett, "The Effect of Meteors on Radio Transmission through the Kennelly-Heaviside Layer," *Physical Review* 37 (1931): 1668; Skellett, "The Ionizing Effect of Meteors," *Proceedings of the Institute of Radio Engineers* 23 (1935): 132–149. Skellett was a part-time graduate student in astronomy at Princeton University and an employee of Bell Telephone Laboratories, New York City. The research described in this article came out of a study of the American Telegraph and Telephone Company transatlantic short-wave telephone circuits in 1930–1932, and how they were affected by meteor ionization. DeVorkin, p. 275.

the Air Ministry to study meteors immediately after World War II. They concluded that meteors caused abnormal bursts of ionization as they passed through the ionosphere.[31]

During the war, the military had investigated meteor trails with radar. When the Germans started bombarding London with V2 rockets, the Army's gun-laying radars were hastily pressed into service to detect the radar reflections from the rockets during their flight in order to give some warning of their arrival. In many cases alarms were sounded, but no rockets were aloft. James S. Hey, a physicist with the Operational Research Group, was charged with investigating these mistaken sightings. He believed that the false echoes probably originated in the ionosphere and might be associated with meteors.

Hey began studying the impact of meteors on the ionosphere in October 1944, using Army radar equipment at several locations until the end of the war. The Operational Research Group, Hey, G. S. Stewart (electrical engineer), S. J. Parsons (electrical and mechanical engineer), and J. W. Phillips (mathematician), found a correlation between visual sightings and radar echoes during the Giacobinid meteor shower of October 1946. Moreover, by using an improved photographic technique that better captured the echoes on the radar screen, they were able to determine the velocity of the meteors.

Neither Hey nor Appleton pursued their radar investigations of meteors. During the war, Hey had detected radio emissions from the Sun and the first discrete source of radio emission outside the solar system in the direction of Cygnus. He left the Operational Research Group for the Royal Radar Establishment at Malvern, where he and his colleagues carried on research in radio astronomy. Appleton, by 1946 a Nobel Laureate and Secretary of the Department of Scientific and Industrial Research, also became thoroughly involved in the development of radio astronomy and became a member of the Radio Astronomy Committee of the Royal Astronomical Society in 1949.[32]

Radar astronomy, however, did gain a foothold in Britain at the University of Manchester under A. C. (later Sir) Bernard Lovell, director of the University's Jodrell Bank Experimental Station. During the war, Lovell had been one of many scientists working on microwave radar.[33] His superior, the head of the Physics Department, was Patrick M. S. Blackett, a member of the Committee for the Scientific Survey of Air Defence that approved Watson-Watt's radar memorandum. With the help of Hey and Parsons, Lovell borrowed some Army radar equipment. Finding too much interference in Manchester, he moved to the University's botanical research gardens, which became the Jodrell Bank Experimental Station. Lovell equipped the station with complete war-surplus radar systems, such as a 4.2-meter gun-laying radar and a mobile Park Royal radar. He purchased at rock-bottom prices or borrowed the radars from the Air Ministry, Army, and Navy, which were discarding the equipment down mine shafts.

---

31.   Appleton and R. Naismith, "The Radio Detection of Meteor Trails and Allied Phenomena," *Proceedings of the Physical Society* 59 (1947): 461–473; James S. Hey and G.S. Stewart, "Radar Observations of Meteors," *Proceedings of the Physical Society* 59 (1947): 858; Lovell, *Meteor Astronomy* (Oxford: Clarendon Press, 1954), pp. 23–24.

32.   Hey, *The Evolution of Radio Astronomy* (New York: Science History Publications, 1973), pp. 19–23 and 33–34; Lovell, *The Story of Jodrell Bank* (London: Oxford University Press, 1968), p. 5; Hey, Stewart, and S.J. Parsons, "Radar Observations of the Giacobinid Meteor Shower," *Monthly Notices of the Royal Astronomical Society* 107 (1947): 176–183; Hey and Stewart, "Radar Observations of Meteors," *Proceedings of the Physical Society* 59 (1947): 858–860 and 881–882; Hey, *The Radio Universe* (New York: Pergamon Press, 1971), pp. 131–134; Lovell, *Meteor Astronomy*, pp. 28–29 and 50–52; Peter Robertson, *Beyond Southern Skies: Radio Astronomy and the Parkes Telescope* (New York: Cambridge University Press, 1992), p. 39; Dudley Saward, *Bernard Lovell, a Biography* (London: Robert Hale, 1984), pp. 142–145; David O. Edge and Michael J. Mulkay, *Astronomy Transformed: The Emergence of Radio Astronomy in Britain* (New York: Wiley, 1976), pp. 12–14. For a brief historical overview of the Royal Radar Establishment, see Ernest H. Putley, "History of the RSRE," *RSRE Research Review* 9 (1985): 165–174; and D.H. Tomin, "The RSRE: A Brief History from Earliest Times to Present Day," *IEE Review* 34 (1988): 403–407. This major applied sciene institution deserves a more rigorously researched history.

33.   See Lovell, *Echoes of War: The Story of H₂S Radar* (Bristol: Adam Hilger, 1991). Lovell's wartime records are stored at the Imperial War Museum, Lambeth Road, London.

Figure 3

The Jodrell Bank staff 1951 in front of the 4.2-meter searchlight aerial used in some meteor radar experiments. Sir Bernard Lovell is in the center front. (Courtesy of the Director of the Nuffield Radio Astronomy Laboratories, Jodrell Bank.)

Originally, Lovell wanted to undertake research on cosmic rays, which had been Blackett's interest, too. One of the primary research objectives of the Jodrell Bank facility, as well as one of the fundamental reasons for its founding, was cosmic ray research. Indeed, the interest in cosmic ray research also lay behind the design and construction of the 76-meter (250-ft) Jodrell Bank telescope. The search for cosmic rays never succeeded, however; Blackett and Lovell had introduced a significant error into their initial calculations.

Fortuitously, though, in the course of looking for cosmic rays, Lovell came to realize that they were receiving echoes from meteor ionization trails, and his small group of Jodrell Bank investigators began to concentrate on this more fertile line of research. Nicolai Herlofson, a Norwegian meteorologist who had recently joined the Department of Physics, put Lovell in contact with the director of the Meteor Section of the British Astronomical Association, J. P. Manning Prentice, a lawyer and amateur astronomer with a passion for meteors. Also joining the Jodrell Bank team was John A. Clegg, a physics teacher whom Lovell had known during the war. Clegg was a doctoral candidate at the University of Manchester and an expert in antenna design. He remained at Jodrell Bank until 1951 and eventually landed a position teaching physics in Nigeria. Clegg converted an Army searchlight into a radar antenna for studying meteors.[34]

34.   Lovell 11 January 1994; Lovell, *Jodrell Bank*, pp. 5–8, 10; Lovell, *Meteor Astronomy*, pp. 55–63; Edge and Mulkay, pp. 15–16; Saward, pp. 129–131; R.H. Brown and Lovell, "Large Radio Telescopes and their Use in Radio Astronomy," *Vistas in Astronomy* 1 (1955): 542–560; Blackett and Lovell, "Radio Echoes and Cosmic Ray Showers," *Proceedings of the Royal Society of London* ser. A, vol. 177 (1941): 183–186; and Lovell, "The Blackett-Eckersley-Lovell Correspondence of World War II and the Origin of Jodrell Bank," *Notes and Records of the Royal Society of London* 47 (1993): 119–131. For documents relating to equipment on loan from the Ministry of Aviation, the War Office, the Royal Radar Establishment, the Admiralty, and the Air Ministry as late as the 1960s, see 10/51, "Accounts," JBA.

The small group of professional and amateur scientists began radar observations of the Perseid meteor showers in late July and August 1946. When Prentice spotted a meteor, he shouted. His sightings usually, though not always, correlated with an echo on the radar screen. Lovell thought that the radar echoes that did not correlate with Prentice's sightings might have been ionization trails created by cosmic ray showers. He did not believe, initially, that the radar might be detecting meteors too small to be seen by the human eye.

The next opportunity for a radar study of meteors came on the night of 9 October 1946, when the Earth crossed the orbit of the Giacobini-Zinner comet. Astronomers anticipated a spectacular meteor shower. A motion picture camera captured the radar echoes on film. The shower peaked around 3 A.M.; a radar echo rate of nearly a thousand meteors per hour was recorded. Lovell recalled that "the spectacle was memorable. It was like a great array of rockets coming towards one."[35]

The dramatic correlation of the echo rate with the meteors visible in the sky finally convinced Lovell and everyone else that the radar echoes came from meteor ionization trails, although it was equally obvious that many peculiarities needed to be investigated. The Jodrell Bank researchers learned that the best results were obtained when the aerial was positioned at a right angle to the radiant, the point in the sky from which meteor showers appear to emanate. When the aerial was pointed at the radiant, the echoes on the cathode-ray tube disappeared almost completely.[36]

Next joining the Jodrell Bank meteor group, in December 1946, was a doctoral student from New Zealand, Clifton D. Ellyett, followed in January 1947 by a Cambridge graduate, John G. Davies. Nicolai Herlofson developed a model of meteor trail ionization that Davies and Ellyett used to calculate meteor velocities based on the diffraction pattern produced during the formation of meteor trails. Clegg devised a radar technique for determining their radiant.[37]

At this point, the Jodrell Bank investigators had powerful radar techniques for studying meteors that were unavailable elsewhere, particularly the ability to detect and study previously unknown and unobservable daytime meteor showers. Lovell and his colleagues now became aware of the dispute over the nature of meteors and decided to attempt its resolution with these techniques.[38]

Astronomers specializing in meteors were concerned with the nature of sporadic meteors. One type of meteor enters the atmosphere from what appears to be a single point, the radiant. Most meteors, however, are not part of a shower, but appear to arrive irregularly from all directions and are called sporadic meteors. Most astronomers believed that sporadic meteors came from interstellar space; others argued that they were part of the solar system.

The debate could be resolved by determining the paths of sporadic meteors. If they followed parabolic or elliptical paths, they orbited the Sun; if their orbit were hyperbolic, they had an interstellar origin. The paths of sporadic meteors could be determined by an accurate measurement of both their velocities and radiants, but optical means were insufficiently precise to give unambiguous results. Fred L. Whipple, future director of the

35.   Lovell 11 January 1994; Lovell, *Jodrell Bank*, pp. 7–8, 10.
36.   Lovell 11 January 1994; Lovell, *Jodrell Bank*, pp. 8–10; Lovell, Clegg, and Congreve J. Banwell, "Radio Echo Observations of the Giacobinid Meteors 1946," *Monthly Notices of the Royal Astronomical Society* 107 (1947): 164–175. Banwell was a New Zealand veteran of the Telecommunications Research Establishment wartime radar effort and an expert on receiver electronics.
37.   Saward, p. 137; Herlofson, "The Theory of Meteor Ionization," *Reports on Progress in Physics* 11 (1946–47): 444–454; Ellyett and Davies, "Velocity of Meteors Measured by Diffraction of Radio Waves from Trails during Formation," *Nature* 161 (1948): 596-597; Clegg, "Determination of Meteor Radiants by Observation of Radio Echoes from Meteor Trails," *Philosophical Magazine* ser. 7, vol. 39 (1948): 577-594; Davies and Lovell, "Radio Echo Studies of Meteors," *Vistas in Astronomy 1* (1955): 585-598, provides a summary of meteor research at Jodrell Bank.
38.   Lovell, *Jodrell Bank*, p. 12; Lovell, *Meteor Astronomy*, pp. 358–383.

Harvard College Observatory, a leading center of United States meteor research, attempted state-of-the-art optical studies of meteors with the Super Schmidt camera, but the first one was not operational until May 1951, at Las Cruces, New Mexico.[39]

Radar astronomers thus attempted to accomplish what optical methods had failed to achieve. Such has been the pattern of radar astronomy to the present. Between 1948 and 1950, Lovell, Davies, and Mary Almond, a doctoral student, undertook a long series of sporadic meteor velocity measurements. They found no evidence for a significant hyperbolic velocity component; that is, there was no evidence for sporadic meteors coming from interstellar space. They then extended their work to fainter and smaller meteors with similar results.

The Jodrell Bank radar meteor studies determined unambiguously that meteors form part of the solar system. As Whipple declared in 1955, "We may now accept as proven the fact that bodies moving in hyperbolic orbits about the sun play no important role in producing meteoric phenomena brighter than about the 8th effective magnitude."[40] Astronomers describe the brightness of a body in terms of magnitude; the larger the magnitude, the fainter the body.

The highly convincing evidence of the Jodrell Bank scientists was corroborated by Canadian radar research carried out by researchers of the Radio and Electrical Engineering Division of the National Research Council under Donald W. R. McKinley. McKinley had joined the Council's Radio Section (later Branch) before World War II and, like Lovell, had participated actively in wartime radar work.

McKinley conducted his meteor research with radars built around Ottawa in 1947 and 1948 as part of various National Research Council laboratories, such as the Flight Research Center at Arnprior Airport. Earle L. R. Webb, Radio and Electrical Engineering Division of the National Research Council, supervised the design, construction, and operation of the radar equipment. From as early as the summer of 1947, the Canadian radar studies were undertaken jointly with Peter M. Millman of the Dominion Observatory. They coordinated spectrographic, photographic, radar, and visual observations. The National Research Council investigators employed the Jodrell Bank technique to determine meteor velocities, a benefit of following in the footsteps of the British.[41]

Their first radar observations took place during the Perseid shower of August 1947, as the first radar station reached completion. Later studies collected data from the Geminid shower of December 1947 and the Lyrid shower of April 1948, with more radar stations brought into play as they became available. Following the success of Jodrell Bank,

39.   Ron Doel, "Unpacking a Myth: Interdisciplinary Research and the Growth of Solar System Astronomy, 1920-1958," Ph.D. diss. Princeton University, 1990, pp. 33–35, 42–44 and 108–111; DeVorkin, pp. 96, 273, 278 and 293; Luigi G. Jacchia and Whipple, "The Harvard Photographic Meteor Programme," *Vistas in Astronomy* 2 (1956): 982–994; Whipple, "Meteors and the Earth's Upper Atmosphere," *Reviews of Modern Physics* 15 (1943): 246–264; Whipple, "The Baker Super-Schmidt Meteor Cameras," *The Astronomical Journal* 56 (1951): 144–145, states that the first such camera was installed in New Mexico in May 1951. Determining the origin of meteors was not the primary interest of Harvard research.

40.   Whipple, "Some Problems of Meteor Astronomy," in H. C. Van de Hulst, ed., *Radio Astronomy* (Cambridge: Cambridge University Press, 1957), p. 376; Almond, Davies, and Lovell, "The Velocity Distribution of Sporadic Meteors," *Monthly Notices of the Royal Astronomical Society* 111 (1951): 585–608; 112 (1952): 21–39; 113 (1953): 411–427. The meteor studies at Jodrell Bank were continued into later years. See, for instance, I. C. Browne and T. R. Kaiser, "The Radio Echo from the Head of Meteor Trails," *Journal of Atmospheric and Terrestrial Physics* 4 (1953): 1–4.

41.   W. E. Knowles Middleton, *Radar Development in Canada: The Radio Branch of the National Research Council of Canada, 1939–1946* (Waterloo, Ontario: Wilfred Laurier University Press, 1981), pp. 18, 25, 27, 106–109; Millman and McKinley, "A Note on Four Complex Meteor Radar Echoes," *Journal of the Royal Astronomical Society of Canada* 42 (1948): 122; McKinley and Millman, "A Phenomenological Theory of Radar Echoes from Meteors," *Proceedings of the Institute of Radio Engineers* 37 (1949): 364–375; McKinley and Millman, "Determination of the Elements of Meteor Paths from Radar Observations," *Canadian Journal of Research* A27 (1949): 53–67; McKinley, "Deceleration and Ionizing Efficiency of Radar Meteors," *Journal of Applied Physics* 22 (1951): 203; McKinley, *Meteor Science and Engineering* (New York: McGraw-Hill, 1961), p. 20; Lovell, *Meteor Astronomy*, pp. 52–55.

McKinley's group initiated their own study of sporadic meteors. By 1951, with data on 10,933 sporadic meteors, McKinley's group reached the same conclusion as their British colleagues: meteors were part of the solar system. Soon, radar techniques became an integral part of Canadian meteor research with the establishment in 1957 of the National Research Council Springhill Meteor Observatory outside Ottawa. The Observatory concentrated on scientific meteor research with radar, visual, photographic, and spectroscopic methods.[42]

These meteor studies at Jodrell Bank and the National Research Council, and only at those institutions, arose from the union of radar and astronomy; they were the beginnings of radar astronomy. Radar studies of meteors were not limited to Jodrell Bank and the National Research Council, however. With support from the National Bureau of Standards, in 1957 Harvard College Observatory initiated a radar meteor project under the direction of Fred Whipple. Furthermore, radar continues today as an integral and vital part of worldwide meteor research. Its forte is the ability to determine orbits better than any other technique. In the last five years, a number of recently built radars have studied meteors in Britain (MST Radar, Aberytswyth, Wales), New Zealand (AMOR, Meteor Orbit Radar, Christchurch), and Japan (MU Radar, Shigaraki), not to mention earlier work in Czechoslovakia and Sweden.[43]

Unlike the Jodrell Bank and National Research Council cases, the radar meteor studies started in the United States in the early 1950s were driven by civilian scientists doing ionospheric and communications research and by the military's desire for jam-proof, point-to-point secure communications. While various military laboratories undertook their own research programs, most of the civilian U.S. radar meteor research was carried out at Stanford University and the National Bureau of Standards, where investigators fruitfully cross-fertilized ionospheric and military communications research. The Stanford case is worth examining not only for its later connections to radar astronomy, but also for its pioneering radar study of the Sun that arose out of an interest in ionospheric and radio propagation research.

In contrast to the Stanford work, many radar meteor experiments carried out in the United States in the 1940s were unique events. As early as August and November 1944, for instance, workers in the Federal Communications Commission Engineering Department associated visual observations of meteors and radio bursts. In January 1946, Oliver Perry Ferrell of the Signal Corps reported using a Signal Corps SCR-270B radar to detect meteor ionization trails.[44] The major radar meteor event in the United States and elsewhere,

42.   Millman, McKinley, and M. S. Burland, "Combined Radar, Photographic, and Visual Observations of the 1947 Perseid Meteor Shower," *Nature* 161 (1948): 278–280; McKinley and Millman, "Determination of the Elements," p. 54; Millman and McKinley, "A Note," pp. 121–130; McKinley, "Meteor Velocities Determined by Radio Observations," *The Astrophysics Journal* 113 (1951): 225–267; F. R. Park, "An Observatory for the Study of Meteors," *Engineering Journal* 41 (1958): 68–70.

43.   Whipple, "Recent Harvard-Smithsonian Meteoric Results," *Transactions of the IAU* 10 (1960): 345–350; Jack W. Baggaley and Andrew D. Taylor, "Radar Meteor Orbital Structure of Southern Hemisphere Cometary Dust Streams," pp. 33–36 in Alan W. Harris and Edward Bowell, eds., *Asteroids, Comets, Meteors 1991* (Houston: Lunar and Planetary Institute, 1992); Baggaley, Duncan I. Steel, and Taylor, "A Southern Hemisphere Radar Meteor Orbit Survey," pp. 37–40 in ibidem; William Jones and S. P. Kingsley, "Observations of Meteors by MST Radar," pp. 281–284 in ibidem; Jun-ichi Wattanabe, Tsuko Nakamura, T. Tsuda, M. Tsutsumi, A. Miyashita, and M. Yoshikawa, "Meteor Mapping with MU Radar," pp. 625–627 in ibidem. The MST Radar and the AMOR were newly commissioned in 1990. The MU Radar is intended primarily for atmospheric research.

For the meteor radar research in Sweden and Czechoslovakia, see B. A. Lindblad and M. Simek, "Structure and Activity of Perseid Meteor Stream from Radar Observations, 1956–1978," pp. 431–434 in Claes-Ingva Lagerkvist and Hans Rickman, eds., *Asteroids, Comets, Meteors* (Uppsala: Uppsala University, 1983); A. Hajduk and G. Cevolani, "Variations in Radar Reflections from Meteor Trains and Physical Properties of Meteoroids," pp. 527–530 in Lagerkvist, H. Rickman, Lindblad, and M. Lindgren, *Asteroids, Comets, Meteors III* (Uppsala: Uppsala University, 1989); Simek and Lindblad, "The Activity Curve of the Perseid Meteor Stream as Determined from Short Duration Meteor Radar Echoes," pp. 567–570 in ibidem.

44.   Ferrell, "Meteoric Impact Ionization Observed on Radar Oscilloscopes," *Physical Review* 2d ser., vol. 69 (1946): 32–33; Lovell, *Meteor Astronomy*, p. 28.

however, was the spectacular meteor shower associated with the Giacobini-Zinner comet.

On the night of 9 October 1946, 21 Army radars were aimed toward the sky in order to observe any unusual phenomena. The Signal Corps organized the experiment, which fit nicely with their mission of developing missile detection and ranging capabilities. The equipment was operated by volunteer crews of the Army ground forces, the Army Air Forces, and the Signal Corps located across the country in Idaho, New Mexico, Texas, and New Jersey. For mainly meteorological reasons, only the Signal Corps SCR-270 radar successfully detected meteor ionization trails. No attempt was made to correlate visual observations and radar echoes. A Princeton University undergraduate, Francis B. Shaffer, who had received radar training in the Navy, analyzed photographs of the radar screen echoes at the Signal Corps laboratory in Belmar, New Jersey.

This was the first attempt to utilize microwave radars to detect astronomical objects. The equipment operated at 1,200 MHz (25 cm), 3,000 MHz (10 cm), and 10,000 MHz (3 cm), frequencies in the L, S, and X radar bands that radar astronomy later used. "On the basis of this night's experiments," the Signal Corps experimenters decided, "we cannot conclude that microwave radars do not detect meteor-formed ion clouds."[45]

In contrast to the Signal Corps experiment, radar meteor studies formed part of ongoing research at the National Bureau of Standards. Organized from the Bureau's Radio Section in May 1946 and located at Sterling, Virginia, the Central Radio Propagation Laboratory (CRPL) division had three laboratories, one of which concerned itself exclusively with ionospheric research and radio propagation and was especially interested in the impact of meteors on the ionosphere. In October 1946, Victor C. Pineo and others associated with the CRPL used a borrowed SCR-270-D Signal Corps radar to observe the Giacobinid meteor shower. Over the next five years, Pineo continued research on the effects of meteors on the ionosphere, using a standard ionospheric research instrument called an ionosonde and publishing his results in *Science*.

Pineo's interest was in ionospheric physics, not astronomy. Underwriting his research at the Ionospheric Research Section of the National Bureau of Standards was the Air Force Cambridge Research Center (known later as the Cambridge Research Laboratories and today as Phillips Laboratory). His meteor work did not contribute to knowledge about the origin of meteors, as such work had in Britain and Canada, but it supported efforts to create secure military communications using meteor ionization trails.[46] Also, it related to similar research being carried out concurrently at Stanford University.

The 1946 CRPL experiment, in fact, had been suggested by Robert A. Helliwell of the Stanford Radio Propagation Laboratory (SRPL). Frederick E. Terman, who had headed the Harvard Radio Research Laboratory and its radar countermeasures research during the war, "virtually organized radio and electronic engineering on the West Coast" as

45.   Signal Corps Engineering Laboratories, "Postwar Research and Development Program of the Signal Corps Engineering Laboratories, 1945," (Signal Corps, 1945), "Postwar R&D Program," HL R&D, HAUSACEC; John Q. Stewart, Michael Ference, John J. Slattery, Harold A. Zahl, "Radar Observations of the Draconids," *Sky and Telescope* 6 (March 1947): 35. They reported their earlier results in a paper, "Radar Observations of the Giacobinid Meteors," read before the December 1946 meeting of the American Astronomical Society in Boston. HL Diana 46 (26), HAUSACEC.

46.   Wilbert F. Snyder and Charles L. Bragaw, *Achievement in Radio: Seventy Years of Radio Science, Technology, Standards, and Measurement at the National Bureau of Standards* (Boulder: National Bureau of Standards, 1986), pp. 461–465; Ross Bateman, A. G. McNish, and Pineo, "Radar Observations during Meteor Showers, 9 October 1946," *Science* 104 (1946): 434–435; Pineo, "Relation of Sporadic E Reflection and Meteoric Ionization," *Science* 110 (1949): 280–283; Pineo, "A Comparison of Meteor Activity with Occurrence of Sporadic-E Reflections," *Science* 112 (1950): 5051; Pineo and T. N. Gautier, "The Wave-Frequency Dependence of the Duration of Radar-Type Echoes from Meteor Trails," *Science* 114 (1951): 460–462. Other articles by Pineo on his ionospheric research can be found in Laurence A. Manning, *Bibliography of the Ionosphere: An Annotated Survey through 1960* (Stanford: Stanford University Press, 1962), pp. 421–423.

Stanford Dean of Engineering, according to historian C. Stewart Gillmor. Terman negotiated a contract with the three military services for the funding of a broad range of
research, including the SRPL's long-standing ionospheric research program.[47]

Helliwell, whose career was built on ionospheric research, was joined at the SRPL by
Oswald G. Villard, Jr. Villard had earned his engineering degree during the war for the
design of an ionosphere sounder. As an amateur radio operator in Cambridge,
Massachusetts, he had noted the interference caused by meteor ionizations at shortwave
frequencies called Doppler whistles.[48]

In October 1946, during the Giacobinid meteor shower, Helliwell, Villard, Laurence
A. Manning, and W. E. Evans, Jr., detected meteor ion trails by listening for Doppler whistles with radios operating at 15 MHz (20 meters) and 29 MHz (10 meters). Manning then
developed a method of measuring meteor velocities using the Doppler frequency shift of
a continuous-wave signal reflected from the ionization trail. Manning, Villard, and Allen
M. Peterson then applied Manning's technique to a continuous-wave radio study of the
Perseid meteor shower in August 1948. The initial Stanford technique was significantly
different from that developed at Jodrell Bank; it relied on continuous-wave radio, rather
than pulsed radar, echoes.[49]

One of those conducting meteor studies at Stanford was Von R. Eshleman, a graduate student in electrical engineering who worked under both Manning and Villard. While
serving in the Navy during World War II, Eshleman had studied, then taught, radar at the
Navy's radar electronics school in Washington, DC. In 1946, while returning from the war
on the U.S.S. Missouri, Eshleman unsuccessfully attempted to bounce radar waves off the
Moon using the ship's radar. Support for his graduate research at Stanford came through
contracts between the University and both the Office of Naval Research and the Air Force.

Eshleman's dissertation considered the theory of detecting meteor ionization trails
and its application in actual experiments. Unlike the British and Canadian meteor studies, the primary research interest of Eshleman, Manning, Villard, and the other Stanford
investigators was information about the winds and turbulence in the upper atmosphere.
Their investigations of meteor velocities, the length of ionized meteor trails, and the fading and polarization of meteor echoes were part of that larger research interest, while
Eshleman's dissertation was an integral part of the meteor research program.

Eshleman also considered the use of meteor ionization trails for secure military communications. His dissertation did not explicitly state that application, which he took up
after completing the thesis. The Air Force supported the Stanford meteor research mainly to use meteor ionization trails for secure, point-to-point communications. The Stanford
meteor research thus served a variety of scientific and military purposes simultaneously.[50]

---

47.   Gillmor, "Federal Funding and Knowledge Growth in Ionospheric Physics, 1945–1981," *Social Studies of Science* 16 (1986): 124.

48.   Oswald G. Villard, Jr., "Listening in on the Stars," *QST* 30 (January, 1946): 59–60, 120 and 122; Helliwell, *Whistlers and Related Ionospheric Phenomena* (Stanford: Stanford University Press, 1965), pp. 11–23; Leslie, p. 58; Gillmor, "Federal Funding," p. 129.

49.   Manning, Helliwell, Villard, and Evans, "On the Detection of Meteors by Radio," *Physical Review* 70 (1946): 767–768; Manning, "The Theory of the Radio Detection of Meteors," *Journal of Applied Physics* 19 (1948): 689–699: Manning, Villard, and Peterson, "Radio Doppler Investigation of Meteoric Heights and Velocities," *Journal of Applied Physics* 20 (1949): 475–479; Von R. Eshleman, "The Effect of Radar Wavelength on Meteor Echo Rate," *Transactions of the Institute of Radio Engineers* 1 (1953): 37–42. DeVorkin, pp. 287–288, points out that, when given an opportunity to make radio observations in coordination with rocket flights, Stanford declined.

50.   Eshleman 9 May 1994; Eshleman, "The Mechanism of Radio Reflections from Meteoric Ionization," Ph.D. diss., Stanford University, 1952; Eshleman, *The Mechanism of Radio Reflections from Meteoric Ionization*, Technical Report no. 49 (Stanford: Stanford Electronics Research Laboratory, 15 July 1952), pp. ii–iii and 3; Manning, "Meteoric Radio Echoes," *Transactions of the Institute of Radio Engineers* 2 (1954): 82–90; Manning and Eshleman, "Meteors in the Ionosphere," *Proceedings of the Institute of Radio Engineers* 47 (1959): 186–199.

The meteor research carried out at Stanford had nontrivial consequences. Eshleman's dissertation has continued to provide the theoretical foundation of modern meteor burst communications, a communication mode that promises to function even after a nuclear holocaust has rendered useless all normal wireless communications. The pioneering work at Stanford, the National Bureau of Standards, and the Air Force Cambridge Research Laboratories received new attention in the 1980s, when the Space Defense Initiative ("Star Wars") revitalized interest in using meteor ionization trails for classified communications. Non-military applications of meteor burst communications also have arisen in recent years.[51]

Early meteor burst communications research was not limited to Stanford and the National Bureau of Standards. American military funding of early meteor burst communications research extended beyond its shores to Britain. Historians of Jodrell Bank radio astronomy and meteor radar research stated that radio astronomy had surpassed meteor studies at the observatory by 1955. However, that meteor work persisted until 1964 through a contract with the U.S. Air Force, though as a cover for classified military research.[52]

Auroras provided additional radar targets in the 1950s. A major initiator of radar auroral studies was Jodrell Bank. As early as August 1947, while conducting meteor research, the Jodrell Bank scientists Lovell, Clegg, and Ellyett received echoes from an aurora display. Arnold Aspinall and G. S. Hawkins then continued the radar auroral studies at Jodrell Bank in collaboration with W. B. Housman, Director of the Aurora Section of the British Astronomy Association, and the aurora observers of that Section. In Canada, McKinley and Millman also observed an aurora during their meteor research in April 1948.[53]

The problem with bouncing radar waves off an aurora was determining the reflecting point. Researchers in the University of Saskatchewan Physics Department (B. W. Currie, P. A. Forsyth, and F. E. Vawter) initiated a systematic study of auroral radar reflections in 1948, with funding from the Defense Research Board of Canada. Radar equipment was lent by the U.S. Air Force Cambridge Research Center and modified by the Radio and Electrical Engineering Division of the National Research Council. Forsyth had completed a dissertation on auroras at McGill University and was an employee of the Defense Research Board's Telecommunications Establishment on loan to the University of Saskatchewan for the project. The Saskatchewan researchers discovered that the echoes bounced off small, intensely ionized regions in the aurora.[54]

Other aurora researchers, especially in Sweden and Norway, took up radar studies. In Sweden, Götha Hellgren and Johan Meos of the Chalmers University of Technology

51. Robert Desourdis, telephone conversation, 22 September 1994; Donald Spector, telephone conversation, 22 September 1994; Donald L. Schilling, ed., *Meteor Burst Communications: Theory and Practice* (New York: Wiley, 1993); Jacob Z. Schanker, *Meteor Burst Communications* (Boston: Artech House, 1990). For a civilian use of meteor burst communications, see Henry S. Santeford, *Meteor Burst Communication System: Alaska Winter Field Test Program* (Silver Spring, MD: U.S. Dept. of Commerce, National Oceanic and Atmospheric Administration, National Weather Service, Office of Hydrology, 1976).

52. Lovell 11 January 1994; 7 and 8/55, "Accounts," JBA; Lovell, "Astronomer by Chance," typed manuscript, February 1988, p. 376, Lovell materials; Lovell, *Jodrell Bank*, p. 157; G. Nigel Gilbert, "The Development of Science and Scientific Knowledge: The Case of Radar Meteor Research," in Gerard Lemaine, Roy Macleod, Michael Mulkay, and Peter Weingart, eds., *Perspectives on the Emergence of Scientific Disciplines* (Chicago: Aldine, 1976), p. 191; Edge and Mulkay, pp. 330–331.

53. Lovell, Clegg, and Ellyett, "Radio Echoes from the Aurora Borealis," *Nature* 160 (1947): 372; Aspinall and Hawkins, "Radio Echo Reflections from the Aurora Borealis," *Journal of the British Astronomical Association* 60 (1950): 130–135; various materials in File Group "International Geophysical Year," Box 1, File 4, JBA; McKinley and Millman, "Long Duration Echoes from Aurora, Meteors, and Ionospheric Back-Scatter," *Canadian Journal of Physics* 31 (1953): 171–181.

54. Currie, Forsyth, and Vawter, "Radio Reflections from Aurora," *Journal of Geophysical Research* 58 (1953): 179–200.

Research Laboratory of Electronics in Gothenburg decided to conduct radar studies of auroras as part of their ionospheric research program. Beginning in May 1951, the Radio Wave Propagation Laboratory of the Kiruna Geophysical Observatory undertook round-the-clock observations of auroras with a 30.3-MHz (10-meter) radar. In Norway, Leiv Harang, who had observed radar echoes from an aurora as early as 1940, and B. Landmark observed auroras with radars lent by the Norwegian Defense Research Establishment and installed at Oslo (Kjeller) and Tromsö, where a permanent center for radar investigation of auroras was created later.[55]

These and subsequent radar investigations changed the way scientists studied auroras, which had been almost entirely by visual means up to about 1950. Permanent auroral observatories located at high latitudes, such as those at Oslo and Tromsö in Norway, at Kiruna in Sweden, and at Saskatoon in Saskatchewan, integrated radar into a spectrum of research instruments that included spectroscopy, photography, balloons, and sounding rockets. The International Geophysical Year, 1957–1958, was appropriately timed to further radar auroral research; it coincided with extremely high sunspot and auroral activity, such as the displays visible from Mexico in September 1957 and the "Great Red Aurora" of 10 February 1958. Among those participating in the radar aurora and meteor studies associated with the International Geophysical Year activities were three Jodrell Bank students and staff who joined the Royal Society expedition to Halley Bay, Antarctica.[56]

To the Moon Again

The auroral and meteor radar studies carried out in the wake of the lunar radar experiments of DeWitt and Bay were, in essence, ionospheric studies. While the causes of auroras and meteor ionization trails arise outside the Earth's atmosphere, the phenomena themselves are essentially ionospheric. At Jodrell Bank, meteor and auroral studies provided the initial impetus, but certainly not the sustaining force, for the creation of an ongoing radar astronomy program. That sustaining force came from lunar studies. However, like so much of early radar astronomy, those lunar studies were never far from ionospheric research. Indeed, the trailblazing efforts of DeWitt and Bay opened up new vistas of ionospheric and communications research using radar echoes from the Moon.

Historically, scientists had been limited to the underside and lower portion of the ionosphere. The discovery of "cosmic noise" by Bell Telephone researcher Karl Jansky in 1932 suggested that higher frequencies could penetrate the ionosphere. The experiments of DeWitt and Bay suggested radar as a means of penetrating the lower regions of the ionosphere. DeWitt, moreover, had observed unexpected fluctuations in signal strength that lasted several minutes, which he attributed to anomalous ionospheric refraction.[57] His observations invited further investigation of the question.

The search for a better explanation of those fluctuations was taken up by a group of ionosphericists in the Division of Radiophysics of the Australian Council for Scientific and Industrial Research: Frank J. Kerr, C. Alex Shain, and Charles S. Higgins. In 1946, Kerr and Shain explored the possibility of obtaining radar echoes from meteors, following the

55.   Hellgren and Meos, "Localization of Aurorae with 10m High Power Radar Technique, using a Rotating Antenna," *Tellus* 3 (1952): 249–261; Harang and Landmark, "Radio Echoes Observed during Aurorae and Geomagnetic Storms using 35 and 74 Mc/s Waves Simultaneously," *Journal of Atmospheric and Terrestrial Physics* 4 (1954): 322–338; ibidem *Nature* 171 (1953): 1017–1018; Harang and J. Tröim, "Studies of Auroral Echoes," *Planetary and Space Science* 5 (1961): 33–45 and 105–108.
56.   Jean Van Bladel, *Les applications du radar à l'astronomie et à la météorologie* (Paris: Gauthier-Villars, 1955), pp. 78–80; Neil Bone, *The Aurora: Sun-Earth Interactions* (New York: Ellis Horwood, 1991), pp. 36, 45–49; Alistair Vallance Jones, *Aurora* (Boston: D. Reidel Publishing Company, 1974), pp. 9, 11 and 27; Lovell, "Astronomer by Chance," manuscript, February 1988, p. 201, Lovell materials.
57.   DeWitt and Stodola, p. 239.

example of Lovell in Britain, but Project Diana turned their attention toward the Moon. In order to study the fluctuations in signal strength that DeWitt had observed, Kerr, Shain, and Higgins put together a rather singular experiment.

For a transmitter, they used the 20-MHz (15-meter) Radio Australia station, located in Shepparton, Victoria, when it was not in use for regular programming to the United States and Canada. The receiver was located at the Radiophysics Laboratory, Hornsby, New South Wales, a distance of 600 km from the transmitter. Use of this unique system was limited to days when three conditions could be met all at the same time: the Moon was passing through the station's antenna beams; the transmitter was available; and atmospheric conditions were favorable. In short, the system was workable about twenty days a year.[58]

Kerr, Shain, and Higgins obtained lunar echoes on thirteen out of fifteen attempts. The amplitude of the echoes fluctuated considerably over the entire run of tests as well as within a single test. Researchers at ITT's Federal Telecommunications Laboratories in New York City accounted for the fluctuations observed by DeWitt by positing the existence of smooth spots that served as "bounce points" for the reflected energy. Another possibility they imagined was the existence of an ionosphere around the Moon.[59] The Australians disagreed with the explanations offered by DeWitt and the ITT researchers, but they were initially cautious: "It cannot yet be said whether the reductions in intensity and the long-period variations are due to ionospheric, lunar or inter-planetary causes."[60]

During a visit to the United States in 1948, J. L. Pawsey, a radio astronomy enthusiast also with the Council for Scientific and Industrial Research's Division of Radiophysics, arranged a cooperative experiment with the Americans. A number of U.S. organizations with an interest in radio, the National Bureau of Standards CRPL, the Radio Corporation of America (Riverhead, New York), and the University of Illinois (Urbana), attempted to receive Moon echoes simultaneously from Australia, beginning 30 July 1948. Ross Bateman (CRPL) acted as American coordinator. The experiment was not a great success. The times of the tests (limited by transmitter availability) were all in the middle of the day at the receiving points. Echoes were received in America on two occasions, 1 August and 28 October, and only for short periods in each case.

Meanwhile, Kerr and Shain continued to study lunar echo fading with the Radio Australia transmitter. Based on thirty experiments (with echoes received in twenty-four of them) conducted over a year, they now distinguished rapid and slow fading. Kerr and Shain proposed that each type of fading had a different cause. Rapid fading resulted from the Moon's libration, a slow wobbling motion of the Moon. Irregular movement in the ionosphere, they originally suggested, caused the slower fading.[61] Everyone agreed that the rapid fading of lunar radar echoes originated in the lunar libration, but the cause of slow fading was not so obvious.

The problem of slow fading was taken up at Jodrell Bank by William A. S. Murray and J. K. Hargreaves, who sought an explanation in the ionosphere. Although Lovell had proposed undertaking lunar radar observations as early as 1946, the first worthwhile results were not obtained until the fall of 1953. Hargreaves and Murray photographed and analyzed some 50,000 lunar radar echoes at the Jodrell Bank radar telescope in October and November 1953 to determine the origin of slow fading.

---

58.   Kerr, Shain, and Higgins, "Moon Echoes and Penetration of the Ionosphere," *Nature* 163 (1949): 310; Kerr and Shain, "Moon Echoes and Transmission through the Ionosphere," *Proceedings of the IRE* 39 (1951): 230; Kerr, "Early Days in Radio and Radar Astronomy in Australia," pp. 136–137 in Sullivan. Kerr and Shain, pp. 230–232, contains a better description of the system. See also Kerr, "Radio Superrefraction in the Coastal Regions of Australia," *Australian Journal of Scientific Research*, ser. A, vol. 1 (1948): 443–463.

59.   D. D. Grieg, S. Metzger, and R. Waer, "Considerations of Moon-Relay Communication," *Proceedings of the IRE* 36 (1948): 660.

60.   Kerr, Shain, and Higgins, p. 311.

61.   Kerr and Shain, pp. 230–242.

With rare exceptions, nighttime runs showed a steady signal amplitude, while daytime runs, especially those within a few hours of sunrise, were marked by severe fading. The high correlation between fading and solar activity strongly suggested an ionospheric origin. However, Hargreaves and Murray believed that irregularities in the ionosphere could not account for slow fading over periods lasting up to an hour. They suggested instead that slow fading resulted from Faraday rotation, in which the plane of polarization of the radio waves rotated, as they passed through the ionosphere in the presence of the Earth's magnetic field.

Hargreaves and Murray carried out a series of experiments to test their hypothesis in March 1954. The transmitter had a horizontally polarized antenna, while the primary feed of the receiving antenna consisted of two dipoles mounted at right angles. They switched the receiver at short intervals between the vertical and horizontal feeds so that echoes would be received in both planes of polarization, a technique that is a standard planetary radar practice today.

As the plane of polarization of the radar waves rotated in the ionosphere, stronger echo amplitudes were received by the vertical feed than by the horizontal feed. If no Faraday rotation had taken place, both the transmitted and received planes of polarization would be the same, that is, horizontal. But Faraday rotation of the plane of polarization in the ionosphere had rotated the plane of polarization so that the vertical feed received more echo power than the horizontal feed. The results confirmed that slow fading was caused, at least in part, by a change in the plane of polarization of the received lunar echo.[62]

Murray and Hargreaves soon took positions elsewhere, yet Jodrell Bank continued to feature radar astronomy through the persistence of Bernard Lovell. Lovell became entangled in administrative affairs and the construction of a giant radio telescope, while John V. Evans, a research student of Lovell, took over the radar astronomy program. Evans had a B.Sc. in physics and had had an interest in electronics engineering since childhood. He chose the University of Manchester Physics Department for his doctoral degree, because the department, through Lovell, oversaw the Jodrell Bank facility. The facility's heavy involvement in radio and radar astronomy, when Evans arrived there on his bicycle in the summer of 1954, assured Evans that his interest in electronics engineering would be sated.

With the approval and full support of Lovell, Evans renewed the studies of lunar radar echoes, but first he rebuilt the lunar radar equipment. It was a "poor instrument," Evans later recalled, "and barely got echoes from the Moon." After he increased the power output from 1 to 10 kilowatts and improved the sensitivity of the receiver by rebuilding the front end, Evans took the lunar studies in a new direction. Unlike the majority of Jodrell Bank research, Evans's lunar work was underwritten through a contract with the U.S. Air Force, which was interested in using the Moon as part of a long-distance communications system.

With his improved radar apparatus, Evans discovered that the Moon overall was a relatively smooth reflector of radar waves at the wavelength he used (120 MHz; 2.5 meters). Later, from the way that the Moon appeared to scatter back radar waves, Evans speculated that the lunar surface was covered with small, round objects such as rocks and stones. Hargreaves proposed that radar observations at shorter wavelengths should be able to give interesting statistical information about the features of the lunar surface.[63] That idea was

62.   Murray and Hargreaves, "Lunar Radio Echoes and the Faraday Effect in the Ionosphere," *Nature* 173 (1954): 944–945; Browne, Evans, Hargreaves, and Murray, p. 901; 1/17 "Correspondence Series 7," JBA; Lovell, "Astronomer by Chance," p. 183.

63.   Evans 9 September 1993; Hargreaves, "Radio Observations of the Lunar Surface," *Proceedings of the Physical Society* 73 (1959): 536–537; Evans, "Research on Moon Echo Phenomena," Technical (Final) Report, 1 May 1956, and earlier reports in 1/4 "Correspondence Series 2," JBA.

the starting point for the creation of planetary radar techniques that would reveal the surface characteristics of planets and other moons.

Experimenters prior to Evans had assumed that the Moon reflected radar waves from the whole of its illuminated surface, like light waves. They debated whether the power returned to the Earth was reflected from the entire visible disk or from a smaller region. The question was important to radar astronomers at Jodrell Bank as well as to military and civilian researchers developing Moon-relay communications.

In March 1957, Evans obtained a series of lunar radar echoes. He photographed both the transmitted pulses and their echoes so that he could make a direct comparison between the two. Evans also made range measurements of the echoes at the same time. In each case, the range of the observed echo was consistent with that of the front edge of the Moon. The echoes came not from the entire visible disk but from a smaller portion of the lunar surface, that closest to the Earth and known as the subradar point.[64] This discovery became fundamental to radar astronomy research.

Because radar waves reflected off only the foremost edge of the Moon, Evans and John H. Thomson (a radio astronomer who had transferred from Cambridge in 1959) undertook a series of experiments on the use of the Moon as a passive communication relay. Although initial results were "not intelligible," because FM and AM broadcasts tended to fade, Lovell bounced Evans' "hello" off the Moon with a Jodrell Bank transmitter and receiver during his BBC Reith Lecture of 1958. Several years later, in collaboration with the Pye firm, a leading British manufacturer of electronic equipment headquartered in Cambridge, and with underwriting from the U.S. Air Force, a Pye transmitter at Jodrell Bank was used to send speech and music via the Moon to the Sagamore Hill Radio Astronomy Observatory of the Air Force Cambridge Research Center, at Hamilton, Massachusetts. The U.S. Air Force thus obtained a successful lunar bounce communication experiment at Jodrell Bank for a far smaller sum than that spent by the Naval Research Laboratory.[65]

The Moon Bounce

The lunar communication studies at Jodrell Bank illustrate that astronomy was not behind all radar studies of the Moon. Much of the lunar radar work, especially in the United States, was performed to test long-distance communication systems in which the Moon would serve as a relay. Thus, the experiments of DeWitt and Bay may be said to have begun the era of satellite communications. Research on Moon-relay communications systems by both military and civilian laboratories eventually drew those institutions into the early organizational activities of radar astronomers. After all, both communication research and radar astronomy shared an interest in the behavior of radio waves at the lunar surface. Hence, a brief look at that research would be informative.

Before the advent of satellites, wireless communication over long distances was achieved by reflecting radio waves off the ionosphere. As transmission frequency increased, the ionosphere was penetrated. Long-distance wireless communication at high frequencies had to depend on a network of relays, which were expensive and technically complex. Using the Moon as a relay appeared to be a low-cost alternative.[66]

---

64.   Evans 9 September 1993; Evans, "The Scattering of Radio Waves by the Moon," *Proceedings of the Physical Society* B70 (1957): 1105–1112.

65.   Evans 9 September 1993; Edge and Mulkay, p. 298; Materials in 1/4 "Correspondence Series 2," and 2/53 "Accounts," JBA. With NASA funding, Jodrell Bank later participated in the Echo balloon project.

66.   Harold Sobol, "Microwave Communications: An Historical Perspective," *IEEE Transactions on Microwave Theory and Techniques* MTT-32 (1984): 1170–1181.

Reacting to the successes of DeWitt and Bay, researchers at the ITT Federal Telecommunications Laboratories, Inc., New York City, planned a lunar relay telecommunication system operating at UHF frequencies (around 50 MHz; 6 meters) to provide radio telephone communications between New York and Paris. If such a system could be made to work, it would provide ITT with a means to compete with transatlantic cable carriers dominated by rival AT&T. What the Federal Telecommunications Laboratories had imagined, the Collins Radio Company, Cedar Rapids, Iowa, and the National Bureau of Standards CRPL, accomplished.

On 28 October and 8 November 1951, Peter G. Sulzer and G. Franklin Montgomery, CRPL, and Irvin H. Gerks, Collins Radio, sent a continuous-wave 418-MHz (72-cm) radio signal from Cedar Rapids to Sterling, Virginia, via the Moon. On 8 November, a slowly hand-keyed telegraph message was sent over the circuit several times. The message was the same sent by Samuel Morse over the first U.S. public telegraph line: "What hath God wrought?"[67]

Unbeknownst to the CRPL/Collins team, the first use of the Moon as a relay in a communication circuit was achieved only a few days earlier by military researchers at the Naval Research Laboratory (NRL). The Navy was interested in satellite communications, and the Moon offered itself as a free (if distant and rough) satellite in the years before an artificial satellite could be launched. In order to undertake lunar communication studies, the NRL built what was then the world's largest parabolic antenna in the summer of 1951. The dish covered over an entire acre (67 by 80 meters; 220 by 263 ft) and had been cut into the earth by road-building machinery at Stump Neck, Maryland. The one-megawatt transmitter operated at 198 MHz (1.5 meters). The NRL first used the Moon as a relay in a radio communication circuit on 21 October 1951. After sending the first voice transmission via the Moon on 24 July 1954, the NRL demonstrated transcontinental satellite teleprinter communication from Washington, DC, to San Diego, CA, at 301 MHz (1 meter) on 29 November 1955 and transoceanic satellite communication, from Washington, DC, to Wahiawa, Oahu, Hawaii, on 23 January 1956.[68]

Later in 1956, the NRL's Radio Astronomy Branch started a radar program under Benjamin S. Yaplee to determine the feasibility of bouncing microwaves off the Moon and to accurately measure both the Moon's radius and the distances to different reflecting areas during the lunar libration cycle. Aside from the scientific value of that research, the information would help the Navy to determine relative positions on the Earth's surface. The first NRL radar contact with the Moon at a microwave frequency took place at 2860 MHz (10-cm) and was accomplished with the Branch's 15-meter (50-ft) radio telescope.[69]

Although interest in bouncing radio and radar waves off the Moon drew military and civilian researchers to early radar astronomy conferences, lunar communication schemes failed to provide either a theoretical or a funding framework within which radar astronomy could develop. The rapidly growing field of ionospheric research, on the other hand, provided both theoretical and financial support for radar experiments on meteors and the Moon. Despite the remarkable variety of radar experiments carried out in the years following World War II, radar achieved a wider and more permanent place in ionospheric research (especially meteors and auroras) than in astronomy.

---

67. Grieg, Metzger, and Waer, pp. 652–663; "Via the Moon: Relay Station to Transoceanic Communication," *Newsweek* 27 (11 February 1946): 64; Sulzer, Montgomery, and Gerks, "An U-H-F Moon Relay," *Proceedings of the IRE* 40 (1952): 361. A few years later, three amateur radio operators, "hams" who enjoyed detecting long-distance transmissions (DXing), succeeded in bouncing 144-Mhz radio waves off the Moon, on 23 and 27 January 1953. E. P. T., "Lunar DX on 144 Mc!" *QST* 37 (1953): 11–12 and 116.

68. Gebhard, pp. 115–116; James H. Trexler, "Lunar Radio Echoes," *Proceedings of the IRE* 46 (1958): 286–288.

69. NRL, "The Space Science Division and E. O. Hulburt Center for Space Research, Program Review," 1968, NRLHRC; Yaplee, R. H. Bruton, K. J. Craig, and Nancy G. Roman, "Radar Echoes from the Moon at a Wavelength of 10 cm," *Proceedings of the IRE* 46 (1958): 293–297; Gebhard, p. 118.

All that changed with the start of the U.S./U.S.S.R. Space Race and the announcement of the first planetary radar experiment in 1958. That experiment was made possible
by the rivalries of the Cold War, which fostered a concentration of expertise and financial,
personnel, and material resources that paralleled, and in many ways exceeded, that of
World War II. The new Big Science of the Cold War and the Space Race, often indistinguishable from each other, gave rise to the radar astronomy of planets.

The Sputnik and Lunik missions were not just surprising demonstrations of Soviet
achievements in science and technology. Those probes had been propelled off the Earth
by ICBMs, and an ICBM capable of putting a dog in Earth-orbit or sending a probe to the
Moon was equally capable of delivering a nuclear bomb from Moscow to New York City.
Behind the Space Race lay the specter of the Cold War and World War III, or to para-
phrase Clausewitz, the Space Race was the Cold War by other means. Just as the vulnerability of Britain to air attacks had led to the creation of the Chain Home radar warning
network, the defenselessness of the United States against aircraft and ICBM attacks with
nuclear bombs and warheads led to the creation of a network of defensive radars. The
development of that network in turn provided the instrument with which planetary radar
astronomy, driven by the availability of technology, would begin in the United States.

Chapter Two

Fickle Venus

In 1958, MIT's Lincoln Laboratory announced that it had bounced radar waves off Venus. That apparent success was followed by another, but in England, during Venus' next inferior conjunction. In September 1959, investigators at Jodrell Bank announced that they had validated the 1958 results, yet Lincoln Laboratory failed to duplicate them. All uncertainty was swept aside, when the Jet Propulsion Laboratory (JPL) obtained the first unambiguous detection of echoes from Venus in 1961.

As we saw in the case of radar studies of meteors and the Moon in the 1940s and 1950s, planetary radar astronomy was driven by technology. The availability of military apparatus made possible the rise of radar astronomy in Britain in the 1940s. Just as the threat of airborne invasion gave rise to the Chain Home radar, the Cold War and its scientific counterpart, the Space Race, demanded the creation of a new generation of defensive radars, and those radars made possible the first planetary radar experiments. Even British and Soviet planetary radar astronomy were not free of the sway of military and space efforts. Thus, the Big Science efforts brought into being by the Cold War and the Space Race provided the material resources necessary for the emergence of planetary radar astronomy.

The initial radar detections of Venus signaled a benchmark in radar capacity that separated a new generation of radars from their predecessors. High-speed digital computers linked to more powerful transmitters and more sensitive receivers utilizing state-of-the-art masers and parametric amplifiers provided the new capacity. As we saw in Chapter One, initial radar astronomy targets were either ionospheric phenomena, like meteors and auroras, or the Moon, whose mean distance from Earth is about 384,000 kilometers. The new radars reached beyond the Moon to Venus, about 42 million kilometers distant at its closest approach to Earth.

Radar detections of the planets, while sterling technical achievements, were incapable of demonstrating the value of planetary radar as an ongoing scientific activity. As radar astronomy already had achieved with meteor studies, planetary radar became a scientific activity by solving problems left unsolved or unsatisfactorily solved by optical means.

As they made their first detections of Venus, planetary radar astronomers found and solved two such problems. One was the rotation of Venus, the determination of which was prevented by the planet's optically impenetrable atmosphere. The other problem was the astronomical unit, the mean radius of the Earth's orbit around the Sun. Astronomers express the distances of the planets from the Sun in terms of the astronomical unit, but agreement on its exact value was lacking. Radar observations of Venus provided an exact value, which the International Astronomical Union adopted, and revealed the planet's retrograde rotation.

While the astronomical unit and the rotation of Venus interested astronomers, they also held potential benefit for the nascent space program. In many respects, the problems solved by the first planetary radar experiments needed solutions because of the Space Race. By February 1958, when Lincoln Laboratory first tried to bounce radar waves off Venus, Sputnik 1 and the Earth-orbiting dog Laika were yesterday's news. The Space Race was hot, and so was the competition between the United States and the Soviet Union.

Planetary radar astronomy rode the cresting waves of Big Science (the Space Race) and the Cold War well into the 1970s.

From the Rad Lab to Millstone Hill

Scientists and engineers at MIT's Lincoln Laboratory attempted to reach Venus by radar in 1958, because they had access to a radar of unprecedented capability. The radar existed because MIT, as it had since the days of the Radiation Laboratory, conducted military electronics research. Lincoln Laboratory did not emerge directly from the Radiation Laboratory but through its direct descendant, the Research Laboratory of Electronics (RLE).

The RLE, a joint laboratory of the Physics and Electrical Engineering Departments, continued much of the fundamental electronic research of the Radiation Laboratory. The Signal Corps, Air Force, and the Office of Naval Research jointly funded the new laboratory, with the Signal Corps overseeing the arrangement. Former Radiation Laboratory employees filled research positions at the RLE, which occupied a temporary structure on the MIT campus erected earlier for the Radiation Laboratory. The two leaders of the Lincoln Laboratory Venus radar experiment, Robert Price and Paul E. Green, Jr., were both student employees of the RLE. Price also had an Industrial Fellowship in Electronics from Sperry. Among the other early RLE fellowship sponsors were the General Radio Company, RCA, ITT, and the Socony-Vacuum Oil Company.

In September 1949, the Soviet Union detonated its first nuclear bomb; within months civil war exploded in Korea. The need for a United States air defense capable of coping with a nuclear attack was urgent. Project Charles, a group of military and civilian experts, studied the problems of air defense. Its findings led directly to the creation of Lincoln Laboratory in the Autumn of 1951.[1]

MIT was, in the words of Hoyt S. Vandenberg, U.S. Air Force chief of staff, "uniquely qualified to serve as contractor to the Air Force for the establishment of the proposed [Lincoln] laboratory. Its experience in managing the Radiation Laboratory of World War II, the participation in the work of ADSEC [Air Defense Systems Engineering Committee] by Professor [George E.] Valley and other members of the MIT staff, its proximity to AFCRL [Air Force Cambridge Research Laboratories], and its demonstrated competence in this sort of activity have convinced us that we should be fortunate to secure the services of MIT in the present connection."[2]

Lincoln Laboratory was to design and develop what became known as SAGE (Semi-Automatic Ground Environment), a digital, integrated computerized North-American network of air defense. SAGE involved a diversity of applied research in digital computing and data processing, long-range radar, and digital communications. The Army, Navy, and Air Force jointly underwrote Lincoln Laboratory through an Air Force prime contract. The Air Force provided nearly 90 percent of the funding. In 1954, Lincoln Laboratory moved out of its Radiation Laboratory buildings on the MIT campus and into a newly constructed facility at Hanscom Field, in Lexington, Massachusetts, next to the Air Force Cambridge Research Center.

---

1.   "President's Report Issue," *MIT Bulletin* vol. 82, no. 1 (1946): 133–136; ibid., vol. 83, no. 1 (1947): 154–157; ibid., vol. 86, no. 1 (1950): 209; "Government Supported Research at MIT: An Historical Survey Beginning with World War II: The Origins of the Instrumentation and Lincoln Laboratories," May 1969, typed manuscript, pp. 15–19 and 30–31, MITA; George E. Valley, Jr., rough draft, untitled four page manuscript, 13 October 1953, 6/135/AC 4, and MIT Review Panel on Special Laboratories, "Final Report," pp. 132–133, MITA. James R. Killian, Jr., *The Education of a College President: A Memoir* (Cambridge: The MIT Press, 1985), pp. 71–76, recounts the founding of Lincoln Laboratory, too.

2.   Vandenberg to James R. Killian, Jr., 15 December 1950, 3/136/AC 4, MITA. A portion of the quote also appears in Killian, p. 71.

Lincoln Laboratory quickly began work on the Distant Early Warning (DEW) Line in the arctic region of North America. The first experimental DEW-line radar units were in place near Barter Island, Alaska, by the end of 1953. The radar antennas were enclosed by a special structure called a radome, which protected them from arctic winds and cold.

InterContinental Ballistic Missiles (ICBMs) challenged the DEW Line and the North American coordinated defense network, which had been designed to warn against airplane attacks. ICBMs could carry nuclear warheads above the ionosphere, higher than any pilot could fly; existing warning radars were useless. In order to detect and track ICBMs, radars would have to recognize targets smaller than airplanes at altitudes several hundred kilometers above the Earth and at ranges of several thousand kilometers. The new radars would have to distinguish between targets and auroras, meteors, and other ionospheric disturbances, which experience already had shown were capable of crippling military communications and radars.[3]

In 1954, Lincoln Laboratory began initial studies of Anti-InterContinental Ballistic Missile (AICBM) systems and the creation of the Ballistic Missile Early Warning System (BMEWS). By the spring of 1956, the construction of an experimental prototype BMEWS radar was underway. Its location, atop Millstone Hill in Westford, Massachusetts, was well away from air routes and television transmitters and close to MIT and Lincoln Laboratory. The Air Force owned and financed the radar, while Lincoln Laboratory managed it under Air Force contract through the adjacent Air Force Cambridge Research Center.

Herbert G. Weiss was in charge of designing and building Millstone. After graduating from MIT in 1936 with a BS in electrical engineering, Weiss conducted microwave research for the Civil Aviation Authority in Indianapolis and worked in the MIT Radiation Laboratory. After the war, Weiss worked at Los Alamos, then at Raytheon, before returning to MIT to work on the DEW radars.

Millstone embodied a new generation of radars capable of detecting smaller objects at farther ranges. Thanks to specially designed, 3-meter-tall (11-feet-tall) klystron tubes, Millstone was intended to have an unprecedented amount of peak transmitting power, 1.25 megawatts from each klystron (2.5 megawatts total). Its frequency was 440 MHz (68 cm). The antenna, a steerable parabolic dish 26 meters (84-feet) from rim to rim, stood on a 27-meter-high (88-foot-high) tower of concrete and steel. Millstone began operating in October 1957, just in time to skin track the first Sputnik.

3.   Valley; "Final Report," pp. 133–137; "Government Supported," p. 33; C. L. Strong, Information Department, Western Electric Company, press release, 1 October 1953, 6/135/AC 4, MITA; Carl F. J. Overhage to Lt. Gen. Roscoe C. Wilson, 15 October 1959, and brochure, "Haystack Family Day, 10 October 1964," 1/24/AC 134, MITA; F. W. Loomis to Killian, 17 April 1952, 4/135/AC 4, MITA; various documents in 2/136/AC 4 and 7/135/AC 4, MITA; Overhage, "Reaching into Space with Radar," paper read at MIT Club of Rochester, 25 February 1960, pp. 6–7, LLLA. For a popular introduction to the DEW Line, see Richard Morenus, *Dew Line: Distant Early Warning, The Miracle of America's First Line of Defense* (New York: Rand McNally, 1957).

Figure 4
The Lincoln Laboratory Millstone Hill Radar Observatory, ca. 1958. (Courtesy of MIT Lincoln Laboratory, Lexington, Massachusetts, photo no. P489-128.)

Millstone furnished valuable scientific and technological information to the military, while advancing ionospheric and lunar radar research. In addition to testing and evaluating new defense radar techniques and components, its scientific missions included measuring the ionosphere and its influence on radar signals (such as Faraday rotation), observing satellites and missiles, and performing radar studies of auroras, meteors, and the Moon, all of which were potential sources of false alarm for BMEWS radars.[4]

The Lunchtime Conversazione

The idea of using the Millstone Hill radar to bounce signals off Venus arose during one of the customary lunchtime discussions between Bob Price and Paul Green. As MIT doctoral students and later as Lincoln Laboratory engineers, Price and Green worked closely together under Wilbur B. Davenport, Jr., their laboratory supervisor and dissertation director. They worked on different aspects of NOMAC (NOise Modulation And Correlation), a high-frequency communication system (known by the Army Signal Corps production name F9C) that used pseudonoise sequences, and on Rake, a receiver that

4.   Weiss 29 September 1993; "Final Report," pp. 136 and 138; Overhage, "Reaching into Space," p. 2; Overhage to Wilson, 30 June 1961, 1/24/AC 134, MITA; Allen S. Richmond, "Background Information on Millstone Hill Radar of MIT Lincoln Laboratory," 5 November 1958, typed manuscript, LLLA; Weiss, *Space Radar Trackers and Radar Astronomy Systems*, JA-1740-22 (Lexington: Lincoln Laboratory, June 1961), pp. 21–23, 29, 44 and 64; Price, "The Venus Radar Experiment," in E. D. Johann, ed., *Data Handling Seminar, Aachen, Germany, September 21, 1959* (London: Pergamon Press, 1960), p. 81; Price, P. Green, Thomas J. Goblick, Jr., Robert H. Kingston, Leon G. Kraft, Jr., Gordon H. Pettengill, Roland Silver, William B. Smith, "Radar Echoes from Venus," *Science* 129 (1959): 753; "Missile Radar Probes Arctic," *Electronics* 30 (1957): 19; Pettengill 28 September 1993.

solved NOMAC multipath propagation problems. Later, what Lincoln Laboratory called NOMAC came to be called spread spectrum.

Their work was vital to maintaining military communications in the face of enemy jamming. One of their units went to Berlin in 1959 in anticipation of a blockade to provide essential communications in case of jamming. The Soviet Union already had demonstrated its jamming expertise against the Voice of America. Conceivably, all NATO communications could be jammed in time of war. The Lincoln Laboratory anti-jamming project was a direct response to that threat.[5]

Radio astronomy, which influenced the rise of planetary radar astronomy during the 1960s, played a small role in the Lincoln Laboratory Venus experiment. Price actually had worked at the University of Sydney under radio astronomer Gordon Stanley and met such pioneers as Pawsey, Taffy Bowen, Paul Wild, Bernie Mills, and Chris Christiansen. A recently published book on radio astronomy by the Australian scientists J. L. Pawsey and Ronald N. Bracewell was the subject of lunch conversation between Green and Price in the Lincoln Laboratory cafeteria. The chapter on radar astronomy predicted that one day man would bounce radar waves off the planets. But radio astronomy did not give rise to the decision to attempt a radar detection of Venus.[6]

What *did* trigger the decision was the completion of the Millstone facility. Green and Price wondered if it was powerful enough to bounce radar signals off Venus. Gordon Pettengill, a junior member of the team, joined the lunchtime discussions. Trained in physics at MIT and an alumnus of Los Alamos, Pettengill had an office at Millstone. After making calculations on a paper napkin, though, they estimated that Millstone did not have enough detectability for the experiment, even if one assumed that Venus was perfectly reflective.

The lunchtime conversazione went nowhere, until Robert H. Kingston, who had a joint MIT and Lincoln Laboratory appointment, joined the discussions. Kingston had just built a maser. "Within an hour," Green recalled, "we had the whole damn thing mapped out."[7] The maser gave the radar receiver the sensitivity necessary to carry out the experiment.

The maser, an acronym for Microwave Amplification by Stimulated Emission of Radiation, was a new type of solid-state microwave amplifying device vaunted by one author as "the greatest single technological step in radio physics for many years, with the possible exception of the transistor, comparable say with the development of the cavity magnetron during the Second World War." The maser was at the heart of the low-noise microwave amplifiers used in radio astronomy. The first radio-astronomy maser application, a joint effort by Columbia University and the Naval Research Laboratory, occurred in April 1958. The first use of a maser in radar astronomy, however, preceded that application by two months, in February 1958, at Millstone. While most masers

5.   William W. Ward, "The NOMAC and Rake Systems," *The Lincoln Laboratory Journal* vol. 5, no. 3 (1992): 351–365; Green 20 September 1993; Price 27 September 1993. Green and Price acknowledged each other in their dissertations. Green, "Correlation Detection using Stored Signals" D.Sc. diss., MIT, 1953, and Price, "Statistical Theory Applied to Communication through Multipath Disturbances," D.Sc. diss., MIT, 1953.

A history of the subject, R. A. Scholtz, "The Origins of Spread-Spectrum Communications," *IEEE Transactions on Communications* COM-30 (1982): 822–854, is reproduced in Marvin K. Simon, Jim K. Omura, Scholtz, and Barry K. Levitt, eds., *Spread Spectrum Communications* (Rockville, Md.: Computer Science Press, Inc., 1985), Volume 1, Chapter 2, "The Historical Origins of Spread-Spectrum Communications," pp. 39–134. Price, "Further Notes and Anecdotes on Spread-Spectrum Origins," *IEEE Transactions on Communications* COM-31 (January 1983): 85–97, provides an absorbing anecdotal sequel to Scholtz.

6.   Pawsey and Bracewell, *Radio Astronomy* (Oxford: Clarendon Press, 1955); Green 20 September 1993; Price 27 September 1993.

7.   Green 20 September 1993; Pettengill 28 September 1993. For a description of the maser, see Kingston, *A UHF Solid State Maser,* Group Report M35-79 (Lexington: Lincoln Laboratory, 1957); and Kingston, *A UHF Solid State Maser,* Group Report M35-84A (Lexington: Lincoln Laboratory, 1958).

functioned above 1,000 MHz, Kingston's operated in the UHF region, around 440 MHz, and reduced overall system noise temperature to an impressive 170 K.[8]

Despite the maser's low noise level, Price and Green knew that they would have to raise the level of the Venus echoes above that of the noise. Their NOMAC anti-jamming work had prepared them for this problem. They chose to integrate the return pulses over time, as Zoltán Bay had done in 1946. In theory, the signals buried in the noise reinforced each other through addition, while the noise averaged out by reason of its random nature.[9]

A digital computer, as well as additional digital data processing equipment, linked to the Millstone radar system performed the integration and analysis of the Venusian echoes. An analog-to-digital convertor, initially developed for ionospheric research by William B. Smith, digitized information on each radar echo. That information simultaneously was recorded on magnetic tape and fed to a solid-state digital computer. The experiment was innovative in digital-signal processing and marked one of the earliest uses of digital tape recorders.[10]

Venus or Bust

Kingston's maser was installed at Millstone Hill just in time for the inferior conjunction of Venus. However, a klystron failure left only 265 kilowatts of transmitter power available for the experiment. On 10 and 12 February 1958, the radar was pointed to detect Venus, then some 45 million kilometers (28 million miles) away. The radar signals took about five minutes to travel the round-trip distance. In contrast, John DeWitt's signals went to the Moon and back to Fort Monmouth, NJ, in only about 2.5 seconds.

Of the five runs made, only four of the digital recordings had few enough tape blemishes that they could be easily edited and run through the computer. Two of the four runs, one from each day, showed no evidence of radar returns. The others had one peak each. Price recalled, "When we saw the peaks, we felt very blessed."[11] It was not absolutely clear, however, that the two peaks were really echoes.

Green explained: "We looked into our soul about whether we dared to go public with this news. Bob was the only guy who really stayed with it to the end. He had convinced himself that he had seen it, and he had convinced me that he had seen it. Management asked us to have a consultant look at our results, and we did." Thomas Gold of Cornell University looked at the peaks and said "Yes, I think you should publish this." Green and Price then published their findings in the 20 March 1959 issue of *Science*, the journal of

8.   J. V. Jelley, "The Potentialities and Present Status of Masers and Parametric Amplifiers in Radio Astronomy," *Proceedings of the IEEE* 51 (1963): 31 and 36, esp. 30; J. W. Meyer, *The Solid State Maser—Principles, Applications, and Potential*, Technical Report ESD-TR-68-261 (Lexington: Lincoln Laboratory, 1960), pp. 14–16; J. A. Giordmaine, L. E. Alsop, C. H. Mayer, and C. H. Townes, "A Maser Amplifier for Radio Astronomy at X-band," *Proceedings of the IRE* 47 (1959): 1062–1070; Pettengill and Price, "Radar Echoes from Venus and a New Determination of the Solar Parallax," *Planetary and Space Science* 5 (1961): 73. For Townes and the invention of the maser, see Paul Forman, "Inventing the Maser in Postwar America," *Osiris* ser. 2, vol. 7 (1992): 105–134.

9.   Price, p. 70; Price et al, p. 751. Later, Price acknowledged the pioneering integration work of Zoltán Bay in 1946. Price, p. 73. Kerr, "On the Possibility of Obtaining Radar Echoes from the Sun and Planets," *Proceedings of the IRE* 40 (1952): 660–666, specifically recommended long-period integration for radar observation of Venus.

10.   Smith graduated MIT in 1955 with a master's degree in electrical engineering and worked with Price and Green on the F9C in Davenport's group. Smith 29 September 1993; Green 20 September 1993; Price 27 September 1993; Price, p. 72; Price et al, p. 751; Scholtz, p. 838; Weiss, *Space Radar Trackers*, pp. 53, 59, 61 and 63–64; "Biographical data, MIT Lincoln Laboratory," 18 March 1959, LLLA.

11.   Price 27 September 1993; Weiss, *Space Radar Trackers*, pp. 29 and 44; Price, pp. 71 and 76; Price et al, p. 751.

the American Association for the Advancement of Science, 13 months after their observations in February 1958.[12]

By then, despite the unsuccessful Lunik I Moon shot, the Soviet Union had achieved a number of successful satellite launches. The United States space effort still was marked by repeated failures. All of the four Pioneer Moon launches of 1958 ended in failure. There was a desperate need for good news; the Lincoln Laboratory publicity department gave the Venus radar experiment full treatment. In addition to a press conference, Green and Price quickly found themselves on national television and on the front page of the *New York Times*. President Eisenhower sent a special congratulatory telegram calling the experiment a "notable achievement in our peaceful ventures into outer space."[13]

Once Price and Green accepted the validity of the two peaks, the next step was to determine the distance the radar waves travelled to Venus and to calculate a value for the astronomical unit. They estimated a value of 149,467,000 kilometers and concluded, moreover, that it did not differ enough from those found in the astronomical literature to warrant a re-evaluation of the astronomical unit.[14]

The Lincoln Laboratory 1958 Venus experiment launched planetary radar astronomy; Millstone Hill was the prototype planetary radar. Its digital electronics, recording of data on magnetic tape for subsequent analysis, use of a maser (or other low-noise microwave amplifier) and a digital computer, and long-period integration all became standard equipment and practice. As with any experiment, scientists must be able to duplicate results. The next inferior conjunction provided an opportunity for scientists at Jodrell Bank to attempt Venus, too.

Jodrell Bank had a new, 76-meter (250-ft) radio telescope, the largest of its type in the world. Although planned as early as 1951, the telescope did not detect its first radio waves until 1957 as a consequence of a long, nightmarish struggle with financial and construction difficulties. The civilian Department of Scientific and Industrial Research and the Nuffield Foundation underwrote its design and construction. Success in detecting Soviet and American rocket launches brought visits from Prince Philip and Princess Margaret and fame. Fame in turn brought solvency and a name (the Nuffield Radio Astronomy Laboratories, Jodrell Bank).

Although the design and construction of the large dish was unquestionably an enterprise carried out with civilian funding, radar research at Jodrell Bank owed a debt to the United States armed forces; however, that military research was limited to meteor studies carried out with the smaller antennas, not the 76-meter (250-ft) dish. The U.S. Air Force and the Office of Naval Research supplied additional money for tracking rocket launches, while the European Office of the U.S. Air Force Research and Development Command (EOARDC) funded general electronics research at a modest level. During the Cuban missile crisis, the 76-meter (250-ft) radio telescope served to detect missiles that might be launched from the Soviet Union. From intelligence sources, the locations of such missiles directed against London were known, and the telescope was aimed accordingly. No U.S. equipment or funding were engaged in this effort, though.[15]

---

12.   Green 20 September 1993; Gold 14 December 1993; Price et al, pp. 751–753.

13.   Green 20 September 1993; Price 27 September 1993; Pettengill 28 September 1993; Overhage to Wilson, 24 March 1959, 1/24/AC 134, MITA; "Venus is Reached by Radar Signals," *New York Times*, vol. 108 (20 March 1959), pp. 1 and 11.

14.   For their calculation of the astronomical unit, see Pettengill and Price, "Radar Echoes from Venus and a New Determination of the Solar Parallax," *Planetary and Space Science* 5 (1961): 71–74.

15.   Lovell, 11 January 1994; Lovell, *Jodrell Bank*, passim, but especially pp. 220–222, 224, 242, 225. On the Foundation, see Ronald William Clark, *A Biography of the Nuffield Foundation* (London: Longman, 1972). Created in 1962, EOARDC was essentially a military operation headquartered in Brussels. It underwrote a wide range of European scientific research, though more money went into electronics research than any other field. Howard J. Lewis, "How our Air Force Supports Basic Research in Europe," *Science* 131 (1960): 15–20. From

Figure 5

The Jodrell Bank 250-foot (76-meter) telescope in June 1961. The control room is partially visible bottom left. The 1962 and 1964 Jodrell Bank Venus radar experiments were carried out using a U.S.-supplied continuous-wave radar mounted on this telescope. (Courtesy of the Director of the Nuffield Radio Astronomy Laboratories, Jodrell Bank.)

Preparation for the 1959 Venus experiment began in 1957, as the dish was reaching completion. The telescope, however, was not yet ready for radar work. John Evans recognized that its transmitter power and operating frequency would have to be raised in order to achieve critical extra gain for the Venus experiment. The 100-MHz (3-meter), 10-kilowatt Moon radar was not powerful enough. The University of Manchester Physics Department had developed a 400-MHz (75-cm), 100-kilowatt klystron. "It was a real kludge," Evans later recalled, "because it was basically a Physics Department experiment. It was continuously pumped; it sat on top of vacuum pumps, which required liquid nitrogen for cooling."[16]

Lovell had the General Electric Company of Britain supply a modulator for the klystron. Evans was responsible for designing and building the rest of the equipment. As the 1958 Venus inferior conjunction approached, "we simply were not ready, and Lovell was quite upset," Evans explained. Out of desperation, Evans employed the 100-MHz Moon radar enhanced with a computer integration scheme, but the equipment failed to detect echoes. When Lincoln Laboratory announced its success, Evans recalled, "We shrugged and felt we were beaten to the punch."

The 1958 Jodrell Bank failure put all that much more pressure on Evans to produce results during the next inferior conjunction of September 1959. The transmitter was more

---

August 1957, when Jodrell Bank began preliminary calibration measurements to August 1970, the telescope gathered results for 68,538 hours. Of those, 4,877 hours (7.1% of operational time) represented "miscellaneous use." Of that "miscellaneous use," 2,498 hours (3.6% of operational time) were directly concerned with the space programs of the United States and the Soviet Union. Lovell, *Out of the Zenith: Jodrell Bank, 1957–1970* (New York: Harper & Row, 1973), p. 2.

16.   Evans 9 September 1993.

or less ready. The klystron was mounted in one of the telescope towers. "It was a royal pain," Evans remembered, "because we had to take liquid nitrogen up the elevator and then a vertical ladder to get to this darn thing." As if that were not enough, a water pump burned up, and the connectors on the coaxial cable carrying power to the dish burned out every ten or fifteen minutes. While still struggling with the connector problem, Evans made several runs on Venus.

Evans was a junior scientist, having just received his Ph.D. in 1957. He felt he was under great pressure to produce positive results. Lovell was anxious to know if they had found an echo; the Duke of Edinburgh was about to visit. Evans looked at his data, taken from the first few minutes of each run, when he thought the apparatus was working. He had what looked like a return, but it could have been noise. Evans decided, "Well, I think we have an echo." The Venus detection was announced in the 31 October 1959 issue of *Nature.* The Duke of Edinburgh visited Jodrell Bank on 11 November 1959; he received an explanation and a demonstration of the technique, using the Moon as a target.

Despite the patchwork equipment, the 50-kilowatt, 408-MHz (74-cm) radar obtained a total of 58 and three quarters hours of useful operating data, before Venus passed beyond its range. As expected, none of the echoes were stronger than the receiver noise level; integration techniques increased the strength of the echoes.[17] The Jodrell Bank signal processing equipment was rather limited in its ability to search. Without accurate range or Doppler correction information, Evans had to make assumptions; he chose the Lincoln Laboratory 1958 published value. Not surprisingly, the value Jodrell Bank derived for the astronomical unit agreed with that determined at Lincoln Laboratory. The Jodrell Bank confirmation of the Lincoln Laboratory results placed them on solid scientific ground, that is, until Lincoln Laboratory repeated the experiment.

Fickle Venus

Bob Price and his fellow Lincoln Laboratory investigators were highly optimistic about verifying their 1958 results. Millstone now had a peak transmitter power of 500 kilowatts, almost twice the 1958 level. In addition to using a higher pulse repetition rate, which improved signal detectability, Price's team replaced the maser with a parametric amplifier. Like the maser, the parametric amplifier was a solid-state microwave amplifier. Parametric amplifiers were simpler, smaller, cheaper, and lighter than masers, and they did not require cryogenic fluids to keep them cool. Although masers generally were less noisy, the Millstone parametric amplifier was, Pettengill and Price reported, "gratifyingly stable and reliable in its operation."[18]

Over a four-week period around the inferior conjunction of Venus, the Lincoln Laboratory team made two types of radar observations. On 66 runs, they recorded the echoes digitally for subsequent computer processing, as they had done in 1958. The second approach, used on 117 runs, involved initial analog processing in a series of electronic circuits, followed by digitization and integration in real time by the site's computer. It was their first attempt at a real-time planetary detection by radar. Of all the runs, only one displayed a peak sufficiently above the noise level to be statistically significant. When subjected to detailed analysis, though, the peak turned out to be only noise. Price and

17.   Evans 9 September 1993; Jodrell Bank, *Moon and Venus Radar Passive Satellite Observations: Technical (Final) Report, October 1958–December 1960,* AFCRL Report 1129 (Macclesfield: Nuffield Radio Astronomy Laboratories, 1961), p. 22; Evans and G. N. Taylor, "Radio Echo Observations of Venus," *Nature* 184 (1959): 1358–1359; Lovell, *Out of the Zenith,* p. 193. The noise figure was 4.6 db. The frequency of the lunar radar was lowered from 120 MHz to 100 MHz, when it was found to interfere with operations at nearby Manchester Airport.

18.   Pettengill and Price, p. 73.

Pettengill concluded that "none of the individual runs show strong evidence of Venus echoes."[19]

Jodrell Bank had corroborated the 1958 results; yet with an improved radar, Lincoln Laboratory could not confirm them. The disparity between the results was perplexing—and bothersome. "It is difficult to explain the disparity between the results obtained at the two Venus conjunctions. Our current feeling," wrote Green and Pettengill, "is that the planet's reflectivity may be highly variable with time, and that the two successes in 1958 were observations made on very favorable occasions."[20]

At the Jet Propulsion Laboratory (JPL), the Lincoln Laboratory and Jodrell Bank experiments were viewed with disbelief. As an internal report stated in 1961, "It is not known at the present time with certainty that a radio signal has ever been reflected from the surface of Venus and successfully detected."[21] JPL investigators intended to obtain the first unambiguous detection of radar echoes from the Venusian surface.

The Jet Propulsion Laboratory

JPL began modestly in Pasadena, California, in 1936 as the Guggenheim Aeronautical Laboratory, California Institute of Technology (GALCIT), rocket project, led by Hungarian-born professor Theodore von Kàrmàn and financed by Harry Guggenheim. Starting in 1940, with backing from the Army Air Corps, the GALCIT group turned into a vital rocket research, development, and testing facility. A 1944 contract signed by GALCIT, the Army Air Force, and the California Institute of Technology (Caltech) transformed it into a large permanent laboratory called the Jet Propulsion Laboratory, whose major responsibility was research, development, and testing of missile technology, including the country's first tactical nuclear missiles, the Corporal and Sergeant, for the Army.

JPL electronics arose out of the need for missile guidance and tracking systems. William Pickering, a Caltech electrical engineering professor with a Ph.D. in physics, became the director of JPL in 1954 and remained in that position until 1976. His specialization was electronics, not propulsion. Under Pickering's aegis, electronics grew in prominence at JPL and came to the forefront in 1958, when JPL became a NASA laboratory and started work on a worldwide, civilian satellite communications network known today as the Deep Space Network (DSN).[22]

The communications network, known originally as the Deep Space Instrumentation Facility (DSIF), was the home of planetary radar at JPL. The three leaders of the Venus radar experiment were engineers involved in its design, Eberhardt Rechtin, Robertson Stevens, and Walter K. Victor. Rechtin, the architect of the DSIF, had a Ph.D. in electrical engineering from Caltech. He also was an inventor, with Richard Jaffe (also at JPL), of CODORAC (COded DOppler, Ranging, And Command), a radio communication system

---

19.   Pettengill and Price, p. 73; Green and Pettengill, "Exploring the Solar System by Radar," *Sky and Telescope* 20 (1960): 12–13; Jelley, pp. 30 and 35. During the 1959 Lincoln Laboratory Venus experiment, over 150 runs were made, yet no echoes as strong as those of 1958 were observed. Overall system noise temperature rose from 170 Kelvins in 1958 to 185 Kelvins with the parametric amplifier. For a discussion of parametric amplifiers, see Karl Heinz Locherer, *Parametric Electronics: An Introduction* (New York: Springer-Verlag, 1981), pp. 276–286.

20.   Green and Pettengill, p. 13.

21.   JPL, *Research Summary No. 36–7, Volume 1, for the period December 1, 1960 to February 1, 1961* (Pasadena: JPL, 1961), pp. 68 and 70.

22.   "Jet" was a broader term than rocket and avoided any stigma still attached to that word. Clayton R. Koppes, *JPL and the American Space Program: A History of the Jet Propulsion Laboratory* (New Haven: Yale University Press, 1982), pp. ix, 4–5, 10–17, 20, 38, 45 and 65.

that detected and tracked narrow band signals in the presence of wideband noise. CODORAC, whose electronics in many ways resembled Lincoln Laboratory's NOMAC, became the basis for much of the DSIF's electronics. Bob Stevens had an M.S. in electrical engineering from the University of California at Berkeley, and Walt Victor, who assisted Rechtin in developing CODORAC, had a B.S. in mechanical engineering from the University of Texas.

JPL located its share of the DSIF antennas in the Mojave Desert, about 160 kilometers from JPL, on the Fort Irwin firing range near Goldstone Dry Lake, where GALCIT earlier had tested Army rockets.[23] The two antennas on which JPL investigators performed their Venus experiment in 1961 were artifacts of the funding and research agendas of both the military and NASA. The first was a 26-meter-diameter (85-feet-diameter) dish named the HA-DEC antenna, because its axes were arranged to measure angles in terms of local hour angle (HA) and declination (DEC). JPL installed it at Goldstone during the second half of 1958 to track and receive telemetry from the military's Pioneer probes.[24]

Figure 6

JPL Goldstone 26-meter HA-DEC antenna erected in late 1958 to track and receive telemetry from the military's Pioneer probes. It was used with the 26-meter AZ-EL antenna to detect radar echoes from Venus in 1961. (Courtesy of Jet Propulsion Laboratory, photo no. 333-5968AC.)

23. Rechtin, telephone conversation with author, 13 September 1993; Stevens 14 September 1993; Nicholas A. Renzetti, ed., *A History of the Deep Space Network from Inception to January 1, 1969*, vol. 1, Technical Report 32–1533 (Pasadena: JPL, 1 September 1971), pp. 6–7 and 11; William R. Corliss, *A History of the Deep Space Network*, CR-151915 (Washington: NASA, 1976), pp. 3–4 and 16; Craig B. Waff, "The Road to the Deep Space Network," *IEEE Spectrum* (April 1993): 53; Scholtz, pp. 841–843; additional background material supplied from oral history collection, JPLA.

24. Dish diameters have been expressed in meters only recently. Initially, they were measured in feet. For the sake of consistency, diameters are given in both feet and meters throughout the text. Victor, "General System Description," p. 6 in Victor, Stevens, and Solomon W. Golomb, eds., *Radar Exploration of Venus: Goldstone Observatory Report for March–May 1961*, Technical Report No. 32–132 (Pasadena: JPL, 1961); Corliss, *Deep Space Network*, pp. 16–17 and 20–25.

JPL erected the second antenna for Project Echo. Echo, a large balloon in Earth orbit, tested the feasibility of long-range satellite communications. As such, it was heir to the lunar-repeater communication tests discussed in Chapter One. Originally funded by NASA's predecessor, the National Advisory Committee for Aeronautics (NACA), and the Defense Department's space research organization, the Advanced Research Projects Agency (ARPA), Project Echo became a JPL, NASA, and Bell Telephone Laboratories undertaking in an agreement signed in January 1959.

The Echo experiments used the existing HA-DEC antenna to receive as part of a satellite circuit running from east to west. The west-to-east circuit, however, required the construction of an antenna capable of transmitting. Therefore, JPL installed a second 26-meter-diameter (85-feet-diameter) dish at Goldstone about a year after the HA-DEC antenna for Project Echo. The axes of the second antenna measured angles in terms of azimuth (AZ) and elevation (EL); hence, it was referred to as the AZ-EL antenna.[25]

Figure 7
Jet Propulsion Laboratory Goldstone 26-meter AZ-EL antenna built for Project Echo and used with the 26-meter HA-DEC antenna to detect echoes from Venus in 1961. (Courtesy of Jet Propulsion Laboratory, photo no. 332-168.)

25. Victor, "General System Description," in Victor, Stevens, and Golomb, p. 6; Corliss, *Deep Space Network*, pp. 25–27; Donald C. Elder, III, "Out From Behind the Eight Ball: Echo I and the Emergence of the American Space Program, 1957–1960," Ph.D. diss., University of California at San Diego, 1989, passim. For a history of ARPA, see Richard J. Barber Associates, Inc., *The Advanced Research Projects Agency, 1958–1974* (Washington, D.C.: National Technical Information Service, 1975). For the story of JPL and Project Echo, see Stevens and Victor, eds., *The Goldstone Station Communications and Tracking System for Project Echo*, Technical Report 32–59 (Pasadena: JPL, 1960); Victor and Stevens, "The Role of the Jet Propulsion Laboratory in Project Echo," *IRE Transactions on Space Electronics and Telemetry* SET-7 (1961): 20–28.

By August 1960, as Goldstone prepared to participate in Project Echo, the Lincoln Laboratory and Jodrell Bank Venus experiments already had taken place. Solomon Golomb, assistant chief of the Communications System Research Section under Walt Victor, asked his employee, Richard Goldstein, to design a space experiment to feed the rivalry between Eb Rechtin, JPL program director for the DSIF, and Al Hibbs, who was in charge of space science at JPL. Goldstein suggested the Venus radar experiment. Victor, JPL project engineer for the Echo program and recently promoted to chief of the Communications System Research Section, and Bob Stevens, head of the Communications Elements Research Section, became the project managers.[26]

Rechtin, Victor, and Stevens organized the Venus experiment as a drill of the DSIF and its technical staff. The functional, organizational, and budgetary status of planetary radar astronomy as a test of the DSIF originated in their conception of the 1961 Venus experiment and defined planetary radar at JPL for over two decades. At the time, the laboratory was preparing for the first Mariner missions. Consequently, as Rechtin pointed out, JPL had "a particular interest in an accurate determination of the distance to Venus in order that we might guide our space probes to that target."[27]

The NASA Office of Space Science approved the Mariner 1 and 2 missions in July 1960. Goldstone was to provide communications with them. The task would be more challenging than communicating with a Ranger Moon probe. While a Ranger mission required three days, the Mariner missions would involve months of round-the-clock, high-level technical performance. In June 1960, even before final approval of the Mariner probes, Rechtin proposed the radar experiment to NASA, emphasizing not its scientific value, but the "practical, purely project point of view."[28]

In order to perform the Venus experiment, JPL had to modify the Echo equipment. Venus was a much farther object than the Earth-orbiting Echo balloon, and both differed radically as radar targets. Victor and Stevens, moreover, wanted to avoid long-term integration and after-the-fact data reduction and analysis, that is, the Lincoln Laboratory and Jodrell Bank approach. Instead, JPL attempted a real-time radar detection of Venus.

The JPL antennas were unlike those of Lincoln Laboratory and Jodrell Bank in many ways. They operated in tandem, the AZ-EL transmitting and the HA-DEC receiving. This bistatic mode, as it is called, offered advantages over the Millstone and Jodrell Bank monostatic mode, in which a single instrument both sent and received. Monostatic radars have to stop transmitting half the time in order to receive, while bistatic radars can operate continuously, gathering twice the data in the same period of time. The Goldstone radars also operated at a higher frequency (S-band v. UHF) and sent a continuous wave, whereas the Lincoln Laboratory and Jodrell Bank radars transmitted discrete pulses.

JPL also boosted the transmitting power and receiver sensitivity of the two radars. The normal output of the AZ-EL transmitter klystron tube was 10 kilowatts at 2388 MHz (12.6 cm), but engineers coaxed a nominal average power output of 13 kilowatts out of it.

26.   Golomb, "The First Touch of Venus," paper presented at the Symposium Celebrating the Thirtieth Anniversary of Planetary Radar Astronomy, Pasadena, October 1991, Renzetti materials; Goldstein 7 April 1993; Goldstein 14 September 1993; Goldstein 19 September 1991; Stevens 14 September 1993; biographical material and JPL Press Release, 23 May 1961, 3–15, Historical File, JPLA.

27.   Rechtin, "Informal Remarks on the Venus Radar Experiment," in Armin J. Deutsch and Wolfgang B. Klemperer, eds., *Space Age Astronomy* (New York: Academic Press, 1962), p. 365; Golomb, "Introduction," in Victor, Stevens, and Golomb, pp. 1–2; Rechtin, telephone conversation, 13 September 1993; Goldstein 19 September 1991.

28.   Golomb, "Introduction," p. 1; JPL, *Research Summary No. 36–7*, p. 70; Rechtin, telephone conversation, 13 September 1993; Waff, "A History of the Deep Space Network," manuscript furnished to author, ch. 6, pp. 22 and 24. Because the manuscript is not paginated sequentially, both chapter and page references are provided.

Raising the sensitivity of the HA-DEC receiver was a daunting challenge; the total receiver system noise temperature on Project Echo had been 1570 K![29]

The technical solution was a maser and a parametric amplifier in tandem on the HA-DEC antenna. Charles T. Stelzried and Takoshi Sato created a 2388-MHz maser specifically for the Venus radar experiment and suitable for Goldstone's tough desert ambient temperatures (from -12° to 43°C; 10° to 110°F) and climate (rain, dust, and snow). The maser and 2388-MHz parametric amplifier combined gave an overall average system noise temperature of about 64 K during the two months of the Venus experiment, considerably lower than the best achieved at Millstone in 1958 (170 K). As Victor and Stevens proclaimed, "This is believed to be the most sensitive operational receiving system in the world."[30]

"No Echo, No Thesis"

Besides testing the personnel and materiel of the Goldstone facility, the JPL Venus experiment also was the doctoral thesis topic of two employees in Walt Victor's section, Duane Muhleman and Richard Goldstein. Muhleman graduated from the University of Toledo with a BS in physics in 1953, then worked two years at the NACA Edwards Air Force Base High-Speed Flight Station as an aeronautical research engineer, before joining JPL. As part of his duties at JPL, Muhleman tested the Venus radar system and its components during January, February, and March 1961, using the Moon as a target. For the Venus experiment, Muhleman contributed an instrument to measure Doppler spreading.[31]

Goldstein was a Caltech graduate student in electrical engineering. His task on the Venus radar experiment was to build a spectrum measuring instrument. It recorded what the spectrum looked like during reception of an echo and what it looked like when the receiver saw only noise. JPL hired his brother, Samuel Goldstein, a JPL alumnus and radio astronomer at Harvard College Observatory, as a consultant on the Venus experiment; Samuel also helped his brother with some of the radio techniques.

Dick Goldstein wanted to use the Venus radar experiment as his thesis topic at Caltech, but his advisor, Hardy Martel, was highly skeptical. The inability of Lincoln Laboratory to detect Venus was widely known. Although he thought the task indisputably impossible, Martel finally agreed to accept the topic, but with a firm admonition: "No echo, no thesis."[32]

---

29.   Rechtin, p. 366; Victor, "General System Description," pp. 6-7; Stevens and Victor, "Summary and Conclusions," p. 95; Victor and Stevens, "The 1961 JPL Venus Radar Experiment," *IRE Transactions on Space Electronics and Telemetry* SET-8 (1962): 85–90; Charles T. Stelzried, "System Capability and Critical Components: System Temperature Results," in Victor, Stevens, and Golomb, pp. 28–29. For a general description of the radar system, see M. H. Brockman, Leonard R. Malling, and H. R. Buchanan, "Venus Radar Experiment," in JPL, *Research Summary No. 36–8, Volume 1, for the period February 1, 1961 to April 1, 1961* (Pasadena: JPL, 1961), pp. 65–73; Victor and Stevens, "Exploration of Venus by Radar," *Science* 134 (1961): 46. The Jodrell Bank transmitter had a peak power of 50 kilowatts; Millstone's peak power was 265 kilowatts in 1958 and 500 kilowatts in 1959. However, comparing the peak power ratings of pulse and continuous-wave radars is the electronic equivalent of comparing apples and oranges. One must compare their average power outputs.

30.   Stevens and Victor, "Summary and Conclusions," p. 95; Sato, "System Capability and Critical Components: Maser Amplifier," in Victor, Stevens, and Golomb, p. 17; Stelzried, "System Capability and Critical Components: System Temperature Results," pp. 28–29; H. R. Buchanan, "System Capability and Critical Components: Parametric Amplifier," in Victor, Stevens, and Golomb, pp. 22–25; Walter H. Higa, *A Maser System for Radar Astronomy*, Technical Report 32–103 (Pasadena: JPL, 1961); Higa, "A Maser System for Radar Astronomy," in K. Endresen, *Low Noise Electronics* (New York: Pergamon Press, 1962), pp. 296–304.

31.   Muhleman 8 April 1993; Muhleman 19 May 1994; Muhleman 27 May 1994; Goldstein 19 September 1991; Stevens 14 September 1993; Golomb, "Introduction," p. 3; Stevens, "Additional Experiments: Resume," in Victor, Stevens, and Golomb, p. 70. Muhleman's dissertation was "Radar Investigations of Venus," Ph.D. diss., Harvard University, 1963.

32.   Goldstein 7 April 1993; Goldstein 19 September 1991; Goldstein 14 September 1993.

On 10 March 1961, a month before inferior conjunction, the Goldstone radars were pointed at Venus. The first signals completed the round-trip of 113 million kilometers in about six and a half minutes. During the 68 seconds of electronic signal integration time, 1 of 7 recording styluses on Goldstein's instrument deviated significantly from its zero level and remained at the new level.

To verify that the deflection came from Venus and was not leakage from the transmitter or an instability in the receiver, the transmitter antenna was deliberately allowed to drift off target. Six and a half minutes later, the recording stylus on Goldstein's instrument returned to its zero setting. The experiment was immediately repeated with the same result. JPL had achieved the first real-time detection of a radar signal from Venus. And Dick Goldstein had his dissertation topic.[33]

On 16 March, Eb Rechtin telexed Paul Green: "HAVE BEEN OBTAINING REAL TIME RADAR REFLECTED SIGNALS FROM VENUS SINCE MARCH 10 USING 10 KW CW AT 2388 MC AT A SYSTEM TEMPERATURE OF 55 DEGREES." The following day, Green, John Evans (then at Lincoln Laboratory), Pettengill, and Price telexed back: "HEARTIEST CONGRATULATIONS ON YOUR SUCCESS WITH THE FICKLE LADY. MILLSTONE IS ON WITH THE USUAL MODE OF OPERATION BUT HAS HAD NO SUCH LUCK AS YET. PRESENT PARAMETERS 2.4 MEGAWATTS PEAK FOR 2 MILLISECONDS EVERY 33 MILLISECONDS 190 DEGREES KELVIN."[34]

Following the initial contact, JPL conducted additional radar experiments almost daily from 10 March to 10 May 1961, collecting 238 hours of recorded radar data about Venus.[35] No previous Venus radar experiment, nor any others carried out in 1961, collected as many hours of data as the JPL experiment.

The JPL experiment succeeded, because it did not depend on knowing the range to Venus, specifically; it did not depend on prior knowledge of the precise value of the astronomical unit. On the other hand, Lincoln Laboratory, as well as Jodrell Bank, had based its experiment on an assumed, yet commonly accepted, value for the astronomical unit, and, consequently, for the distance between Earth and Venus during inferior conjunction.

"We Were Wrong."

The results obtained by Lincoln and other laboratories in 1961 agreed with those obtained by JPL. That agreement led Gordon Pettengill to discern the error of the 1958 Lincoln Laboratory observations. "In view of the generally excellent agreement among the various observations made at several wavelengths [in 1961]," Pettengill and his colleagues concluded, "it seems likely that the results reported from observations of the 1958 inferior conjunction are in error, although no explanation has been found."[36]

Green recalled: "It was sort of devastating, when the next conjunction of Venus came around, and we learned that we were wrong. We had the wrong value of the astronomical unit. It wasn't over here; it was way over there someplace. In fact, it wasn't even easy to go back and look at the original data and conclude that it was really over there. The original

33. JPL Press Release, 23 May 1961, 3-15, Historical File, JPLA; Malling and Golomb, "Radar Measurements of the Planet Venus," *Journal of the British Institution of Radio Engineers* 22 (1961): 298; Victor and Stevens, "The 1961 JPL Venus Radar Experiment," *IRE Transactions on Space Electronics and Telemetry* SET-8 (1962): 90–91. Goldstein's dissertation was "Radar Exploration of Venus," Ph.D. diss., California Institute of Technology, 1962.

34. 3–15, Historical File, JPLA.

35. Victor and Stevens, "1961 JPL Venus Radar Experiment," p. 91.

36. Pettengill, Briscoe, Evans, Gehrels, Hyde, Kraft, Price, and Smith, "A Radar Investigation of Venus," *The Astronomical Journal* 67 (1962): 186.

data just had turned out to be too noisy....It was a chastening experience for us."[37] Price remembered someone entering his office with "a rather long look on his face" and saying, "Bob, I think we've been found to be wrong." It was an embarrassing moment.

Price re-examined the Lincoln Laboratory 1958 tapes. "I wanted to be sure that we hadn't detected it. I really mean that. I wanted to make sure that we had a negative result and that by accident we didn't have two wrongs making a right, that is, false processing of the 1958 data led to a false result, so the proper processing of the 1958 data would agree with JPL. I wanted to prove that that was *not* the case. So I went back and found the peaks, just as I had done before. I made a meticulous measurement of their position, which is the whole thing that the false echo hinged on. I developed with magnetic powder over and over again those tapes, and I inspected them until my eyes were sore. I reran the Fortran programs and checked all the programs, because you could create a timing error in the program."

The experience reminded Price of his work in Australia. Every day, his group had made ink-pen recordings of the radio sky over the antenna, usually recording only random lines, but a peak appeared on two successive days. Did the peak mean a detection of deuterium? They decided that it was a fluke and published their negative results. "If we had behaved the same way at Millstone," Price reflected, "we might have saved ourselves some embarrassment. But that is hindsight." The two Venus pulses arrived 2.2 milliseconds apart. "We just turned our back on it," Price admitted, "did a little wishful thinking, and said, 'That's the same pulse.'...I just pulled them together, ignored the 2.2-millisecond difference, and sat one on top of the other."[38]

Whatever the cause of the 1958 false readings, JPL was unquestionably the first to detect radar waves reflected off Venus. The literature contains two earlier, but after-the-fact detections. Only months after acknowledging JPL's priority, Lincoln Laboratory found on their data tapes a detection of Venus on 6 March 1961, a few days prior to that of JPL. Later, in 1963, Lincoln Laboratory electrical engineer Bill Smith re-examined the 1959 data tapes and found that an echo had been recorded on 14 September 1959.[39] Such after-the-fact discoveries are not uncommon in the history of science, and radar astronomers from both JPL and MIT thirty years later commemorated JPL's uncontested priority in detecting radar waves reflected off Venus.

Once JPL unambiguously detected echoes from Venus, the key question planetary radar astronomers addressed was the size of the astronomical unit. In order to determine more precisely the Earth-to-Venus distance, JPL ran ranging experiments between 18 April and 5 May 1961. In the July 1961 issue of *Science*, Victor and Stevens announced a preliminary value for the astronomical unit of 149,599,000 kilometers with an accuracy of ± 1500 kilometers.[40] That value was over 100,000 kilometers larger than the false radar value determined by Lincoln Laboratory in 1958 and confirmed by Jodrell Bank in 1959, 149,467,000 kilometers. Values obtained from preliminary analyses of radar data at Lincoln Laboratory and elsewhere in 1961 agreed closely with that of JPL (Table 1).

When Lincoln Laboratory undertook its 1961 Venus radar experiment, Gordon Pettengill, joined by John Evans, took over Bob Price's leadership role. Evans had left Jodrell Bank for Lincoln Laboratory during the previous summer, after being courted by the National Bureau of Standards and Stanford. At Jodrell Bank, Evans had had one

---

37.   Green 20 September 1993.
38.   Price 27 September 1993.
39.   Smith 29 September 1993; Smith, "Radar Observations of Venus, 1961 and 1959," *The Astronomical Journal* 68 (1963): 17; Pettengill et al, "A Radar Investigation of Venus," p. 183.
40.   Rechtin, p. 367; Victor, "General System Description," p. 7; Victor and Stevens, "1961 JPL Venus Radar Experiment," p. 88; Victor and Stevens, "Exploration of Venus by Radar," p. 46.

Table 1
Radar Values for the Astronomical Unit, 1961–1964

	Error of Measurement (in kilometers)	Value of Astronomical Unit (in kilometers)
Optical Values		
Spencer Jones	±17,000	149,675,000
Eugene Rabe	±10,000	149,530,000
1961 Conjunction		
Jet Propulsion Laboratory		
July 1961 (1)	±1,500	149,599,000
August 1961 (2)	±500	149,598,500
Muhleman (3)	±250	149,598,845
Lincoln Laboratory		
May 1961 (4)	±1,500	149,597,700
Corrected value (5)	±400	149,597,850
Jodrell Bank (6)	±5,000	149,601,000
RCA/Flower and Cook Observatory (7)	±200	149,596,000
Soviet Union		
Pravda value (8)	±130,000P	149,457,000
November 1961 (9)	±3,300	149,598,000
Revised Value (10)	±2,000	149,599,300
Space Technology Laboratories (11)	±13,700	149,544,360
1962 Conjunction		
Jodrell Bank (12)	±900	149,596,600
Soviet Union (13)	±270	149,597,900
Jet Propulsion Laboratory Muhleman (14)	±670	149,598,900
1964 Conjunction		
Lincoln Laboratory (15)	±100	149,598,000
Jet Propulsion Laboratory (16)	±100	149,598,000
Soviet Union (17)	±400	149,598,000
IAU Value		149,600,000

Sources

1. W.K. Victor and R. Stevens, "Exploration of Venus by Radar," *Science* 134 (July 1961): 46–48.
2. D.O. Muhleman, D.B. Holdridge, and N. Block, "Determination of the Astronomical Unit from Velocity, Range, and Integrated Velocity Data, and the Venus-Earth Ephemeris," pp. 83–92 in W.K. Victor, R. Stevens, and S.W. Golomb, eds., *Radar Exploration of Venus: Goldstone Observatory Report for March–May 1961*, Technical Report 32–132 (Pasadena: Jet Propulsion Laboratory, 1 August 1961).
3. D.O. Muhleman, D.B. Holdridge, and N. Block, "The Astronomical Unit Determined by Radar Reflections from Venus," *The Astronomical Journal* 67 (1962): 191–203.
4. Staff, Millstone Radar Observatory, Lincoln Laboratory, "The Scale of the Solar System," *Nature* 190 (13 May 1961): 592.
5. G.H. Pettengill, H.W. Briscoe, J.V. Evans, E. Gehrels, G.M. Hyde, L.G. Kraft, R. Price, and W.B. Smith, "A Radar Investigation of Venus," *The Astronomical Journal* 67 (1962): 181–190.
6. J.H. Thomson, J.E.B. Ponsonby, G.N. Taylor, and R.S. Roger, "A New Determination of the Solar Parallax by Means of Radar Echoes from Venus," *Nature* 190 (1961): 519–520.
7. I. Maron, G. Luchak, and W. Blitzstein, "Radar Observation of Venus," *Science* 134 (1961): 1419–1421.
8. V.A. Kotelnikov, "Radar Contact with Venus," *Journal of the British Institution of Radio Engineers* 22 (1961): 293–295.
9. V.A. Kotelnikov, V.M. Dubrovin, V.A. Morozov, G.M. Petrov, O.N. Rzhiga, Z.G. Trunova, and A.M. Shakhovoskoy, "Results of Radar Contact with Venus in 1961," *Radio Engineering and Electronics Physics* 11 (November 1961): 1722–1733.
10. V.A. Kotelnikov, B.A. Dubinskii, M.D. Kislik, and D.M. Tsvetkov, "Refinement of the Astronomical Unit on the Basis of the Results of Radar Observations of the Planet Venus in 1961," NASA TT F-8532, October 1963.
11. J.B. McGuire, E.R. Spangler, and L. Wong, "The Size of the Solar System," *Scientific American* vol. 204, no. 4 (1961): 64–72.
12. J.E.B. Ponsonby, J.H. Thomson, and K.S. Imrie, "Radar Observations of Venus and a Determination of the Astronomical Unit," *Monthly Notices of the Royal Astronomical Society* 128 (1964): 1–17.
13. V.A. Kotelnikov, V.M. Dubrovin, V.A. Dubinskii, M.D. Kislik, B.I. Kusnetsov, I.V. Lishin, V.A. Morosov, G.M. Petrov, O.N. Rzhiga, G.A. Sytsko, and A.M. Shakhovskoi, "Radar Observations of Venus in the Soviet Union in 1962," *Soviet Physics-Doklady* 8 (1964): 642–645.
14. D.O. Muhleman, *Relationship Between the system of Astronomical Constants and the Radar determinations of the Astronomical Unit*, Technical Report 32–477 (Pasadena: Jet Propulsion Laboratory, 15 January 1964).
15. J.C. Pecker, ed., Proceedings of the Twelfth General Assembly (New York: Academic Press, 1966), p. 602.
16. J.C. Pecker, ed., Proceedings of the Twelfth General Assembly (New York: Academic Press, 1966), p. 603.
17. V.A. Kotelnikov, Yu. N. Aleksandrov, L.V. Apraksin, V.M. Dubrovin, M.D. Kislik, B.I. Kuznetsov, G.M. Petrov, O.N. Rzhiga, A.V. Frantsesson, and A.M. Shakhovskoi, "Radar Observations of Venus in the Soviet Union in 1964," *Soviet Physics-Doklady* 10 (1966): 578–580.

technical assistant; but at Lincoln Laboratory, as Bernard Lovell pointed out, he had "an army of engineers and technicians together with a transmitter vastly superior to the one at Jodrell Bank."

Evans' departure from Jodrell Bank could not have come at a worse time, in the opinion of Lovell. "For me it was the beginning of a distressing series of losses of the brilliant young men who had been with me throughout the crisis of the telescope and whose devotion and skill had been a determining factor in the immediate success of the instrument. But who could expect a young man to resist a lavish red carpet reception and an offer of a salary many times greater than any sum which we could possibly offer him?"[41]

During the 1961 Venus experiment, the Millstone Hill radar ran at peak transmitting power, 2.5 megawatts. The increased transmitter power overcame the higher overall receiver noise temperature (240 K) to make the telescope a far more capable instrument. Pettengill and his colleagues aimed their radar at Venus on 6 March 1961, again using a technique to provide real-time detection. No echoes appeared until 24 March. Preliminary analysis yielded a value for the astronomical unit of 149,597,700 ± 1,500 kilometers in May 1961.[42] That agreed closely with JPL's preliminary value, 149,599,000 kilometers. Despite considerable obstacles, and chastened by their 1959 false detection, Jodrell Bank investigators also found a value for the astronomical unit that agreed with the JPL value.

In 1959, John H. Thomson took over the planetary radar program, and in the autumn of 1960, Lovell added John E. B. Ponsonby, who had come to Jodrell Bank to work on a doctorate after graduating in electrical engineering from Imperial College, London. Ponsonby had experience in meteor radar through his high school teacher and one-time member of the Jodrell Bank group, Ian C. Browne.[43]

Working from notes and memoranda left by Evans, the new team, which included G. N. Taylor and R. S. Roger, put together a radar system that "yielded a clear-cut and decisive answer after only a few 5 minute integration periods."[44] The first thing they did, however, was to abandon the atrocious klystron. With most of the problems that plagued the 1959 experiment overcome, with a more sensitive receiver, and with peak power output boosted from 50 to 60 kilowatts, the 76-meter (250-ft) Jodrell Bank telescope detected Venus beginning 8 April 1961, a few weeks after both JPL and Lincoln Laboratory had started their experiments, and ending 25 April 1961.

Jodrell Bank calculated a value for the astronomical unit, 149,600,000 ± 5000 kilometers,[45] close to the preliminary values of JPL (149,599,000 kilometers) and Lincoln

---

41.   Lovell, *Out of the Zenith*, pp. 192 and 195; Evans 9 September 1993; Green 20 September 1993; Smith 29 September 1993; Pettengill 28 September 1993.

42.   The Staff, Millstone Radar Observatory, Lincoln Laboratory, "The Scale of the Solar System," *Nature* 190 (1961): 592; Pettengill et al, "A Radar Investigation of Venus," pp. 182–183; Pettengill and Price, p. 73; Pettengill, "Radar Measurements of Venus," in Wolfgang Priester, ed., *Space Research III, Proceedings of the Third International Space Science Symposium* (New York: Interscience Publishers Division, John Wiley and Sons, 1963), p. 874; Overhage to Wilson, 22 May 1961, 1/24/AC 134, MITA.

43.   Ponsonby 11 January 1994; I. C. Browne and T. R. Kaiser, "The Radio Echo from the Head of Meteor Trails," *Journal of Atmospheric and Terrestrial Physics* 4 (1953): 1–4.

44.   Evans 9 September 1993; Lovell, *Out of the Zenith*, pp. 198–199; Thomson, Ponsonby, Taylor, and Roger, "A New Determination of the Solar Parallax by Means of Radar Echoes from Venus," *Nature* 190 (1961): 519–520. The Jodrell Bank experiment was funded by Air Force contract no. AF61(052)-172. John Evans, then of Lincoln Laboratory, privately had communicated the laboratory's results to Thomson at Jodrell Bank.

45.   I have calculated this value from the information provided in Thomson, Ponsonby, Taylor, and Roger, pp. 519–520. While the authors concern themselves with the solar parallax, they also provide a figure for the light-time of the astronomical unit, 499,011 ±0.017 seconds, which represents the time taken by radar waves to travel the distance of one astronomical unit, and another for the speed of light, 299,792.5 kilometers per second, which is the same as the speed of electromagnetic waves. By multiplying the two figures, I obtained a product of 149,599,750 kilometers.

The first published value of the astronomical unit I have found was in the comments given by Thomson following a presentation by Malling and Golomb at a convention in Oxford that took place 5–8 July 1961. The date of publication was October 1961. Malling and Golomb, p. 302.

Laboratory (149,597,700 kilometers), but with a far greater possible error of measurement. Similar results came from an unexpected source. RCA's Missile and Surface Radar Division in Moorestown, New Jersey, carried out its first and last planetary radar experiment in 1961. The Division performed radar research for the Army Signal Corps and the Navy, and in 1960, the Division performed solar radio experiments using a missile-tracking radar. On their Venus radar experiment, RCA investigators collaborated with the Flower and Cook Observatory of the University of Pennsylvania. Between 12 March and 8 April 1961, RCA tracked Venus with a BMEWS experimental radar in order to measure the astronomical unit. In over six hours of transmitted signals, they found only four peaks from which they calculated a value for the astronomical unit of 149,596,000 ± 200 kilometers,[46] only 3,000 kilometers less than the JPL value. Not all Venus radar results agreed with those of JPL, however.

In the Soviet Union, planetary radar was fundamental to the space program. One of the main objectives of the Crimean Venus experiment was to calculate a more precise value for the astronomical unit for use in launching planetary probes. The calculation of the orbit of the Mars-1 probe, in November 1962, utilized a radar-based value for the astronomical unit. The Institute of Radio Engineering and Electronics (IREE) of the U.S.S.R. Academy of Sciences, in association with other unnamed (but presumably military and intelligence) organizations and under the direction of Vladimir A. Kotelnikov, of the Soviet Academy of Sciences, designed and built planetary radar equipment that was installed at the Long-Distance Space Communication Center, located near Yevpatoriya in the Crimea. The IREE installation had nothing to do with the radar work carried out in the Soviet Union in 1946 on meteors or between 1954 and 1957 on the Moon.

The IREE planetary radar was a monostatic pulse 700-MHz (43-cm) system. For the receiver, the IREE expressly designed both a parametric and a paramagnetic amplifier, another form of solid-state, low-noise microwave amplifier. The noise temperature of the entire receiver (without antenna) was claimed to be 20 ± 10 K. The antenna was an array of eight 16-meter dishes, unlike any design ever used in the United States or Britain for planetary radar astronomy.[47]

Kotelnikov and his colleagues observed Venus between 18 and 26 April 1961. Their preliminary analysis of the data yielded an estimate of the astronomical unit, 149,457,000 kilometers, which appeared in the newspapers *Pravda* and *Izvestiia* on 12 May 1961. Over 100,000 kilometers less than the JPL and other values, the Soviet astronomical unit measurement was so incredibly incongruous, that Solomon Golomb told a conference of astronomers, "we should congratulate our Russian colleagues on the discovery of a new

46.   W. O. Mehuron, "Passive Radar Measurements at C-Band using the Sun as a Noise Source," *The Microwave Journal* 5 (April, 1962): 87–94; David K. Barton, "The Future of Pulse Radar for Missile and Space Range Instrumentation," *IRE Transactions on Military Electronics* MIL-5, no. 4 (October, 1961): 330–351; Irving Maron, George Luchak, and William Blitzstein, "Radar Observation of Venus," *Science* 134 (1961): 1419–1420.

47.   B. I. Kuznetsov and I. V. Lishin, "Radar Investigations of the Solar System Planets," in Air Force Systems Command, *Radio Seventy Years* (Wright-Patterson AFB, Ohio: Air Force Systems Command, 1967), pp. 187–188, 190 and 201; Vladimir A. Kotelnikov, "Radar Contact with Venus," *Journal of the British Institution of Radio Engineers* 22 (1961): 293; Kotelnikov, L. V. Apraksin, V. O. Voytov, M. G. Golubtsov, V. M. Dubrovin, N. M. Zaytsev, E. B. Korenberg, V. P. Minashin, V. A. Morozov, N. I. Nikitskiy, G. M. Petrov, O. N. Rzhiga, and A. M. Shakhovskoy, "Radar System Employed during Radar Contact with Venus in 1961," *Radio Engineering and Electronic Physics* 11 (1962): 1715–1716. For a brief history of the IREE, see Y. V. Gulyaev, "40 Years of the Institute of Radioengineering and Electronics of the Russian Academy of Sciences," *Radiotekhnika Elektronika* vol. 38, no. 10 (October 1993): 1729–1733. Soviet investigators performed radar studies of meteors in 1946 and of the Moon in 1954–1957, according to A. E. Solomonovich, "The First Steps of Soviet Radio Astronomy," pp. 284–285 in Sullivan. Although radar astronomers recently have used the arrayed dishes of the Very Large Array in bistatic experiments, dish arrays have not been used as transmitting antennas.

planet. It surely wasn't Venus!" Retrospectively, Kotelnikov explained that "random real-izations of noise were taken for reflected signals."[48]

The cause of the Soviet error might have been rooted in Cold War competition, which placed Soviet scientists under great pressure to produce results quickly for political reasons. The *Pravda* and *Izvestiia* announcements appeared on 12 May 1961, six days after the Jodrell Bank, but before the Lincoln Laboratory, announcements. If published sources had guided Kotelnikov and his colleagues, they would have been the erroneous Lincoln Laboratory and Jodrell Bank results of 1958 and 1959, with which the *Izvestiia* value agreed closely (within 10,000 kilometers).

The Cold War prevented communication and cooperation among planetary radar investigators. The Space Race in 1961 was still an extension of the Cold War; informal communications did not exist. Lincoln Laboratory did secret military research; JPL was a sensitive space research center with connections to ARPA, a military research agency. Jodrell Bank did not yet have ties with their Soviet counterparts. While Lincoln Laboratory, JPL, and Jodrell Bank personnel exchanged data, such informal links with Soviet scientists did not and could not exist.

Kotelnikov and his associates at the IREE, after realizing their error, turned their attention to a complete analysis of the raw radar data recorded on magnetic tape with the help of a special analyzer. Their new value, 149,598,000 ± 3300 kilometers, agreed closely with those of the United States and Britain.[49] Although the Soviet and British errors of measurement were greater than those of the American laboratories, they were far less than the values obtained by optical methods. The accuracy of the radar over the optical method and the general agreement among the preliminary results obtained in the United States, Britain, and the Soviet Union were the basis for a re-evaluation of the astronomi-cal unit by the International Astronomical Union (IAU).

Redefining the Astronomical Unit

The re-evaluation of the astronomical unit was part of a general movement within the IAU to reform the entire system of astronomical constants conventionally used to com-pute ephemerides. On 21 August 1961, shortly after JPL, Lincoln Laboratory, and Jodrell Bank announced their first estimations of the astronomical unit, the IAU executive com-mittee decided to organize a symposium on the system of astronomical constants. That sys-tem rested upon observations made in the nineteenth century and values adopted at international conferences held in Paris in 1896 and 1911.[50]

By 1950, two competing optical methods provided more accurate values for the astro-nomical unit. Harold Spencer Jones, Astronomer Royal of Great Britain from 1933 to 1955, used a trigonometric approach based on the triangulation of Eros. The orbit of the

48.   Kotelnikov et al, "Radar System," pp. 1715 and 1721; Kotelnikov, "Radar Contact," p. 294; Malling and Golomb, p. 300; Kotelnikov, "Radar Observations of the Planet Venus in the Soviet Union in April, 1961," typed manuscript, 27 February 1963, anonymous translation of a technical report of the Soviet Institute of Radio Engineering and Electronics, DTIC report number AD-401137, pp. 41–42, Renzetti materials. The Soviet publi-cation venue and aberrant astronomical unit value raise serious doubts about the veracity of their announce-ment.
49.   Kotelnikov et al, "Radar System," p. 1721; Kuznetsov and Lishin, p. 188; Kotelnikov, "Radar Observations," p. 2; Kotelnikov, Dubrovin, Morozov, Petrov, Rzhiga, Z. G. Trunova, and Shakhovoskoy, "Results of Radar Contact with Venus in 1961," *Radio Engineering and Electronics Physics* 11 (1962): 1722 and 1725. For a discussion of the integration technique, see V. I. Bunimovich and Morozov, "Small-Signal Reception by the Method of Binary Integration," *ibid.*, pp. 1734–1740.
50.   Jean Kovalevsky, ed., *The System of Astronomical Constants* (Paris: Gauthier-Villars and Cie., 1965), p. 1; Walter Fricke, "Arguments in Favor of the Revision of the Conventional System of Astronomical Constants," in J. C. Pecker, ed., *Proceedings of the Twelfth General Assembly* (New York: Academic Press, 1966), p. 604.

asteroid, discovered in 1898 by Berlin astronomer Gustav Witt, approaches Earth at regular intervals. As president of the IAU Solar Parallax Commission, Spencer Jones oversaw a worldwide operation to record photographic observations of Eros during its closest approach to Earth in 1930 and 1931. Through a complicated analysis of nearly 3,000 photographs, Spencer Jones estimated the astronomical unit to be 149,675,000 ± 17,000 kilometers. Eugene Rabe, an astronomer at the Cincinnati Observatory, applied the so-called dynamic method to observations of Eros between 1926 and 1945. He took into account the gravitational effects of the Earth, Mars, Mercury, and Venus on the orbit of Eros, and arrived at a value of 149,530,000 ± 10,000 kilometers.[51]

In addition, investigators at the Space Technology Laboratories (STL), a wholly-owned subsidiary of Ramo-Wooldridge (later TRW), computed a value from data acquired during the Pioneer 5 mission. In figuring the probe's trajectory, STL chose Rabe's value over that of Lincoln Laboratory in 1958. Not surprisingly, STL found a value for the astronomical unit, 149,544,360 ± 13,700 kilometers, in agreement with Rabe, but with a greater error of measurement. The STL value hardly challenged the more accurate ground-based radar measurements. Its "published accuracy," Walter Fricke, astronomer and professor at the Heidelberg Astronomisches Rechen-Institut, judged, "does not yet indicate any advantage over the traditional methods."[52] The Pioneer 5 value did not play any part in the IAU's revision of the astronomical unit.

The organizing committee of the IAU symposium on astronomical constants brought together astronomers from the United States and Europe who were responsible for drawing up the ephemerides. COSPAR (the Committee on Space Research) named an ad hoc committee to participate in the symposium, and additional astronomers from the United States, Britain, France, West Germany, Portugal, the Soviet Union, and South Africa took part. The members of the organizing committee included Eb Rechtin, the JPL manager of the DSIF; Dirk Brouwer, director of the Yale Observatory; and Gerald M. Clemence, scientific director of the U.S. Naval Observatory in Washington. Both Brouwer and Clemence had helped JPL with the Venus radar experiment ephemerides. Among the additional astronomers participating in organizing committee activities were two radar astronomers, Dewey Muhleman and Irwin I. Shapiro.[53]

Soon after the 1961 Venus experiment, Muhleman left JPL for the Harvard Astronomy Department. There, under Fred Whipple, A. Edward Lilley, and William Liller, he completed a doctoral dissertation based on Venus radar data collected at Goldstone in June 1963. After returning to JPL, Muhleman took a teaching position in the Cornell Astronomy Department in 1965. Shapiro had a Ph.D. in physics from Harvard and had worked on the detection of objects with radar in a clutter environment and on ballistic missile defense systems, before joining the team conducting radar experiments on Venus as the "guru" who calculated the ephemerides for Lincoln Laboratory planetary radar research.[54]

---

51.   Spencer Jones, "The Solar Parallax and the Mass of the Moon from Observations of Eros at the Opposition of 1931," *Memoirs of the Royal Astronomical Society* 66 (1938–1941): 11–66; Rabe, "Derivation of Fundamental Astronomical Constants from the Observations of Eros during 1926-1945," *The Astronomical Journal* 55 (1950): 112–126; Fricke, "Inaugural Address Delivered at the IAU-Symposium No. 21," in Kovalevsky, pp. 12-13.

52.   Fricke, "Inaugural Address," p. 13; James B. McGuire, Eugene R. Spangler, and Lem Wong, "The Size of the Solar System," *Scientific American* vol. 204, no. 4 (1961): 64–72. The value given in the article is 92,925,100 ±8,500 miles, which I have converted into kilometers for consistency.

53.   Rechtin, p. 368; Muhleman, D. Holdridge, and N. Block, "Determination of the Astronomical Unit from Velocity, Range and Integrated Velocity Data, and the Venus-Earth Ephemeris," in Victor, Stevens, and Golomb, pp. 83–92. Kovalevsky, p. 1, provides a list of their names.

54.   Muhleman 8 April 1993; Muhleman 19 May 1994; Shapiro 30 September 1993; Evans 9 September 1993.

The IAU symposium took place at the Paris Observatory between 27 and 31 May 1963. By then, Lincoln Laboratory and JPL had refined the accuracy of their calculations even further, to ± 400 and ± 250 kilometers respectively. In his inaugural address, Walter Fricke lauded the accuracy and general agreement of the radar measurements. As far as Fricke and other symposium participants were concerned, the real debate was between the radar and dynamic methods. Spencer Jones' trigonometric method contained too many inherent sources of systematic error. In an attempt to reconcile the dynamic and radar methods, Brian G. Marsden, an astronomer at the Yale University Observatory, concluded in favor of the radar measurements. Rabe defended his method in person, arguing that the radar observations were inconsistent with the observed orbit of Eros and with gravitational theory.[55]

Muhleman and Shapiro supported the radar method and explained the basis on which JPL and Lincoln Laboratory had obtained their results. Additional support for the radar method came from Britain. D. H. Sadler, Superintendent of H. M. Nautical Almanac Office at the Royal Greenwich Observatory, read a paper on the results of the Jodrell Bank 1962 Venus experiment.

Lest it appear that there was unanimous approval of the radar method, COSPAR raised the question of the discrepancy between the radar observations of 1958 and 1959 and those of 1961. Both Muhleman and Shapiro insisted that a discussion of the 1958 data, which they both labelled "manifestly wrong," would be too difficult and serve no purpose. They explained that the 1958 technology was highly inadequate and stressed the harmonious agreement among the 1961 measurements.[56]

The participants unanimously adopted Resolution Six, which recommended that the astronomical constants be studied by both existing and new methods, so that the results might be compared. The IAU Executive Committee then translated Resolution Six into Resolution Four, which recommended that a working group study the system of astronomical constants, including the astronomical unit expressed in meters. Next, the IAU Executive Committee named the Working Group on astronomical constants: Dirk Brouwer, Jean Kovalevsky (Bureau of Longitudes, Paris), Walter Fricke (chairman), Aleksandr A. Mikhailov (director of the Pulkovo Observatory, Soviet Union), and George A. Wilkins (Royal Observatory of Greenwich; Secretary). The Working Group sent a circular letter and copies of the Paris resolutions to all persons, some 80 in number, who were thought to be likely to be able to help the Group or who might be affected by the introduction of new constants. The Working Group met in January 1964, at the Royal Greenwich Observatory, Herstmonceux Castle, and drew up a list of constants, including the astronomical unit, for consideration by the IAU general assembly, which met in Hamburg later that year.[57]

The Working Group met again during the Hamburg meeting on 27 August. Muhleman and Pettengill, who read Shapiro's paper in his place, reviewed the latest radar determinations of the astronomical unit by JPL and Lincoln Laboratory from new observations made in 1964. Pettengill reported that preliminary analysis of the new data confirmed a value of 149,598,000 kilometers, while Muhleman disclosed the JPL value of

55. Kovalevsky, p. 3; Fricke, "Inaugural Address," pp. 12–13; Fricke, "Arguments in Favor of the Revision of the Conventional System of Astronomical Constants," in Pecker, p. 606; Marsden, "An Attempt to Reconcile the Dynamical and Radar Determinations of the Astronomical Unit," in Kovalevsky, pp. 225–236; Rabe, "On the compatibility of the Recent Solar Parallax Results from Radar Echoes of Venus with the Motion of Eros," in Kovalevsky, pp. 219–223.

56. Shapiro, "Radar Determination of the Astronomical Unit," in Kovalevsky, pp. 177–215, and Muhleman, "Relationship between the System of Astronomical Constants and the Radar Determinations of the Astronomical Unit," in ibid., pp. 153–175; Kovalevsky, pp. 298 and 311.

57. Kovalevsky, pp. 314 and 323; "Joint Discussion on the Report of the Working Group on the IAU System of Astronomical Constants," in Pecker, p. 600.

149,598,500 kilometers. The error of measurement reported by both laboratories, ± 100 kilometers, was the smallest yet.[58]

Walter Fricke, chair of the Working Group, had misgivings about the radar method: "One could argue that the radar results are still too fresh to deserve full confidence. My personal distrust of them in so far as it originates in their newness has a counterpart in my distrust of the dynamical [Rabe] result obtained from the discussion of the observations of Eros."[59]

Without any discussion of the dynamic method, however, the Working Group recommended adoption of a value expressed in meters and based on radar observations. The IAU general assembly then adopted the recommended value, 149,600 X 10^6 meters (149,600,000 kilometers).[60] It was now a matter of incorporating the new value into the various national almanacs and ephemerides.

The Rotation of Venus

The establishment of a highly accurate value for the astronomical unit and its adoption by the IAU was but one way that planetary radar demonstrated its value as a problem-solving scientific activity. The distance from Earth to Venus as measured by JPL radar also proved essential in keeping the 1962 Mariner 2 Venus probe on target. Early in its flight, Mariner 2 went off course. The Pioneer and Echo antennas sent midcourse commands, and a 34-minute maneuver put Mariner 2 on course. Had Rabe's value for the astronomical unit been used in place of the radar value, Mariner 2 would have passed Venus without acquiring any useful data.[61]

Valuable insight into the rotation of Venus further demonstrated the problem-solving scientific merit of planetary radar. Optical and spectrographic methods failed to reveal the planet's period or direction of rotation, because Venus' thick, opaque cloud layer hid all evidence of its motion. Astronomers could only infer and imagine. Radar waves, on the other hand, were quite capable of penetrating the Venusian atmosphere; yet determining the planet's rotation by radar was still not easy. The key was methodical and meticulous attention to the shape of the echo spectra. Although JPL, Lincoln Laboratory, Jodrell Bank, and the Soviet Yevpatoriya facility calculated rotational rates for Venus, only JPL and Lincoln Laboratory found its "locked" orbit and retrograde motion.[62]

Evans and Taylor at Jodrell Bank published the first estimate of the planet's rotational period, about 20 days, using their erroneous 1959 data. In 1964, John Thomson reckoned a slow rotational rate, "probably" somewhere between 225 days and a similar retrograde period. After seeming to be on the brink of discovery, Thomson pulled back, concluding, "Future observations of the change of spectral width with time should enable the rotation rate and rotation axis to be determined." "Retrograde rotation," he held, was "physically unlikely."[63]

---

58. "Joint Discussion," pp. 591, 599 and 602–603; Shapiro, "Radar Determinations," in Pecker, pp. 615–623.

59. "Joint Discussion," p. 606.

60. Ibid., p. 606; "Report to the Executive Committee of the Working Group on the System of Astronomical Constants," in Pecker, p. 594.

61. Renzetti 17 April 1992; Renzetti, *A History*, pp. 20 and 31; Renzetti, *Tracking and Data Acquisition Support for the Mariner Venus 1962 Mission*, Technical Memorandum 33–212 (Pasadena: JPL, 1 July 1965), pp. 9, 17 and 75–76.

62. RCA did not hesitate a guess on the rotation rate or direction. Maron, Luchak, and Blitzstein, pp. 1419–1421.

63. Evans and Taylor, p. 1359; Ponsonby, Thomson, and Imrie, "Radar Observations of Venus and a Determination of the Astronomical Unit," *Monthly Notices of the Royal Astronomical Society* 128 (1964): 14–16.

As close as Jodrell Bank came to discovering Venus' retrograde motion, the Soviets were that far away. Looking at frequency shifts in their 1961 data, Kotelnikov's group persistently estimated the planet's rotational period as 11 days, if not 9 or 10 days. They entirely missed the planet's retrograde motion. The Soviet error arose from their finding that the spectrum had a wide base, at least 400 hertz wide, indicating rapid motion. All British and United States workers agreed that the spectrum was far narrower. Lincoln Laboratory, for example, found a narrow spectrum of only 0.6 hertz. After their 1962 radar study of Venus, Kotelnikov and his colleagues re-evaluated their data and concluded a retrograde rotational period of 200 to 300 days.[64]

By then, though, JPL and Lincoln Laboratory already had discovered Venus' retrograde motion. Finding it was not easy. Along the way, both laboratories concluded that the Venusian day was as long as its year, about 225 days. Venus was "locked" in its orbit, turning one face always toward the Earth at the moment of inferior conjunction. However, these initial reports failed to note the planet's retrograde motion.[65]

The investigators who found it did not follow the same path of discovery. Just as the availability of technology had made planetary radar astronomy possible, the limits of that technology shaped the paths of discovery. JPL harvested the benefits of a powerful, low-noise continuous-wave radar in their 1962 and 1964 Venus experiment, while Lincoln Laboratory reaped the rewards of their computer and signal processing skills.

The Goldstone radar permitted Roland L. Carpenter to find the retrograde motion of Venus in a rather novel fashion. Carpenter actually had a BA in psychology from California State University at Los Angeles, but he had been interested in astronomy since childhood, and he had worked at Griffith Observatory as a guide. Finding very little work available in psychology, Carpenter found a job at Collins Radio as an electrician thanks to his friend, astronomer George Abell (known for Abell's clusters of galaxies), who had a summer job there. Carpenter gradually worked his way up to electronics engineer, simply through his work experience at Collins Radio. Then, when JPL began hiring people with experience in radio communications for the Deep Space Network, Carpenter jumped at the opportunity. Carpenter worked with Dewey Muhleman in Walt Victor's group and took advantage of JPL's employee benefits program by pursuing an advanced degree in astronomy at UCLA, while working full-time at JPL. His doctoral dissertation, "The Study of Venus by CW Radar," written under Lawrence Aller and completed in 1966, used data from the 1964 JPL Venus radar experiment.[66] By then, however, Carpenter already had published his discovery of the retrograde rotation of Venus.[67]

64.   Kuznetsov and Lishin, pp. 199-201; Kotelnikov, "Radar Contact with Venus," *Journal of the British Institution of Radio Engineers* 22 (1961): 295; Kotelnikov et al, "Results of Radar Contact," p. 1732; Kotelnikov, Dubrovin, M. D. Kislik, Korenberg, Minashin, Morozov, Nikitskiy, Petrov, Rzhiga, and Shakhovskoy, "Radar Observations of the Planet Venus," *Soviet Physics—Doklady* 7 (1963): 728–731; Kotelnikov, Dubrovin, V. A. Dubinskii, Kislik, Kusnetsov, Lishin, Morozov, Petrov, Rzhiga, G. A. Sytsko, and Shakhovskoy, "Radar Observations of Venus in the Soviet Union in 1962," *Soviet Physics—Doklady* 8 (1964): 644; Smith, p. 15. Rzhiga, "Radar Observations of Venus in the Soviet Union in 1962," in M. Florkin and A. Dollfus, eds. *Life Sciences and Space Research II* (New York: Interscience Publishers, 1964), pp. 178–189, states 300 days but still misses the retrograde motion.
    65.   Pettengill et al, "A Radar Investigation of Venus," pp. 189–190; Pettengill, "Radar Measurement of Venus," in Priester, pp. 880–883. The range given was between 115 and 500 days, that is, 225 (+275, -110) days. The first JPL external announcement of that finding was made in a paper read by Solomon Golomb and Leonard R. Malling at a convention on radio techniques and space research held at Oxford in July 1961. Malling and Golomb, pp. 297–303. The paper was not published until October 1961 and was preceded in print by the internal report, Victor and Stevens, "Summary and Conclusions," pp. 94–95. See also Victor and Stevens, "Exploration of Venus by Radar," pp. 46–47; Muhleman, "Early Results of the 1961 JPL Venus Radar Experiment," *The Astronomical Journal* 66 (1961): 292; Victor and Stevens, "The 1961 JPL Venus Radar Experiment," p. 94.
    66.   Carpenter, telephone conversation, 14 September 1993.
    67.   Carpenter, "An Analysis of the Narrow-Band Spectra of Venus," in *JPL Research Summary No. 36-14 for the Period February 1, 1962 to April 1, 1962* (Pasadena: JPL, 1 May 1962), pp. 56–59.

His first announcement of the planet's retrograde motion appeared in a JPL internal report dated 1 May 1962 and was based on the 1961 Venus experiment. Carpenter suggested a retrograde rotational period of about 150 days, but backed off from insisting on his discovery. "Unfortunately," Carpenter concluded, "a definitive answer cannot be given for the rotation period of Venus based on the present data."

Carpenter hesitated until he had the results of the Goldstone 1962 Venus experiment. Between 1 October and 17 December 1962, when Venus was closest to Earth, Goldstone made nearly daily radar observations of the planet with a 13-kilowatt continuous-wave transmitter operating at 2388 MHz (12.6 cm). Equipped with a maser and a parametric amplifier, the system's total noise temperature was only 40 K, better than the 64 K achieved in 1961.[68]

The Goldstone radar was sufficiently powerful and sensitive that a large feature on the planet's surface showed up as an irregularity or "detail" on the power spectrum. The surface feature scattered back to the radar antenna more energy than the surrounding area. Normally, most spectral irregularities resulted from random fluctuations produced by noise. The power and sensitivity of the Goldstone radar made all the difference.

"On close examination," Carpenter wrote, "one irregularity was found to persist from day to day and to change its position slowly....The relative permanence of the detail strongly suggests that it was caused by an actual physiographic feature on the surface of Venus and that its motion was the result of the planet's rotation. The true nature of the feature can only be guessed at; however, it is not unreasonable to assume that it is a particularly rough region of rather large extent."

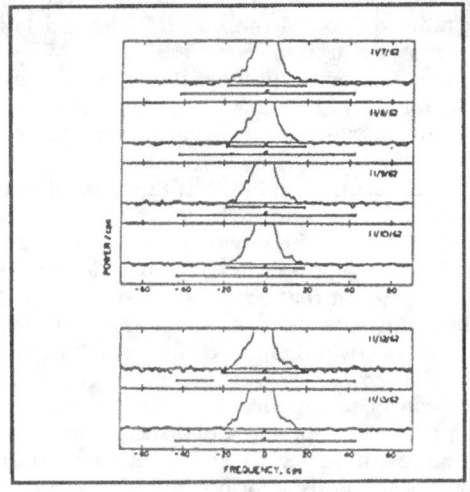

Figure 8
Lower portion of the spectra obtained by Roland Carpenter during the week prior to the 1962 conjunction of Venus. Note the persistent detail on the left side of each spectrum. Carpenter followed that detail to determine the retrograde motion of Venus. (Courtesy of Jet Propulsion Laboratory.)

68.   Carpenter, telephone conversation, 14 September 1993; Goldstein and Carpenter, "Rotation of Venus: Period Estimated from Radar Measurements," *Science* 139 (1963): 910; Carpenter, "Study of Venus by CW Radar," *The Astronomical Journal* 69 (1964): 2. Details of the 1962 JPL Venus radar experiment are given in Goldstein, Stevens, and Victor, eds., *Radar Exploration of Venus: Goldstone Observatory Report for October–December 1962*, Technical Report 32–396 (Pasadena: JPL, 1 March 1965).

Carpenter then followed the movement of this "detail" in order to deduce the planet's rotational period. He calculated that Venus had either a forward period of about 1200 days or a retrograde period of 230 days from one conjunction to the other. Next, he measured the bandwidth of the lower portion of the spectra; their widths were incompatible with a 1200-day forward rotation. The base bandwidth measurements, however, did "strongly suggest that the sidereal rotation period of Venus is not synchronous, but rather 250 ± 40 days retrograde."[69]

Millstone lacked the power and sensitivity of Goldstone. The discovery of Venus' retrograde motion at Lincoln Laboratory by William B. Smith relied instead on his computer and signal analyzing skills. Although Smith preceded Carpenter in announcing the retrograde motion of Venus in a publication, he did not achieve recognition as its discoverer.

Smith looked at the spectral bandwidths of radar returns on 11 separate days between 2 April and 8 June 1961. Like Carpenter, he failed to verify a synchronous rotation; however, Smith came to realize that the way the signal bandwidth changed over time could be explained only by retrograde motion. He wrote up his findings and submitted them to his supervisor, Paul Green, for approval. Smith wanted to feature the planet's retrograde motion in his paper, but Green remembered an earlier episode, when "we had been badly burned." That was the embarrassment of 1958.

Green hesitated. Uranus was the only planet then known to have a retrograde period, "but that one is way the hell out, and who would have thought that the next planet to the Earth would have had that kind of anomalous behavior?" Green admitted, "I guess I was working more on psychological factors than on anything else. So I had Bill tone it down." The published article's abstract read: "The (relatively weak) result implies a very slow or possibly retrograde rotation of the planet." The article itself contained no statement of the planet's retrograde motion.[70]

The watered down version made all the difference. Carpenter published his explicit and unequivocal results jointly with fellow JPL radar astronomer Dick Goldstein in the 8 March 1963 issue of *Science*, while the February 1963 issue of *The Astronomical Journal* carried Smith's suggestive abstract.[71]

Green regretted his decision. "Bill Smith is the man who discovered that Venus has retrograde spin, and he should go down in the history books. Due to me he didn't, because his paper didn't feature it the way it should have. If I hadn't sat on it, it would have featured it, but as it came out, it didn't. The people that look at the fine print realize that he had that message, that that was what his data showed, but it didn't make the big splash and give him the career achievement that he deserved."[72] Fellow Lincoln Laboratory radar astronomer Irwin Shapiro concurred: "I felt he [Smith] got a raw deal, because he made a major discovery for which he never got credit."[73]

The detection of Venus, the measurement of the size of the astronomical unit, and the determination of the rotational period and direction of Venus formed the foundation on which planetary radar astronomy was laid. Planetary radar advanced by solving problems left unresolved or at best unsatisfactorily resolved by optical methods. Deliberately or not, the problems solved supported the NASA mission to explore the solar system. Driving the new scientific activity was the availability of a new generation of radars built for military defense (at Lincoln Laboratory) and for space exploration (at JPL). The limits of that technology shaped the paths of discovery.

---

69.   Carpenter, "Study of Venus by CW Radar," pp. 4–6; Carpenter, telephone conversation, 14 September 1993.
70.   Green 20 September 1993; Smith 29 September 1993; Smith, pp. 15–21.
71.   Goldstein and Carpenter, pp. 910-911; Smith, pp. 15–21. Internal evidence indicates that *Science* received the paper on 15 January 1963.
72.   Green 20 September 1993.
73.   Shapiro 30 September 1993.

Without technology and without funding, planetary radar astronomy was impossible. The emergence of planetary radar coincided with the creation of a national, civilian space agency, NASA, a national, civilian agency to fund scientific research, the National Science Foundation (NSF), and a national, military space research agency, ARPA. It also paralleled the rise of American radio astronomy and the age of the Big Dish. Standing at the intersection of civilian and military research into space, the ionosphere, the Moon, and the Sun, planetary radar offered much to potential patrons. It was a wonderful and unique time to organize a new scientific activity.

Chapter Three
Sturm und Drang

The period between 1958 and 1964 saw the explosive growth of planetary radar astronomy in terms of the number of active facilities and investigators. Investigators in three countries (the United States, Britain, and the Soviet Union) attempted to detect Venus in 1961, and three facilities in the United States alone (Lincoln Laboratory, JPL, and RCA) succeeded. During the 1962 conjunction, the Jicamarca Radar Observatory, a National Bureau of Standards ionospheric facility in Peru, made radar observations of Venus at 50 MHz (6 meters). At the same time, the Lincoln Laboratory solar radar facility at El Campo, Texas, completed in the summer of 1960, observed Venus at 38 MHz (8 meters).[1] Thus, by 1964, five American facilities had performed radar experiments on Venus.

The creation of radar astronomy courses, a textbook, and a conference dedicated solely to radar astronomy also signalled the emergence of a new and rapidly growing scientific field. As it had in carrying out planetary radar experiments, Lincoln Laboratory took the lead in shaping the new field. In addition to organizing radar astronomy courses and a textbook, Lincoln Laboratory sponsored the first, and only, radar astronomy conference and undertook, in association with the Cambridge astronomical community, a campaign to design and build a new radar research instrument.

MIT routinely offered summer courses and asked Lincoln Laboratory to propose some. As John Evans explained, "Radar astronomy was in vogue, we were just entering the Space Age, and Sputnik had been launched." So Lincoln Laboratory agreed to run a summer school in radar astronomy beginning in August 1960. In all, about twenty people gave lectures. Evans talked about lunar radar astronomy. Jack Harrington, head of the Radio Physics Division of Lincoln Laboratory and in charge of the summer course, promised lecturers that the talks would be organized into a book. As it turned out, Evans recalled, "the lecture notes weren't that good. We were all asked to rewrite them."[2]

In August 1961, Harrington and Evans ran the radar astronomy summer course again. The topics and lecturers were somewhat different; the course of 15 lectures lasted only one week. Among the lecturers were Paul Green, Bob Kingston (who had designed the maser for the 1958 Venus experiment), Gordon Pettengill, Bob Price, Herb Weiss (who had built Millstone), and Victor Pineo (formerly of the National Bureau of Standards). Von Eshleman (Stanford), a guest lecturer, discussed solar radar experiments. The week ended with a two-hour tour of the Millstone Hill Radar Observatory led by Pettengill, Pineo, and Evans "to observe firsthand a modern space radar facility and to witness a representative experiment in radar astronomy."[3]

1.   W. K. Klemperer, G. R. Ochs, and Kenneth L. Bowles, "Radar Echoes from Venus at 50 Mc/sec," *The Astronomical Journal* 69 (1964): 22–28; Overhage to Lt. Gen. James Ferguson, 28 March 1963, MITA; Jesse C. James, Richard P. Ingalls, and Louis P. Rainville, "Radar Echoes from Venus at 38 Mc/sec," *The Astronomical Journal* 72 (1967): 1047–1050.
2.   Evans 9 September 1993. MITA does not have a copy of the 1960 summer course lecture notes.
3.   Brochure, MIT, *Radar Astronomy: Summer Session 1961 August 14–18* (Cambridge: MIT, 1961), LLLA; MIT, *Radar Astronomy: Summer Session MIT, August 14-18, 1961, Lectures 1–15*, 3 vols. (Cambridge: MIT, 1961), MITA.

The radar astronomy summer course was not given again, "largely because the people concerned have been occupied with other commitments," Evans later wrote.[4] Price and Green were no longer involved in radar astronomy, and Pettengill had left Lincoln Laboratory. Harrington himself became Director of the MIT Center for Space Research, which he founded with funding from NASA in 1963.

At the end of the 1961 summer course, the lecture notes were assembled into a three-volume tome. Yet, as Evans explained, "We didn't have a good set of course notes that would constitute a book."[5] Paul Green became irritated with the lack of progress on the project, announced that he would no longer contribute any material to the book, and nominated Evans to take over the project from Harrington. Evans found himself in an awkward situation; Harrington was his boss. Fortunately, Wilbur B. Davenport, Jr., one of the Assistant Directors of Lincoln Laboratory, had an interest in radar astronomy and pressured Harrington to get the book done quickly.

Evans recalled: "So my arm got twisted very hard by Davenport. I really didn't want to do it. I was quite busy, and I didn't want to take over Jack's project, so I resisted. I eventually capitulated after enough pressure on the condition that a) I had somebody to help me, and b) I had a secretary assigned to do typing and nothing else, because part of the problem was just getting material out of rough draft form and into typed form. They agreed to both of those conditions." Tor Hagfors, a graduate of Scandinavian technical schools and the Stanford University electrical engineering program, edited the book with Evans.

Next, the project met difficulty at the publisher. The McGraw-Hill editor who had been handling the project left, but no one at Lincoln Laboratory knew. "The manuscript sat in his drawer for almost two years," Evans related. "Meanwhile, we were thinking that the manuscript was going through proofing and so on. Finally, we got a letter from some guy who had inherited this desk and found this manuscript. He got it printed fairly quickly, but in sort of photo-offset form rather than nice copy. At least it came out, belatedly."

Once McGraw-Hill published *Radar Astronomy* in 1968, radar astronomy had a textbook, parts of which are still used to teach radar astronomy. Nonetheless, neither MIT nor Lincoln Laboratory (which is not a teaching institution) offered a course in radar astronomy until 1970.[6] Although the Evans-Hagfors textbook and the MIT summer course might have served to train a generation of radar astronomers, they did not. Planetary radar astronomy was the child of a research center (Lincoln Laboratory), not an educational institution (MIT). As a result, Lincoln Laboratory radar astronomers did not reproduce themselves in a traditional academic fashion through graduate education, but through employment.

Three radar astronomers came to Lincoln Laboratory during the 1960s through employment: Stanley H. Zisk, Richard P. Ingalls, and Alan E. E. Rogers. Zisk, who created lunar radar images for NASA in support of the Apollo program, and Haystack Associate Director Dick Ingalls, who had been a Lincoln Laboratory employee since 1953, both had degrees in electrical engineering. Alan Rogers, born in Salisbury, Rhodesia (now Zimbabwe), earned a Ph.D. in electrical engineering from MIT in 1967, and was trained in radio astronomy, before carrying out radar astronomy experiments.[7]

As far as defining the field of radar astronomy, and particularly in terms of defining actual and potential patrons, the most important step taken by Lincoln Laboratory was

4. Evans and Tor Hagfors, eds., *Radar Astronomy* (New York: McGraw-Hill Book Company, 1968), p. viii.
5. Evans 9 September 1993.
6. Campbell 9 December 1993; E-mail, Pettengill to author, 29 September 1994; Rogers 5 May 1994.
7. Pettengill 28 September 1993; Rogers 5 May 1994; NEROC, "Technical Proposal: Radar Studies of the Moon (Topography)," 12 November 1971, SEBRING.

the organization of a conference on radar astronomy. Never again did another such conference take place, mainly because radar astronomers located themselves within existing professional organizations. Moreover, the small number of radar astronomers never justified the creation of a separate society or journal.

The conference underscored the Big Science environment in which radar astronomy was evolving. Only a few attempts at Venus had been made by Lincoln Laboratory and Jodrell Bank when the conference convened; lunar, meteor, and ionospheric radar studies were well established. Those radar studies were part of growing civilian and military programs in ionospheric and communication research. More importantly for planetary radar, a new civilian space agency, NASA, had been created only the year before. Its creation, and the prospect of participating in space research, eventually shaped the new field of planetary radar astronomy more than any other Big Science patron.

The Conference on Radar Astronomy

The National Academy of Sciences, through its Space Science Board, underwrote the radar astronomy conference. Established in 1958, the Space Science Board maintained liaisons with the National Science Foundation, NASA, ARPA, the Office of the Science Advisor to the President, and other federal agencies participating in the country's space program. The Space Science Board solicited the opinions of scientists through discussions and summer studies and recommended space programs to federal agencies.[8]

Bruno B. Rossi, a member of the Space Science Board and a leading MIT physics professor, organized the radar astronomy conference. Rossi had undertaken experimental research on cosmic rays in the 1930s, before working at Los Alamos Laboratory during World War II. He joined MIT in 1946. In 1958, coincidentally with the creation of NASA, Rossi began to consider the potential value of direct measurement of the ionized interplanetary gas by space probes.[9]

Thomas Gold, recently hired to head Cornell's Center for Radiophysics and Space Research, the parent organization for its radio and radar telescope, and MIT's Philip Morrison, both members of the Space Science Board, assisted Rossi in organizing the conference; however, the brunt of the actual work fell on Rossi's shoulders. He reserved MIT's Endicott House in Dedham, Massachusetts, for 15 and 16 October 1959. Endicott House had a dining area, meeting rooms, large gardens, and accommodations for 8 people; the remainder were lodged at a nearby hotel.

Rossi saw the conference as a small group meeting to develop concrete recommendations for consideration by the Space Science Board at its October meeting. The original conference title, "Reflections and Scattering of Radar Signals Beyond Several Earth Radii," by definition excluded ionospheric radar. However, the revised name, "Conference on Radar Astronomy," was less unwieldy and did not appear to exclude those interested in ionospheric research.[10]

Holding a different vision of the conference was Stanford professor of electrical engineering Von R. Eshleman. Seeking to exploit the creation of NASA, Eshleman proposed radar studies of planetary ionospheres and atmospheres from spacecraft. Such studies were a logical extension of Stanford's ionospheric radio and radar work of the 1950s, which included a pioneering solar radar experiment.

8.   Space Science Board, Proposal for Continuation of Contract NSR 09-012-903, 28 October 1965, "NAS-SSB, 1965," NHO; Joseph N. Tatarewicz, *Space, Technology, and Planetary Astronomy* (Bloomington: Indiana University Press, 1990), p. 38.
9.   Rossi biographical information, MITA; "President's Report Issue," *MIT Bulletin* vol. 82, no. 1 (1946): 137–138.
10.   "Conference on Radar Astronomy Program," n.d., and George A. Derbyshire, Memorandum for the Record, 29 May 1959, "ORG, NAS, 1959 October Space Science Bd., Conferences Radar Astronomy, Dedham," NAS. Hereafter, Conference Program and Derbyshire Memorandum, 29 May 1959, respectively.

In 1959, contemporary with the first radar attempts at Venus, Eshleman and Philip B. Gallagher of Stanford, with Lt. Col. Robert C. Barthle of the U.S. Army Signal Corps, a Stanford graduate student, attempted to bounce radar waves off the solar corona. The Air Force Cambridge Research Center (AFCRC) underwrote the Stanford experiment, and the Office of Naval Research funded the 46-meter (150-ft) dish antenna constructed for ionospheric research under the direction of Oswald Villard. Although Eshleman claimed success, a comparison of his results with those obtained shortly afterward by the El Campo solar radar cast serious doubt about their validity, which some radar astronomers continue to express.[11]

As planning for the radar conference was underway, Eshleman was preparing the solar radar experiment and was on the point of campaigning NASA to underwrite studies of planetary ionospheres from spacecraft. It was a pivotal moment for calling attention to the Stanford radar work. Eshleman saw the conference as a Stanford opportunity. In a letter to Rossi, he claimed that Stanford already "had begun to plan some kind of a meeting to bring together all who are active in this field. However these plans had [sic] not progressed very far." He proposed a larger conference with Stanford and the Stanford Research Institute (SRI) "as co-hosts." If the AFCRC were invited to co-sponsor the conference, Eshleman suggested, part of the travel expenses for foreign visitors might be covered. Conference papers could be published as a group in the *Proceedings* of the Institute of Radio Engineers.[12]

The conference, however, was solely an MIT affair sponsored only by the National Academy of Sciences. The spectrum of United States civilian and military scientific radar research facilities was represented: MIT and Lincoln Laboratory, Stanford and SRI, Cornell University, the NRL, and the National Bureau of Standards CRPL. In addition, radio astronomers were invited from Harvard University, Yale University, the University of Michigan, and the National Radio Astronomy Observatory (NRAO), Green Bank, West Virginia, the country's major radio astronomy center. ARPA and the AFCRC represented the military.

In addition to representatives of the Space Science Board, Rossi invited the National Science Foundation program director for astronomy and NASA Space Science chief Homer E. Newell, Jr. Unable to attend, Newell recommended Nancy G. Roman in his place: "Although we have no program which directly involves radar astronomy, Dr. Roman will be happy to discuss those aspects of our Astronomy and Astrophysics Programs which are related to this field. I am sure that the results of the discussion will be valuable in our program planning."[13] Roman was a felicitous choice; she had carried out lunar radar studies at the NRL.[14]

11.   Eshleman, telephone conversation, 26 January 1993; Eshleman 9 May 1994; Eshleman, Barthle, and Gallagher, "Radar Echoes from the Sun," *Science* 134 (1960): 329–332; Eshleman and Allen M. Peterson, "Radar Astronomy," *Scientific American* 203 (August, 1960): 50–51; Barthle, *The Detection of Radar Echoes from the Sun*, Scientific Report 9 (Stanford: RLSEL, 24 August 1960); Pettengill 28 September 1993.
       The possibility of obtaining radar echoes from the solar corona had been suggested earlier by the Australian ionosphericist Frank Kerr in 1952 and by the Ukrainians F. G. Bass and S. I. Braude in 1957. Kerr, "On the Possibility of Obtaining Radar Echoes from the Sun and Planets," pp. 660–666; Bass and Braude, "[On the Question of Reflecting Radar Signals from the Sun]," *Ukrains'ky Fizychny Zhurnal [Ukrainian Journal of Physics]* 2 (1957): 149–164.
       12.   Eshleman to Rossi, 13 May 1959, "ORG, NAS, 1959 October Space Science Bd., Conferences Radar Astronomy, Dedham," NAS.
       13.   "Preliminary List of Invitees;" "Draft Recommendations of the Conference on Radar Astronomy," Appendix A, "List of Participants;" Newell to Rossi, 18 June 1959; Derbyshire Memorandum, 29 May 1959; and Derbyshire, Memorandum for the Record, 2 June 1959, "ORG, NAS, 1959 October Space Science Bd., Conferences Radar Astronomy, Dedham," NAS.
       14.   For Roman's lunar radar work at the NRL, see, for example, Yaplee, Roman, Craig, and T. F. Scanlan, "A Lunar Radar Study at 10-cm Wavelength," in Bracewell, ed., *Paris Symposium on Radio Astronomy* (Stanford: Stanford University Press, 1959), pp. 19–28, and Ch. 1, note 69.

Invitations to foreign radio and radar investigators went to Jodrell Bank, the Royal Radar Establishment (Malvern, England), the Division of Radiophysics of the Australian Commonwealth Scientific and Industrial Research Organization (CSIRO), the Chalmers University of Technology Research Laboratory of Electronics (Gothenburg, Sweden), and the Canadian Defense Research Board Telecommunications Establishment. No Soviet scientists were invited.

The conference program highlighted the work of Lincoln Laboratory. After a talk by Thomas Gold (Cornell) on the scientific goals of radar astronomy, Jack Harrington (Lincoln Laboratory) explained certain experimental techniques and Herb Weiss (Lincoln Laboratory) spoke on transmitters, receivers, and antennas. Next Paul Green (Lincoln Laboratory) discussed signal detection and processing, and James Chisholm (Lincoln Laboratory) talked about electromagnetic propagation phenomena. In another session, organizations represented at the conference described their research programs. General discussion and the formulation of recommendations took up the second day.[15]

These recommendations defined radar astronomy as a field especially useful to NASA and the rapidly growing space effort. The arguments set forth appeared as attempts to garner the patronage of the new space agency. The first recommendation, for example, spoke directly to NASA and argued the value of radar astronomy for planetary exploration. Launching spacecraft required precise measurements of interplanetary distances and knowledge of planetary surface and atmospheric conditions, all of which radar astronomy was capable of providing. *"The importance of radar astronomy to the efficient development of space science must not be underestimated,"* the recommendation exhorted.

Additional recommendations urged the construction of new radar astronomy facilities operating at a variety of frequencies, as well as the design and construction of large dish and array antennas, high-power high-frequency transmitters, and signal detection and recording techniques. The construction of radar telescopes, the conference recommendations argued, would be far less expensive than building and sending planetary probes.

Conference recommendations also addressed the military and radio astronomy. Planetary radar astronomy at Lincoln Laboratory would not have existed without the construction of the Millstone Hill radar, which the military funded. However, planetary radar experiments officially did not exist; military research was the first priority. Radar astronomy, the recommendations pleaded, needed facilities of its own, where it would receive top priority and be "viewed as pure science."

Conference recommendations also targeted radio astronomers. *"Where large radio telescopes are being planned or built,"* one recommendation proposed, *"serious consideration be given from the beginning to the incorporation of provisions for a high-powered transmitter,* even if a transmitter were not actually installed." The recommendation further suggested specifically that a radar transmitter be installed on the 10-GHz (3-cm) 43-meter (140-ft) NRAO antenna, thereby offering "an excellent opportunity for radar investigations at very high frequencies." While recognizing that the dissimilar needs of radar and radio astronomers often gave rise to conflict, one recommendation stated, compromise could resolve them.[16] As we shall see later, however, those dissimilar needs were beyond compromise.

Bruno Rossi submitted the conference draft recommendations to the Space Science Board at its October 1959 meeting. After some editing and checking that left the recom-

---

15.   Derbyshire Memorandum, 2 June 1959; Conference Program; Rossi to Derbyshire, 10 June 1959, "ORG, NAS, 1959 October Space Science Bd., Conferences Radar Astronomy, Dedham," NAS.

16.   "Draft Recommendations of the Conference on Radar Astronomy," pp. 5–8, "ORG, NAS, 1959 October Space Science Bd., Conferences Radar Astronomy, Dedham," NAS. Emphasis in original text.

mendations unaltered, the Space Science Board endorsed them for distribution to fund-ing agencies and other interested groups.[17] Endicott House was the last conference dedi-cated solely to radar astronomy, though radar astronomers continued to meet under an existing organizational umbrella, one dedicated not to planetary science, since such spe-cialized organizations did not yet exist, but to radio astronomy and electrical engineering.

L'Union Radioscientifique Internationale

Although much of the earliest radar astronomy work grew out of an interest in ionos-pheric questions, ionosphericists and planetary radar astronomers soon went separate ways. Planetary radar astronomers grew closer to their colleagues in radio astronomy, with whom they shared techniques and technologies, such as antennas and low-noise receivers. The shift of planetary radar astronomy from the ionospheric to the radio astronomy com-munity was manifest within the Union Radioscientifique Internationale (URSI), which quickly became the premier forum for planetary radar astronomers.[18]

URSI was an international radio science organization founded in France in 1921 by Gustave Ferrié and other French radio pioneers.[19] Its big tent sheltered a range of fields, including ionospheric and radio astronomy science, united by a common technical inter-est in what might be called radio science. Lacking telescopes committed entirely to their field, planetary radar astronomers worked side-by-side with radio astronomers at the same observatory. As radar astronomer Donald B. Campbell has observed, "There is a tremen-dous amount of cross-fertilization between planetary radar and radio astronomers in terms of techniques and equipment."[20] These shared technical interests and instruments brought planetary radar and radio astronomers together at URSI meetings.

Radio astronomers had had their own commission within URSI since shortly after World War II. In 1946, at its General Assembly meeting in Paris, URSI created a special subcommission on Radio Noise of Extra-Terrestrial Origin, which became Commission 5, Extra-Terrestrial Radio Noise, when URSI revised its commission structure at its 1948 Stockholm meeting. On the proposal of the U.S. National Committee, Commission 5 became the Commission on Radio Astronomy two years later at the General Assembly meeting in Zurich. Commission 5 concerned itself with radio astronomy, as well as obser-vations of meteors and the Moon "by radio techniques," meaning by radar. Thus, for example, at the Paris URSI symposium on radio astronomy held in July 1958, a number of papers featured the latest lunar radar work by U.S. and British investigators.[21]

The first URSI meeting—at which planetary radar astronomers gave papers—took place in San Diego, California, between 19 and 21 October 1959, immediately following the Endicott House Conference on Radar Astronomy. The meeting included a first-of-its-kind symposium on radar astronomy. However, presenting the panel discussion was not Commission 5, but URSI Commission 3, Ionospheric Radio.

17.   Memorandum, E. R. Dyer, Jr., to Participants, Space Science Board Conference on Radar Astronomy, 30 October 1959, and "Report and Recommendations of the Conference on Radar Astronomy," "ORG: NAS, 1959 October Space Science Bd.: Conferences Radar Astronomy: Dedham," NAS.
18.   Pettengill 29 September 1993.
19.   URSI actually dates back to 1913 and the creation of the French Commission Internationale de TSF Scientifique. TSF (Télégraphie Sans Fil) is French for wireless radio. Albert Levasseur, *De la TSF à l'électronique: Histoire des techniques radioélectriques* (Paris: ETSF, 1975), pp. 79 and 87.
20.   Campbell 9 December 1993.
21.   Edge and Mulkay, p. 44; Bracewell, *Paris Symposium*, passim.

The seven panel members, all of whom had participated in the Endicott House conference, were practicing radar astronomers at the NRL, Jodrell Bank, Stanford, Lincoln
Laboratory, Cornell, and the National Bureau of Standards. Von Eshleman was the panel
moderator. The speakers covered lunar, solar, meteor, auroral, and planetary radar, as well
as radar studies of the exosphere and the interplanetary medium. The symposium was of
some historical importance: Paul Green described planetary range-Doppler imaging,
which later became a central planetary radar technique.[22]

By the URSI Tokyo meeting of September 1963, planetary radar astronomy had
moved to the newly renamed Commission 5, Radio and Radar Astronomy. Twenty institutions reported on recent U.S. developments in the two fields. The meeting also brought
together individuals from related areas, such as Commission 7, Radio Electronics, where
investigators reported on parametric amplifiers, masers, and other microwave devices of
interest to planetary radar astronomers.[23]

Although the electronic side of planetary radar astronomy drove it to attend URSI
meetings and to publish in such journals as the *Proceedings of the IRE*, the astronomy side
pulled it toward meetings of the International Astronomical Union (IAU) and the
American Astronomical Society (AAS) and to publication in astronomy and general science journals, primarily *The Astronomical Journal, Science*, and *Nature*. These institutional
and publication forums, though, did not meet the need for specialized discussion of planetary topics.

Sporadic workshops provided only limited forums. For example, the 1962 inferior
conjunction of Venus furnished the occasion for a symposium on radar and radio observations of that planet. Although planetary radio astronomers delivered most of the symposium papers, radar astronomers Roland Carpenter, Dick Goldstein, and Dewey
Muhleman described the latest radar research on Venus.[24] Aside from a preliminary
report by National Bureau of Standards ionospheric researchers on their one-time-only
radar attempt at Venus, the symposium was strictly a JPL affair.

Starting in 1965, the need for a specialized forum for presenting and discussing
radar research began to be met through a joint URSI-IAU Symposium on Planetary
Atmospheres and Surfaces held at Dorado, Puerto Rico, 24–27 May 1965. The Organizing
Committee included radar astronomers John Evans, Dewey Muhleman, and Gordon
Pettengill, while Evans and Pettengill chaired sessions on lunar and planetary radar
astronomy. The latter session brought together practitioners from Lincoln Laboratory,
JPL, Cornell's nearby observatory at Arecibo, and the Soviet Union.[25]

A conference on lunar and planetary science held during the week of 13 September
1965 and organized by Caltech and JPL also had its share of planetary radar papers.
Researchers from JPL, Jodrell Bank, and Cornell's Arecibo Observatory spoke on Venus,
while JPL and Arecibo representatives read papers on Mars. Noticeably absent, however,
were researchers from Lincoln Laboratory, which was still a major planetary radar
research center.[26]

---

22.   Ray L. Leadabrand, "Radar Astronomy Symposium Report," *Journal of Geophysical Research* 65 (April
1960): 1103–1115; Green 20 September 1993. P. Green to author, 21 December 1994, states that Green described
range-Doppler mapping in his earlier talk at the Endicott House conference, but the talk was not published.
23.   "URSI National Committee Report, XIV General Assembly, Tokyo, September, 1963: Commission 5.
Radio and Radar Astronomy," *Journal of Research of the National Bureau of Standards, Section D: Radio Science* 68D
(May 1964): 631–653; "Commission 7. Radio Electronics," ibid., pp. 655–678.
24.   The symposium papers were published in *The Astronomical Journal* 69 (1964): 1–72. *The Astronomical
Journal* is the publication of the American Astronomical Society.
25.   William E. Gordon, "Preface," *Journal of Research of the National Bureau of Standards, Section D: Radio
Science* 69D (July–December 1965): iii. This was a special issue containing the symposium papers.
26.   Harrison Brown, Gordon J. Stanley, Duane O. Muhleman, and Guido Münch, eds., *Proceedings of the
Caltech-JPL Lunar and Planetary Conference* (Pasadena: Caltech and JPL, 15 June 1966).

Planetary radar astronomy is at the convergence of science and engineering. Attendance of radar astronomers at both IAU and URSI meetings during the 1960s reflected the dichotomous nature of radar astronomy, perched between radio engineering (URSI) and astronomical science (IAU). The dichotomy arose from the fact that radar astronomy is a set of techniques (engineering) used to generate data whose interpretation yields answers to scientific questions.

Just as vital to the growth of radar astronomy as meetings and journals was access to instruments, for without them there would be no science to discuss or to publish. The very availability of radar instruments capable of detecting echoes from Venus had given rise to planetary radar astronomy, and the field has remained a technology-driven science to the present. However, radar astronomers did not seek their own instruments. In league with the Cambridge astronomical community, Lincoln Laboratory campaigned to design and build a large new radar and radio astronomy research instrument. It was radio astronomers, not radar astronomers, who performed the entrepreneurial task of promoting the new facility and who carried radar astronomy interests with it. The same radio astronomers also urged opening to outside researchers the Haystack antenna built by Lincoln Laboratory for military communications research.

During the 1960s, radio astronomy underwent the kind of rapid growth rate that typifies Big Science. With fewer facilities and researchers than Australia or Britain, the leading countries in the field, the United States saw radio astronomy balloon into Big Science as funding requests and antenna construction proposals increased in size and number. Radio astronomy thus provided an emerging Big Science onto which radar astronomers piggybacked their search for instruments free of military priorities and where radar astronomy, as recommended at the Endicott House conference, would be "viewed as pure science." The potential rewards of piggybacking were great, but the price of pursuing Big Science patronage was equally great. In the end, the effort proved troublesome and futile.

Needles and a Haystack

The decade of the 1960s was the era of Big Science and the Big Dish in radio astronomy. The period of large telescope construction between 1957, when the Jodrell Bank 76-meter (250-ft) telescope reached completion, and 1971, when the 100-meter (328-ft) radio telescope near Effelsberg (about 40 km from Bonn) began operation, has been dubbed "the age when big was beautiful" in radio astronomy.[27] As the first Venus experiment took place at Lincoln Laboratory in 1958, a host of new radar research instruments of unprecedented size were on the drawing board or under construction thanks chiefly to the largesse of Cold War military spending on scientific research and secondarily to the National Bureau of Standards and NASA.

The NRL was breaking ground on a 183-meter (600-ft) antenna at Sugar Grove, West Virginia. With funding from ARPA, Cornell had completed initial design studies of a 305-meter (1,000-ft) dish. Lincoln Laboratory had plans for a 37-meter (120-ft) antenna at Haystack Hill, Massachusetts, as well as a solar radar facility at El Campo, Texas, both of which were to be built with defense funds.[28] Stanford and SRI were soliciting military backing for a 244-meter (800-ft) antenna.[29] In the civilian sector, the National Bureau of

---

27.   Robertson, pp. 285–291, has a section called "When Big was Beautiful."
28.   The El Campo facility later was transferred from Lincoln Laboratory to the MIT Center for Space Research and was funded by a National Science Foundation grant. MIT, *Radar Studies of the Sun and Venus: Final Report to the National Science Foundation under Grant No. GP-8128* (Cambridge: MIT, June 1969).
29.   Eshleman 9 May 1994; Leadabrand and Eshleman, *A Proposal for an 800-foot Radar Astronomy Telescope* (Stanford: Stanford Research Institute, 9 October 1959), Eshleman materials.

Standards was building a three-station radar at its Long Branch Field Station, Illinois, and a huge array antenna at Jicamarca, Peru, to study the ionosphere. NASA's Jet Propulsion Laboratory started designing a large antenna system for its Deep Space Network. In Europe and Australia, additional large antennas were on the drawing board or under construction.

No less a part of the Big Dish era were the Haystack and CAMROC/NEROC antennas. Lincoln Laboratory designed and built Haystack for military communications research. Cambridge-area astronomers, organized as the Cambridge Radio Observatory Committee (CAMROC), then as the Northeast Radio Observatory Corporation (NEROC), campaigned to open Haystack to outside researchers. CAMROC/NEROC, again in collaboration with Lincoln Laboratory, also sought funding for the design and construction of a new large radio and radar telescope.

Designing and building those big dishes was a nightmarish introduction to Big Science politics for radio astronomers. Bernard Lovell, the veteran planner and builder of several radio telescopes at Jodrell Bank, not to mention one or two never built, in 1983 wrote to Ed Lilley, the Harvard astronomer who headed efforts to build the new CAMROC/NEROC dish and to open Haystack to outside researchers, and asked him to summarize his experience. Lilley replied that the story presented an "excellent example of the mix of politics, power struggles, fiscal problems, technology and dealings with Congress, and, ultimately, defeat from a few scientific luminaries," and that he would "need a cabin overlooking a thunderous sea to stimulate the mood to undertake writing a history of the CAMROC/NEROC campaign."[30]

The campaign began with the construction of the Haystack antenna, which replaced Millstone as the Lincoln Laboratory planetary radar. On 12 April 1962, Millstone stopped operating, so that Lincoln Laboratory could upgrade it to 1,320 MHz (23 cm; L-band) and increase overall system capability, as part of the Space Surveillance Techniques Program. Over the years, Lincoln Laboratory expanded the Millstone location. Near the Millstone planetary radar was the Lincoln Laboratory Communications Site, established in 1957 to test communication equipment. Upon completion of the tests, the antennas were torn down, and the site given over to construction of an X-band transmitting dish for use in Project West Ford, commonly known as Project Needles. A similar X-band station was built at Camp Parks, outside San Francisco.[31]

On 10 May 1963, Project Needles launched nearly 500 million hair-like copper wires into Earth orbit, thereby forming a belt of dipole antennas. Lincoln Laboratory then sent messages coast to coast via the orbiting copper needles between Camp Parks and Millstone at Westford, Massachusetts (hence the name Project West Ford). British radio astronomers, such as Martin Ryle and Lovell, as well as optical astronomers, objected fervently to Project Needles, and the Council of the Royal Astronomical Society formally protested to the U.S. President's Science Advisor.[32] Haystack was intended officially as a state-of-the-art radar for Project Needles.

---

30.   Quoted in Lovell, *The Jodrell Bank Telescopes* (New York: Oxford University Press, 1985), pp. 249–250. Lovell has described his experiences in *Jodrell Bank* and *The Jodrell Bank Telescopes.*

31.   Overhage to Ferguson, 21 May 1962; Overhage to Ferguson, 28 December 1962; Overhage to Roscoe Wilson, 30 June 1961; J. W. Meyer, "The Lincoln Laboratory General Research Program," paper presented at the Joint Services Advisory Committee meeting, 19 April 1962, pp. 5–6; and W. H. Radford to B. A. Schriever, 6 May 1964, 1/24/AC 134, MITA; Lincoln Laboratory, "Millstone Hill Field Station," April 1965, LLLA.

32.   Overhage to Ferguson, 26 June 1963, 1/24/AC 134, MITA; Overhage and Radford, "The Lincoln Laboratory West Ford Program: An Historical Perspective," *Proceedings of the IEEE* 52 (1964): 452–454; Folder "Project West Ford Releases and Reports," LLLA. Much of the *Proceedings of the IEEE* 52 (1964): 452–606, deals exclusively with Project West Ford. For antagonism of radio astronomers to Project Needles, see Lovell, *Astronomer by Chance*, pp. 331–334; Martin Ryle and Lovell, Interference to Radio Astronomy from Belts of Orbiting Dipoles (Needles)," *Quarterly Journal of the Royal Astronomical Society* 3 (1962): 100–108; D. E. Blackwell and R. Wilson, "Interference to Optical Astronomy from Belts of Orbiting Dipoles (Needles)," ibid., pp. 109–117; and H. Bondi, "The West Ford Project," ibid., p. 99.

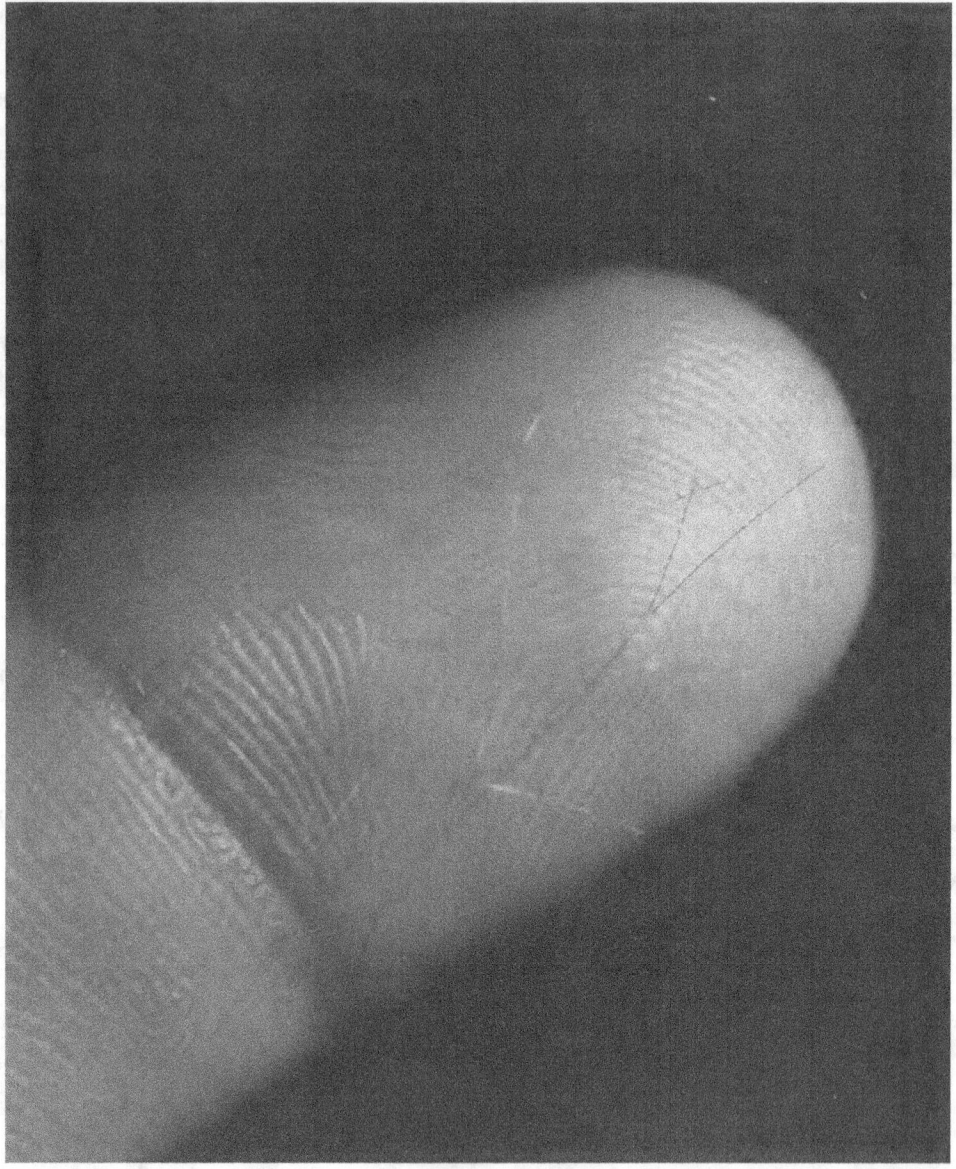

Figure 9
Project Needles planned to launch nearly 500 million hair-like copper wires into Earth orbit, thereby forming a belt of dipole antennas. Haystack Observatory originally was built as part of Project West Ford, which was commonly known as Project Needles. (Courtesy of MIT Lincoln Laboratory, Lexington, Massachusetts, photo no. P201-229.)

Project Needles and the Haystack radar exemplified the new research directions taken by Lincoln Laboratory. The Laboratory had pioneered three major air defense systems: the DEW Line, the SAGE System, and the Ballistic Missile Early Warning System. With the formation of the MITRE Corporation in 1958, Lincoln Laboratory divested itself of manned bomber defense activity and engaged in new research programs that addressed military problems in ballistic missile re-entry systems and ballistic missile defense radars; military satellite communications; and the detection of underground nuclear explosions (Project Vela Uniform). The joint services and ARPA funded this work and supported Lincoln Laboratory's program of general research, which included radar and radio astronomy.[33]

Besides Project Needles, additional applications proposed for Haystack were tracking communication satellites and radar astronomy, the former justified as an adjunct to communications research. The facility's X-band operating frequency ruled out meteor studies. Radio astronomy was also not among the initial proposed uses but emerged later in the earliest funding proposals submitted to the Air Force.[34]

The design of Haystack was an in-house Lincoln Laboratory effort for about a year and a half before the Air Force lent its financial support. The design progressed through several evolutionary stages. The initial March 1958 design called for a 37-meter-diameter (120-ft-diameter) parabolic reflector with a Cassegrainian feed, low-noise maser receivers, and operation in the X-band, all characteristics of the earlier West Ford antennas. The price tag was estimated to be about $5 million, which was too high for Air Force approval.

The problem was to reduce the facility's cost, while designing a reflector that would maintain the high tolerances required for the short X-band wavelength. Exposure to wind and the Sun would warp the dish too much to be effective at X-band. One solution would have been to select a lower frequency range, say S-band, but participation in Project Needles dictated an X-band operating frequency. The solution was to place the antenna inside a radome, which not only protected the antenna from the Sun and wind, but also reduced the weight and power needed to drive the antenna. The radome design was significantly cheaper, too, lowering the estimated cost from $5 million to between $1.5 and $2 million. Adding the radome raised a new design issue, however, because radomes had never been used before at X-band.

Lincoln Laboratory had developed a radome for L-band Millstone-type radars, but it could accommodate a dish no larger than 26 meters (85 ft) in diameter. To enclose the Haystack 37-meter (120-ft) antenna, Lincoln Laboratory engineers raised the radome above ground level and enlarged it from five-eighths to nine-tenths of a complete sphere. Electrical tests carried out in March 1959 determined that a reduction in panel thickness would permit the radome's use at X-band.

In November 1959, Herb Weiss became Haystack project engineer. The following month, the Air Force committed financial support to the project. Lincoln Laboratory took bids on the radar's construction and signed a contract with North American Aviation (Ohio Division) on 1 December 1960. A separate Air Force contract procured the radome and base extension.

Haystack was dedicated on 8 October 1964, at Tyngsboro, Massachusetts, about 30 miles northwest of Boston, but only a half mile up the road from Millstone. Haystack was unique in its use of special plug-in boxes. Each box was 2.4 by 2.4 by 3.7 meters (8 by 8 by 12 ft) and could hold up to 2 tons of equipment. One box contained a 100-kilowatt

33.   Lincoln Laboratory, *The General Research Program*, Report DOR-533 (Lexington: Lincoln Laboratory, 15 June 1967), p. 1.

34.   John Harrington, *The Haystack Hill Station*, Technical Memorandum 78 (Lexington: Lincoln Laboratory, 13 October 1959), pp. 1 and 5–7, LLLA.

Figure 10

Exterior view of the Haystack Observatory in 1964, when the facility was dedicated. There, MIT and Lincoln Laboratory radar astronomers imaged the Moon and Venus and conducted a test of General Relativity. At the time of its dedication, Haystack was one of only three large antennas conducting radar astronomy research on a regular basis. (Courtesy of MIT Lincoln Laboratory, Lexington, Massachusetts, photo no. P10.29-783.)

continuous-wave X-band (7,750 MHz; 4 cm) transmitter, cryogenic low-noise receivers, and associated microwave circuits for planetary radar research.[35]

As Haystack construction was underway, a key meeting of Harvard University astronomers, Donald Menzel, director of the Harvard College Observatory, Leo Goldberg, and Ed Lilley, took place on 24 May 1963. They came together in order to seek access to this new, more sensitive telescope. As a secondary objective, they sought to design and build a larger radio telescope in collaboration with Lincoln Laboratory.

35.   Overhage to Ferguson, 14 November 1962, Overhage to B. A. Schriever, 27 January 1964, and brochure, "Dedication Haystack Microwave Research Facility," 1/24/AC 134, MITA; Memorandum, J. A. Kessler to Radford, 30 September 1964, LLLA; "Millstone Hill Field Station;" Harrington, *Haystack Hill*, pp. 2–3; Weiss 29 September 1993. For a discussion of the design and construction of Haystack, see Weiss, "The Haystack Microwave Research Facility," *IEEE Spectrum* 2 (February 1965): 50–69; Evans, Ingalls, and Pettengill, "The Haystack Planetary Ranging Radar," in L. Efron and C. B. Solloway, eds., *Scientific Applications of Radio and Radar Tracking in the Space Program*, Technical Report 32–1475 (Pasadena: JPL, July 1970), pp. 27–36; and Weiss, W. R. Fanning, F. A. Folino, and R. A. Muldoon, "Design of the Haystack Antenna and Radome," in James W. Mar and Harold Liebowitz, eds., *Structures Technology for Large Radio and Radar Telescope Systems* (Cambridge: MIT Press, 1969), pp. 151–184.

Lincoln Laboratory radar and radio astronomers already enjoyed relatively free access to Haystack, and Lincoln Laboratory radio astronomers often collaborated with their colleagues at Harvard Observatory's Agassiz Station, as well as at the NRAO. The Agassiz Station had been training graduate students in radio astronomy for about ten years under a National Science Foundation grant.

Gaining limited use of Haystack was not difficult. Lilley approached Lincoln Laboratory regarding use of Haystack in July 1964. In September 1965, Lincoln Laboratory and the Air Force reached a mutually agreeable policy on Haystack as well as Millstone. The Air Force encouraged use of the two facilities by scientists outside the Department of Defense and made Lincoln Laboratory responsible for scheduling time. Lincoln Laboratory had to report all outside use of Millstone and Haystack to the Air Force, which had final approval on all requests. Finally, outside agencies would have to pay an hourly fee, to be determined by Lincoln Laboratory, to defray operating and upkeep costs.

At the same Harvard meeting of 24 May 1963, Lilley also suggested that Harvard, MIT (including Lincoln Laboratory), and the Smithsonian Astrophysical Observatory (SAO) jointly undertake a cooperative, regional effort to build a large dish antenna free of military limitations for radio astronomy research. The project sought to marry the strength of Lincoln Laboratory in radar astronomy and the thriving Harvard program in radio astronomy.

The proposed large antenna also would serve the interests of radar astronomers. Although Haystack's greater power and sensitivity outclassed Millstone, Lincoln Laboratory radar astronomers realized that radars then under construction, namely Cornell's 305-meter (1,000-ft) antenna and JPL's 64-meter (210-ft) Mars Station, would outperform Haystack. Lincoln Laboratory radar astronomers therefore sought a telescope with Arecibo's sensitivity, but operating at a higher frequency.[36]

New enthusiasm for the construction of the large telescope ignited upon the release of the Whitford Report, which had endorsed the construction of large dish telescopes for radio astronomy. The Whitford Report grew out of Congressional reaction to the Navy's disastrous attempt to build an enormous steerable dish antenna in West Virginia.

Sugar Grove

The specter that haunted all large radio telescope dish projects was Sugar Grove. In the words of a report of the Comptroller General of the United States to Congress, "The complexity and unique character of the Big Dish [Sugar Grove] were underestimated from the inception of the project."[37] As late as 1965, Harvard astronomer Ed Lilley wrote his colleagues, "International radio scientists still regard the U.S. Navy 600 foot

36. "Ad Hoc Committee on Large Steerable Antenna, Report, 8 July 1963," 5/1/AC 135, Memorandum, Lilley to File, n.d., 10/1/AC 135, Memorandum, Lilley to Sebring and Meyer, 28 July 1964, 11/1/AC 135, Memorandum, 27 September 1965, "A Policy for the Use of the Millstone Hill and Haystack Facilities by Agencies outside the Department of Defense," 6/1/AC 135, and "Ad Hoc Committee on Large Steerable Antenna, Report, 8 July 1963," 5/1/AC 135, MITA; Lincoln Laboratory, *General Research Program*, Report DOR-533, p. 25; MIT Research Laboratory of Electronics, *Annual Research Review and Twentieth Anniversary Program, 10–12 May 1966*, 23 March 1966, pp. 7–8, 13–14, NHOB.

37. Comptroller General, *Report to the Congress of the United States: Unnecessary Costs Incurred for the Naval Radio Research Station Project at Sugar Grove, West Virginia.* (Washington: GPO, April 1964), p. 7. For additional background on the Sugar Grove dish, see Edward F. McClain, Jr., "The 600-foot Radio Telescope," *Scientific American* 202 (January 1960): 45–51; James Bamford, *The Puzzle Palace: A Report on America's Secret Agency* (New York: Penguin, 1983), pp. 218–221; and Daniel S. Greenberg, "Big Dish: How Haste and Secrecy Helped Navy Waste $63 Million in Race To Build Huge Telescope," *Science* 144 (1964): 1111–1112.

paraboloid as a 'radio telescope' fiasco, even though the project had minuscule association with basic research."[38]

As early as 1948, NRL scientists devised a plan for a large steerable telescope for detecting and studying radio sources. By 1956, the NRL had developed an initial proposal which called for a reflector 183 meters (600-ft) in diameter with accurate maneuverability and precision positioning controls. The huge dish would be able to turn a full 360 degrees in the horizon and tilt to any angle of elevation from the zenith to the horizon. If completed, the 183-meter steel-and-aluminum antenna would have stood taller than the Washington Monument, weighed about 22,000 tons (the weight of an ocean liner), and been the largest movable land-based structure ever constructed in the world.

The Navy began breaking ground for the U.S. Naval Radio Research Station, Sugar Grove, West Virginia, telescope in June 1958. As construction got underway, the price tag rose. The initial cost estimate was $20 million, but climbed to $52.2 million in February 1957, when the Department of Defense submitted requests for fiscal 1958 military construction funds to Congress. Later in 1957, coincidental with the launch of Sputnik, the Navy expanded the project concept and included certain (still) classified military surveillance tasks. The nature of those tasks, nonetheless, was an open secret. The Navy planned to listen to Soviet radio communications as they were reflected from the Moon, an idea that grew out of the lunar radar work carried out by Benjamin Yaplee's group at the NRL. Solar, planetary, and ionospheric radar experiments followed.

These new tasks inflated the estimated price tag to $79 million, and the decision to redesign and build the telescope at the same time further ballooned the estimated cost to more than $200 million ($300 million in some estimates), which was the total estimated cost when the Department of Defense canceled the project in July 1962. The fatal decision to design and erect at the same time was an acknowledged "calculated risk" in order to save roughly three or four years of construction time. The emerging new design called for an antenna that was far too heavy for its support structure, which was already under construction. Further complicating the project was an internal turf battle between the Bureau of Yards and Docks and the Naval Research Laboratory. By the time the Department of Defense canceled Sugar Grove, the Navy had spent $42,918,914 on the project, but with the settlement of termination claims included, the secretary of defense estimated that the total expenditure for the telescope amounted to between $63 and $64 million.

An investigation by the comptroller general concluded that the Navy had incurred unnecessary costs in the construction and cancellation of the big dish.[39] The Sugar Grove fiasco raised serious questions about the spending of military research and development dollars. As Senator Hubert H. Humphrey (D-Minn.) pointed out in August 1962, Sugar Grove had "many of the earmarks of other research and development projects which turned out to be 'white elephants.'"[40] The next month, Sugar Grove came under Congressional scrutiny.

---

38.   Memorandum, Lilley, August 1965, "Comments on a Regional Radio and Radar Research Facility for the New England Area," p. 2-1, Box 7, UA V 630.159.10, PAHU.

39.   NRL, *Careers in Space Communications* (Washington: NRL, n.d.), p. 3, NRL, *Radio Astronomy and the 600-foot Dish* (Washington: NRL, n.d.), n. p., and "The Big Dish," typed and edited manuscript, NRLHRC; Comptroller General, pp. 2-4, 6 & 11. Early specifications for Sugar Grove did not include radar experiments. See, for example, *Specifications for the Naval Radio Facility, Sugar Grove, W. Va.* (Washington: NRL, December 1957), and *Specifications for the U.S. Naval Radio Research Station Sugar Grove, W. Va.* (Washington: NRL, September 1959), NRLHRC. Later specifications, though, did indicate radar experiments. P. Green to Robert Page, 14 April 1960, and other documents, Green materials; Eshleman, "Sun Radar Experiment," in MIT, *Radar Astronomy*, vol. 3, lecture 15, p. 10. Fiscal irresponsibility was not the sole factor leading to the termination of the Sugar Grove project; the availability of satellites to perform its espionage functions was certainly another.

40.   *Congressional Record*, 87th Cong., 2d sess., 1962, Vol. 108, pt. 12, pp. 16175-16178.

The Subcommittee on Applications and Tracking and Data Acquisition of the House Committee on Science and Astronautics opened hearings on radio and radar astronomy in September 1962. The Sugar Grove fiasco motivated the hearings, at which radio astronomers defended their telescope projects. Witnesses discussed alternatives to large dishes, such as arrays, in which a number of small antennas electronically linked to each other acted as a single large antenna. Common to the witnesses' testimony was the assertion that the United States lagged behind Australia and Britain in radio astronomy.[41]

American backwardness in radio astronomy was widely accepted in the 1960s by those involved in its funding. For example, in a speech marking the dedication of the NRAO 43-meter (140-ft) radio telescope in 1965, Leland J. Haworth, director of the National Science Foundation, emphasized the Australian, British, and even Dutch lead over the United States in entering the field.[42] While this was neither the first nor the last time that a scientific community would use backwardness to argue for financial support, Cold War competition was not mentioned.

As the federal agency underwriting much of the country's astronomy research, and as the sponsor of the NRAO, the National Science Foundation (NSF) took an avid interest in radio astronomy and its telescopes. In December 1959, well before the Congressional investigation of Sugar Grove, the NSF had appointed an Advisory Panel for Radio Telescopes to appraise current and future needs for radio telescopes. Its report, released in 1961 before the Sugar Grove fiasco was generally realized, did not favor the construction of large dish antennas. Instead, the Panel endorsed arrays using aperture synthesis, a new technique first developed by Martin Ryle in Britain. The endorsement of arrays led immediately to initial design studies of the Very Large Array (VLA), located eventually in New Mexico. The NSF Panel report had more bad news for radar astronomy dishes. Its first resolution stated that antenna requirements for radio and radar astronomy were so different, that radio astronomy antennas "should be primarily designed to meet the needs of passive [radio] astronomy."[43]

The Whitford Report

Radio astronomers clamored for more telescopes. Anyone interested in building a new radio and/or radar telescope dish had to take into consideration the question of parabolic dishes versus arrays, which were still quite experimental and untested, at least in the United States. The NSF was on center stage as the primary civilian funding agency for radio astronomy, and all design concepts and funding requests had to deal with the omnipresent wake of the Sugar Grove disaster. The future of large radio and radar dishes seemed precarious.

Into this situation came the Committee on Government Relations of the National Academy of Sciences. At the suggestion of Harvard astronomer Leo Goldberg, the Committee created the Panel on Astronomical Facilities on 14 October 1963, in order to outline a planned approach to radio and optical telescope construction. Panel membership comprised prominent optical and radio astronomers; Albert E. Whitford of Lick Observatory served as chair.

---

41.   U.S. Congress, House, Committee on Science and Astronautics, Subcommittee on Applications and Tracking and Data Acquisition, *Report on Radio and Radar Astronomy*, 87th Cong., 2d sess., 1962.

42.   "Dedication of new 140-foot radio telescope at the National Radio Astronomy Observatory, Green Bank, West Virginia," remarks by Dr. Leland J. Haworth, 13 October 1965, "Speeches, Leland J. Haworth," NSFHF.

43.   Geoffrey Keller, "Report of the Advisory Panel on Radio Telescopes," *The Astrophysical Journal* 134 (1961): 927–939.

The Panel assembled radio astronomers at a meeting held in Washington on 1 and 2 November 1963 in order to build a consensus. The result of the meeting and the Panel's deliberations was an ambitious, 10-year plan of optical and radio telescope construction. Nonetheless, the result of the Panel's work, known as the Whitford Report, omitted radar and solar astronomy. Solar astronomers protested the neglect in letter after letter.[44]

The Whitford Report specifically rejected solar radar as too costly, but completely neglected planetary radar astronomy. Radar astronomers did not protest. NASA's internal evaluation of the Whitford Report, which Nancy Roman prepared after consulting with those NASA committees and subcommittees responsible for developing the agency's astronomy program, advised NASA to continue its support of radar astronomy. Nonetheless, she wrote, "We do not, at present, foresee NASA support for the construction of new radar facilities, although further experience with radar exploration of the solar system may modify this conclusion." In general, Roman concluded, "Support of astronomy is the province of the National Science Foundation," and the program of telescope construction proposed by the Whitford Report was "within the traditional province of the National Science Foundation which should continue to retain responsibility for them." Although Roman suggested that NASA deep space communications instruments "should incorporate potential use by radio astronomers in their design,"[45] curiously she did not mention lending their use for radar astronomy experiments.

Roman's evaluation summed up what became, for all practical purposes, the NASA position on funding radar astronomy. The construction of ground-based facilities was the responsibility of the NSF; NASA would fund mission-oriented research at existing facilities. The NSF embraced its role as the federal agency with primary responsibility for ground-based astronomy. But full implementation of the Whitford Report construction program required substantial increases in NSF spending on ground-based astronomy, and the Foundation already was the country's major underwriter of ground-based astronomy. In fiscal 1966, of the total federal expenditure of $46.2 million for ground-based astronomy, the NSF share was $21.0 million (46 percent), compared with $9.4 million (20 percent) for NASA, $8.0 million (17 percent) for the Air Force, $4.5 million (10 percent) for the Navy, and $3.3 million (7 percent) for ARPA.[46]

The Whitford Report proposed to spend $224 million (about the cost of Sugar Grove) over 10 years on a number of regional and national facilities. It endorsed 1) a large array as a national facility under the NRAO (the VLA); 2) enlargement of Caltech's Owens Valley Observatory (another array); 3) two fully-steerable 91-meter (300-ft) dishes as regional facilities; 4) a design study of the largest possible steerable dish; and 5) smaller, special purpose instruments.[47]

---

44.   Material in folders "Committees & Boards, Committee on Science and Public Policy, Panels, Astronomical Facilities, 1963," "ADM, C&B, COSPUP, Panels, Astronomical Facilities, Radio Astronomers, Meetings, Agenda, Nov," "Committees & Boards, Committee on Science and Public Policy, Panels, Astronomical Facilities, 1964," and "Committees & Boards, Committee on Science and Public Policy, Panels, Astronomical Facilities, Report, General, 1965," NAS; Gerard F. W. Mulders, "Astronomy Section Annual Report," 25 June 1963, pp. 1–2, and Harold H. Lane, "Astronomy Section Annual Report," 1 July 1964, p. 1, NSFHF; Panel on Astronomical Facilities, *Ground-Based Astronomy: A Ten-Year Program* (Washington: National Academy of Sciences, 1964), p. 57.

45.   Memorandum, Roman to Associate Administrator, Office of Space Science and Applications, 16 March 1965, "ADM, C&B, COSPUP, Astronomical Facilities Rpt Recommendations, Assessment by NSF," NAS.

46.   Haworth to Donald F. Hornig, 5 April 1965, "Committees & Boards, Committee on Science and Public Policy, Panels, Astronomical Facilities, Report, Recommendations, Assessment by NSF, 1965," NAS; "Astronomy Section Annual Report, 1966," p. 1, "MPS Annual Reports," NSFHF.

47.   Harold H. Lane, "Astronomy Section Annual Report," 1 July 1964, p. 2, NSFHF; *Ground-Based Astronomy*, pp. 50–57. In 1955, Caltech began building a radio interferometer consisting of two 90-foot dishes at Owens Valley, California, funded by the U.S. Office of Naval Research. Robertson, pp. 120–121; Marshall H. Cohen, "The Owens Valley Radio Observatory: Early Years," *Engineering and Science* 57 (1994): 8–23.

The Whitford Report favored neither arrays nor dishes, but saw a need for both. As for large dishes, the Report recalled the Sugar Grove fiasco: "The design and evaluation of these solutions are costly and very time-consuming, as has been shown in the unsuccessful attempt at Sugar Grove to build a 600-foot [183-meter] paraboloid." The Report expressed the need for "a thorough-going engineering study" to ensure the construction of large radio telescopes and recommended spending $1 million on design studies for the largest feasible steerable paraboloids "at an early date."[48]

In Dish/Array

The Whitford Report understandably excited both Harvard radio astronomers and Lincoln Laboratory radar astronomers with its endorsements of design studies for large steerable antennas and a regional 91-meter (300-ft) dish. In order to seize the opportunities created by the Whitford Report, Harvard, MIT, and the SAO agreed to undertake a joint study of a large radio and radar telescope, and in August 1965, the group adopted the name Cambridge Radio Observatory Committee and the acronym CAMROC.[49]

In October 1965, when CAMROC drew up a research agenda for the regional telescope, planetary and lunar radar astronomy were featured uses. As Ed Lilley argued: "American radar astronomers have also made major contributions, but in many instances their work has been accomplished by 'borrowing time' on antennas which were mission oriented. In the Cambridge group there are radar scientists who are keenly interested in basic radar astronomy. They, too, need an instrument as powerful and timely as the Palomar 200-inch, where radar astronomy can flourish as a basic science with transmitters and data analysis systems developed for optimum performance on ionospheric, lunar, planetary, and solar problems."[50]

On 29 October 1965, Harvard, MIT, Lincoln Laboratory, and the SAO signed a Memorandum of Agreement, authorizing CAMROC to solicit up to $2.5 million to support design studies for the telescope. MIT would hold, administer, and disburse the funds and act as CAMROC's administrative agent. CAMROC funding was to come from a variety of sources, mostly federal. Of the estimated $2.7 million needed for fiscal 1966 and 1967, the NSF, NASA, and the Smithsonian Institution were to award $1.57 million (58 percent). MIT, Harvard, and private foundations (Kettering and Ford) would provide additional funding.[51]

The NASA money was to come through the Electronics Research Center in Cambridge. Unaware of NASA's evaluation of the Whitford Report, CAMROC submitted a grant proposal to NASA for design studies of the large steerable radio and radar antenna in February 1966. NASA rejected the proposal. As William Brunk, acting chief of Planetary Astronomy, explained, "Support for a project such as this is within the domain of the National Science Foundation and it is recommended that they be approached as a possible source of funding." NASA Deputy Administrator Robert C. Seamans, Jr.,

48.   *Ground-Based Astronomy*, pp. 56 and 75; "Assessment of the recommendations of the Whitford Report, entitled 'Ground-Based Astronomy: A Ten-Year Program,'" Table V, "ADM, C&B, COSPUP, Astronomical Facilities Rpt Recommendations, Assessment by NSF," NAS.

49.   J. A. Stratton to S. Dillon Ripley, 14 May 1965, and Nathan M. Pusey to Stratton, 2 June 1965, 5/1/AC 135, and Minutes of Meeting, 26 August 1965, 14/1/AC 135, MITA.

50.   Lilley, "Comments," p. 2-1, PAHU.

51.   Memorandum, 26 October 1965, "CAMROC Support and Budget," and other documents in 6/1/AC 135 and 12/1/AC 135, MITA.

repeated the message: "The type of effort you proposed is clearly the responsibility of the National Science Foundation."[52]

Despite the clear and consistent reply from NASA, Joel Orlen of MIT and executive officer of the CAMROC Project Office (which was in charge of day-to-day activities) wrote to Jerome Wiesner, MIT provost, "I believe NASA should be pushed on hard to reverse this decision." CAMROC members came to believe that any argument made to NASA had to take into account the risk of offending the advocates of the JPL dish design, that is, the 64-meter (210-ft) Mars Station.

Wiesner wrote to Seamans, requesting that NASA reconsider the rejected proposal; he argued that the technology would be needed in the space effort. Seamans replied that NASA was studying a variety of antenna designs, including arrays, "Because we foresee, in an active and continuing space program, that our ground facilities will be required to support multiple simultaneous flight missions, it may turn out to be more effective to rely on a grouping of antenna systems that can be arrayed together as needed but that can also operate independently for independent missions."[53]

Seamans' reply threw CAMROC plans into disarray. From the beginning, the telescope was to be a large steerable dish. But arrays were gaining popularity and were considered a viable alternative to large radio dishes. The Whitford Report had endorsed both the Owens Valley array and the VLA. In 1955, Caltech began building a pair of 27-meter (90-ft) dishes at Owens Valley, California, with money from the Office of Naval Research; now Caltech proposed expanding the facility. The VLA was to consist of 27 radio telescopes mounted on railroad tracks in a Y formation whose arms were each 21 km long. When completed, each telescope would have a diameter of 25 meters (82 ft).[54] Now, NASA appeared interested in arrays. But were arrays effective in radar astronomy?

Believing that the CAMROC effort would raise questions about the merits of arrays versus dishes, radar astronomer and CAMROC member Gordon Pettengill tackled the question in a memorandum of 9 June 1966. He concluded that arrays had a number of advantages over a single large dish, including the ability to deliver more power to a target. Arrays stretched technology less, promised more reliable capability, and cost less to build. If some array elements were out of service for whatever reason, the deficiency would hardly affect overall performance. Moreover, if full array capability were not needed, the primary array could be divided into several smaller arrays and assigned to different experiments. The major design challenge of arrays, Pettengill pointed out, arose from proving the practicality of phasing a number of elements together. A minor drawback was the need for numerous low-noise receivers and antenna feeds.[55]

Lilley deflected the argument away from the merits of arrays versus dishes by emphasizing the use of the radome. The radome set the design apart from all other radio and radar antenna proposals before the NSF. If the results of the radome tests were satisfactory, Lilley claimed, the CAMROC studies would provide radio and radar astronomy with a "breakthrough in antenna technology," and the CAMROC position would be unique. "Unfortunately," he lamented, "only a small fraction of the radio and radar

52. "Proposal to the National Aeronautics and Space Administration for Support of Design Studies for a Large Steerable Antenna for Radio and Radar Astronomy," February 1966, 55/1/AC 135, and "Project Office Report to CAMROC, Number 2," 30 August 1966, 5/1/AC 135, MITA; William E. Brunk to Director, Grants and Research Contracts, 28 July 1966, NHOB.

53. Memorandum, Joel Orlen to Jerome Wiesner, 7 September 1966, Wiesner to Robert C. Seamans, Jr., 3 October 1966, and Seamans to Wiesner, 15 November 1966, 55/1/AC 135, MITA.

54. For background on the VLA, see David S. Heeschen, "The Very Large Array," *Sky and Telescope* 49 (1975): 344–351; and A. R. Thompson, R. G. Clark, C. M. Wade, and P. J. Napier, "The Very Large Array," *Astrophysical Journal Supplemental Series* 44 (1980): 151–167. The initial theoretical development of arrays is discussed in Bracewell, "Early Work on Imaging Theory in Radio Astronomy," pp. 167–190 in Sullivan. See also P.A.G. Scheuer, "The Development of Aperture Synthesis at Cambridge," pp. 249–265 in ibid.

55. Memorandum, Pettengill to CAMROC Project Office File, 9 June 1966, 18/1/AC 135, MITA.

professional scientists in the United States understand this, and it is unlikely that the National Science Foundation administrators have a clear understanding of the implications of the CAMROC studies."[56]

Although later, in April 1967, the NSF did judge the telescope's unique design feature to be its radome,[57] in the meantime, the ability of the NSF to fund the CAMROC telescope was limited. Lilley foresaw "a dramatic expansion of demand" for federal funding, especially from the NSF, during the summer of 1967 for large radio astronomy telescopes.[58] Nonetheless, the NSF became the largest underwriter of the CAMROC design studies. As of 26 April 1966, total CAMROC funds amounted to $410,000. The largest share, $300,000, came from an NSF grant, with additional money from Harvard ($25,000), the SAO ($20,000), MIT Sloan Funds ($40,000), and the MIT Space Center ($25,000). An earlier attempt to raise money from the Kettering Foundation failed. The Foundation was shifting its funding away from "science" to "education," and the CAMROC telescope was "marginal to their interests." The likelihood of Department of Defense support was equally bleak.[59]

In 1966, the NSF again faced a considerable number of large radio telescope proposals, prompted this time by the large-scale spending proposed by the Whitford Report. In addition to the CAMROC, VLA, and Owens Valley antennas, other projects included "WESTROC," a joint Caltech, Stanford, and University of California at Berkeley telescope. WESTROC was to be a 100-meter (328-ft), fully-steerable S-band radio dish located at the Owens Valley site.

In order to campaign for their telescope, CAMROC held a Conference on Radomes and Large Steerable Antennas on 17 and 18 June 1966. Over 70 persons attended the conference, which dealt exclusively with the proposed CAMROC dish. Participants came from industry (North American Aviation, Rohr Corporation, ESSCO), the NSF, the NASA Electronics Research Center, and the NRL, as well as from MIT, Harvard, the SAO, and Lincoln Laboratory. Lilley also suggested using political pressure.[60] Ultimately, CAMROC did apply political pressure, but not until after employing other tactics, including the expansion of CAMROC into a regional organization.

NEROC

At least as early as February 1966, CAMROC was considering ways of transforming itself into a regional association. The chief reason for the undertaking was to solicit funds for the design, construction, and operation of a regional radio and radar telescope. A regional base, moreover, would be useful in competing for funds against the Very Large Array or WESTROC.[61]

56. Memorandum, Lilley to Edward M. Purcell and Wiesner, 1 August 1966, 22/1/AC 135, MITA.
57. "Report of the Meeting of the Advisory Committee for Mathematical and Physical Sciences," 13–14 April 1967, p. 6, NSFHF.
58. Memorandum, Lilley to Purcell and Wiesner, 1 August 1966, 22/1/AC 135, MITA.
59. CAMROC Funds, 26 April 1966, 7/1/AC 135, Orlen to Wiesner, 24 November 1965, 6/1/AC 135, and various documents in 56/1/AC 135, MITA. NSF Grant GP-5832 was awarded to MIT for the project "Design Studies for a Large Steerable Antenna for Radio and Radar Astronomy." For materials relating to the proposal, see 12/1/AC 135 and 57/1/AC 135, MITA.
60. Memorandum, Lilley to Purcell and Wiesner, 1 August 1966, 22/1/AC 135, and various documents, 49/1/AC 135, MITA. The Institution of Electrical Engineers (London) sponsored a Conference on Large Steerable Aerials for Satellite Communication, Radio Astronomy, and Radar, on 6–28 June 1966. Herb Weiss, William Fanning, and John Ruze from Lincoln Laboratory presented five papers: "Antenna Tolerance Theory: A Review," "Design Considerations for a Large Fully Steerable Radio Telescope," "Performance Measurements on the Haystack Antenna," "Mechanical Design of the Haystack Antenna," "Performance and Design of Metal Space-Frame Radomes." 23/1/AC 135, MITA.
61. Memorandum, Orlen to Wiesner, 8 February 1966, 7/1/AC 135, MITA.

CAMROC reached out to the entire Northeast to establish itself as a regional organization with regional interests, and with justifiable claims to funding for a regional radio and radar telescope. One of the first steps was to choose a name, one which expressed this regional character. The new organization, called the Northeast Radio Observatory Corporation (NEROC), incorporated in Delaware on 26 June 1967. CAMROC also considered a number of corporate arrangements, including the possibility of remaining limited to only Cambridge schools. After lengthy discussion and analysis, CAMROC settled on a corporate structure that combined a "reasonable regional image" with local management. A committee representing qualified users would determine scientific policy, while actual management would remain in the hands of the Cambridge group.[62]

After a detailed study of university astronomy departments in the six New England states, the five adjacent Midatlantic states (New York, New Jersey, Pennsylvania, Maryland, and Delaware), and Washington, DC, NEROC recruited its first members: Boston University, Brandeis University, Brown University, Dartmouth College, Harvard, MIT, the Polytechnic Institute of Brooklyn, the Smithsonian Institution, the State University of New York at Buffalo and Stony Brook, the University of Massachusetts, the University of New Hampshire, and Yale.[63]

Among the universities declining the NEROC invitation was Cornell, which in 1967 managed the world's largest radio and radar antenna at Arecibo, Puerto Rico. Franklin A. Long, vice president for Research and Advanced Studies at Cornell, replied to the MIT invitation to join NEROC on 27 June 1967. Cornell radio astronomers supported the NEROC initiative, he explained, but they did not feel the telescope deserved top priority. The greatest need was for increased resolution, which the VLA promised to deliver. Moreover, they were "still uncertain about the relative advantages of a large steerable dish in the Northeast as compared to the same dish in the Southwest (or Southeast)." Having their own dish as well as an international agreement to use facilities overseas, Cornell was "concerned as to whether formal participation in NEROC would not carry the air of excessive Cornell greediness in this field."[64]

As CAMROC transformed itself into NEROC in 1967, the business of securing additional funding continued. In January 1967, NEROC won a third NSF grant ($675,000) for telescope design studies, bringing the amount of total NSF support to $1,115,000. Nothing guaranteed the continuation of NSF support, however; the Foundation was faced with a multitude of design and construction proposals, and its budget was limited.[65]

In April 1967, the NSF Advisory Committee for Mathematical and Physical Sciences had four radio astronomy projects, including the CAMROC design study, under consideration with a total price tag of $120 million. Funding for all four was not available; the Foundation had to establish which ones to fund. The NSF had no general way to budget for major projects; usually, it treated requests for instrumentation, design studies, or facilities as special cases.[66]

---

62. "Outline of Organization and Management of Radio Observatory," 20 May 1966; untitled document, dated May 1966; and "Alternative Organizational Arrangements," 20 May 1966, 7/1/AC 135; Agenda, CAMROC meeting of 15 June 1967, 8/1/AC 135, and documents in 61/1/AC 135, 66/1/AC 135, and 67/1/AC 135, MITA; NEROC, *Scientific Objectives of the Proposed NEROC Radio-Radar Telescope* (Cambridge: NEROC, 1967), p. 1; Certificate of Incorporation, 22 June 1967, "NEROC," LLLA. The annotated agenda of the first meeting of the NEROC Board of Trustees, the minutes of that meeting, the certificate of incorporation, and the NEROC by-laws are in 11/64/AC 118, MITA.

63. Documents in 8/1/AC 135 and 65/1/AC 135, MITA; Certificate of Incorporation, 22 June 1967, and "Qualifications of Northeastern Institutions for CAMROC Membership," 22 March 1967, "NEROC," LLLA.

64. Long to Wiesner, 27 June 1967, and Wiesner to James A. Perkins, 16 June 1967, 72/1/AC 135, MITA.

65. John T. Wilson to Howard W. Johnson, 17 January 1967, 8/1/AC 135, and Seamans to Wiesner, 15 November 1966, 55/1/AC 135, MITA.

66. "Report of the Meeting of the Advisory Committee for Mathematical and Physical Sciences," 13–14 April 1967, p. 7, NSFHF.

In order to evaluate the four radio telescope proposals, the NSF appointed the Ad Hoc Advisory Panel for Large Radio Astronomy Facilities, called the Dicke Panel after its chair, Robert H. Dicke of Princeton University. By June 1967, when the Panel convened, the NSF had five proposals to consider: the Owens Valley array, the VLA, the Arecibo upgrade, the NEROC antenna, and the WESTROC dish.

The Dicke Panel met in Washington between 24 and 28 July 1967 and listened to technical presentations from members of the proposing institutions. NEROC was asking for $28 million over five years for design and construction of a fully-steerable, radome-enclosed, 440-ft (134-meter) parabolic dish operating at 6,000 MHz (5-cm). Gordon Pettengill wrote the NEROC presentation section on radar astronomy. The NEROC telescope was not the only combined radio and radar astronomy facility looking for money. Thomas Gold, Frank Drake, and Rolf Dyce of Cornell University advocated renovating the Arecibo dish so that it could operate at 3,000 MHz (10-cm) or higher.

Although the Dicke Panel had focused on radio astronomy, it was not blind to radar astronomy. The Panel recognized, for example, that "the use of radar techniques in astronomy has for the first time enabled man to establish direct contact with the planets and to set his own experimental conditions." In contrast to Pettengill's memorandum on radar astronomy arrays, the Dicke Panel judged that "an array cannot be used effectively for spectroscopic work or radar astronomy...without introducing great complications in the electronic system."

Following its deliberations, the Dicke Panel submitted its report to the Director of the NSF on 14 August 1967. The report approved the Owens Valley array, the VLA, and the Arecibo upgrade. To say the least, the Dicke Panel was impressed, perhaps too impressed, by the potential of the spherical Arecibo dish. The Arecibo "type of antenna seems to show great promise for the future and should be considered along with the very large, fully steerable antenna for the next step forward," the Panel ruled. It urged appraisals of Arecibo's performance and suggested that both the WESTROC and NEROC proposals be deferred until more was known of the performance of spherical dishes.[67] As we shall see in the next chapter, the Arecibo antenna was considerably inefficient.

The Dicke Panel report devastated NEROC plans, not to mention planetary radar astronomy at Lincoln Laboratory. The only radar telescope available to Lincoln Laboratory investigators was the Haystack antenna. The Arecibo 305-meter (1,000-ft) dish and JPL's 64-meter (210-ft) Mars Station, moreover, already outclassed Haystack. NEROC tried to salvage its antenna project. MIT physics professor Bernard F. Burke suggested that NEROC consider a smaller, 101-meter (330-ft) dish. "We should not be so beguiled with the idea of being temporarily the master of the world's biggest radio telescope," he wrote, "that we cannot accept an instrument that is only one of the biggest."[68]

Technical reports and symposia papers, though, continued to support the feasibility and desirability of the 134-meter (440-ft) design. The International Symposium on Structures Technology for Large Radio and Radar Telescope Systems, sponsored by MIT and the Office of Naval Research and held at MIT on 18-20 October 1967, saw participants from the United States and six other countries discussing the latest designs for large

---

67. National Science Board, Approved Minutes of the Open Sessions, meeting of 8 September 1967, pp. 113:14–113:15, National Science Board; "Draft of G. Pettengill's material for CAMROC facilities proposal," 21 April 1967, 62/2/AC 135, and NEROC, "A Large Radio-Radar Telescope: Proposal for a Research Facility," June 1967, 61/2/AC 135, MITA; "Report of the Ad Hoc Advisory Panel for Large Radio Astronomy Facilities," 14 August 1967, typed manuscript, pp. 2–4, 9–10 and 13–14, NSFL. The members of the Dicke Panel were Bart J. Bok, Stirling A. Colgate, Rudolph Kompfner, William W. Morgan, Eugene N. Parker, Merle A. Tuve, Gart Westerhout, and Robert H. Dicke.

68. Memorandum, Burke to Lilley, 6 October 1967, 8/2/AC 135, MITA.

telescopes in Europe, the 100-meter (328-ft) Effelsberg antenna and the proposed 122-meter (400-ft) dish at Jodrell Bank.[69]

Design studies for the NEROC radio and radar telescope continued. During an 18-month period in 1966 and 1967, an interim agreement between MIT and the Air Force partially underwrote the studies. Funding at Lincoln Laboratory tightened, however, and Herb Weiss learned that Lincoln Laboratory no longer could pay for personnel doing NEROC studies after 1 January 1968. The design work carried on thanks to modest support from its Cambridge backers. The three original NEROC members, the SAO, MIT, and Harvard, contributed $121,241, of which MIT and Harvard gave 84 percent.[70]

The NEROC project had relied on the technical expertise and financial largesse of Lincoln Laboratory, plus a few not inconsequential NSF grants worth over $1.6 million. At this critical point, as Lincoln Laboratory "soft" money melted and the Dicke Panel advised deferring the NEROC telescope, getting more time on the Haystack telescope became a higher and urgent priority.

HAYROC

In October 1967, Lincoln Laboratory asked NEROC if it were interested in assuming responsibility for Haystack. NEROC was interested and wanted to study costs and use management, but without impairing progress on the design of the 134-meter (440-ft) antenna. As funding for the big dish design studies slowed to a trickle in 1968, NEROC management of Haystack began to look even more desirable. The matter was the first item of business at NEROC's 25 May 1968 meeting. After some discussion, NEROC unanimously voted to begin negotiations with Lincoln Laboratory and to explore sources of financial support to turn Haystack into a regional observatory.[71]

Air Force support of Haystack paid for a single "shift," meaning five eight-hour days a week. NEROC radio astronomers wanted more observing hours, a second and, if possible, a third "shift," that is, additional increments of time averaging forty hours a week. In response to the NEROC interest, Lincoln Laboratory offered a large portion of its current Haystack schedule to NEROC users at no charge, with "overtime" hours at minimal cost beginning January 1969. In stages, NEROC would assume responsibility for antenna management and for securing operating funds, as available observing time increased incrementally toward a maximum schedule of four and a half shifts (three eight-hour shifts each day plus weekends for a total of about 2,000 hours per year for each shift). Lincoln Laboratory still would be an important user of the antenna and would continue to provide substantially to the operating budget.

NEROC established subcommittees responsible for estimating costs, for drawing up mutually agreeable plans between Lincoln Laboratory and its sponsors and between NEROC and its sponsors, for laying out a management structure, and for pursuing funding. Among the funding sources explored were the NASA Electronics Research Center and the state of Massachusetts, both of which encouraged further discussions but cautioned that eventual support, if any, would be in modest amounts. In addition, NEROC

69.   Documents in 62/1/AC 135, MITA. For the Jodrell Bank 440-foot (134-meter) MARK V telescope, see Lovell, *The Jodrell Bank Telescopes*, Chapters 5–6 and 9–11. For the Effelsberg telescope, see Otto Hachenberg, "The 100-meter Telescope of the Max Planck Institute for Radio Astronomy in Bonn," *Proceedings of the IEEE* 61 (1973): 1288–1295, also in Mar and Liebowitz, pp. 13–27, which are the proceedings of the International Symposium on Structures Technology for Large Radio and Radar Telescope Systems.

70.   Weiss to Wiesner, 21 September 1967, 18/2/AC 135, and documents in 63/1/AC 135, MITA. MIT contributed $72,381, Harvard $30,000, and the SAO $18,860; NEROC had received $1,615,000 from the NSF.

71.   "Board of Trustees: Second Meeting of the NEROC Board of Trustees, 10/22/67," 62/1/AC 135, and "Board of Trustees: Third Meeting of the NEROC Board of Trustees, 5/25/68," 63/1/AC 135, MITA.

approached MIT, Harvard, the University of Massachusetts, and the Environmental Science Services Administration (for a very long baseline interferometer with their dish at Boulder). The NSF was not left out of the search.[72]

Meanwhile, the NEROC 134-meter (440-ft) antenna project had languished. Now, though, the Smithsonian Astrophysical Observatory stepped in. The SAO had not contributed technically to the design of the big dish, nor had it contributed significantly to its financial support. But the SAO, through its parent organization, the Smithsonian Institution, could rally political support and make claims for the NEROC/Smithsonian telescope being a national, not a regional, facility.

The possibility of the Smithsonian Institution obtaining Congressional authorization for the NEROC telescope was first summarized in a memorandum to the NEROC Board on 3 September 1967. During the summer of 1968, NEROC and Smithsonian Institution representatives discussed the possibility of the Smithsonian Institution leading the drive to obtain funding for the NEROC telescope. The discussions led to an understanding, which included management of the project during the design, construction, and operational phases of the facility.[73]

As a pivotal preliminary step, the Smithsonian Institution organized a meeting of radio and radar astronomers to marshall agreement on the need to build the NEROC telescope. If the meeting of radar and radio scientists endorsed the NEROC telescope, then the Smithsonian Board of Regents would be asked to approve the attempt to obtain Congressional authorization for it. The meeting took place at the Museum of History and Technology, as it was then called, at Constitution Avenue and 14th Street, NW, on 30 November and 1 December 1968. About three dozen invited radio and radar astronomers and a handful of NSF and NASA officials attended, in response to an invitation from the Secretary of the Smithsonian Institution, S. Dillon Ripley.

After Fred Whipple (Harvard) opened the meeting with a review of the Smithsonian Institution's "historical role" in astronomy, John Findlay (NRAO) explained the purpose and plan of the meeting and pointed out that five years after the Whitford Report, none of the recommended facilities had been built. Talks and discussions covered the gamut of telescope questions, including the Arecibo spherical dish and the issue of using arrays for radar astronomy.

James Bradley, assistant secretary of the Smithsonian Institution, laid out the plan that his institution might follow and assuaged worries about staying on the good side of the NSF. MIT's Edward M. Purcell reviewed the basic design concept: a 134-meter-diameter (440-ft-diameter) dish, enclosed in a 171-meter (560-ft) radome, the whole costing about $35 million. Whipple explained that the telescope would be a national, not a regional, facility, and assured the gathering that the SAO would "absolutely not" dominate the telescope's planning and policy committee.

On the last meeting day, Findlay sought to bring the participants together in agreement around common issues. The formal "Conclusions and Recommendation," by majority vote of the participants, declared that there was "an urgent need for a large filled-aperture radio-radar telescope in the United States to assist in the solution of a wide range

72.   NEROC, Proposal to the National Science Foundation for Programs in Radio and Radar Astronomy at the Haystack Observatory, 8 May 1970, p. V.2, LLLA; "Board of Trustees: Third Meeting of the NEROC Board of Trustees, 5/25/68," 63/1/AC 135; "Board of Trustees: Fourth Meeting of the NEROC Board of Trustees, 1/18/69," 64/1/AC 135; and NEROC, Proposal to the National Science Foundation, for Research Programs in Radio Astronomy Using the Haystack Facility, for the period 1 July 1969 to 30 June 1970, p. 4, 11/64/AC 118, MITA. The proposal can be found in "Research Proposals in Radio Astronomy Using the Haystack Facility, 7/1/69–6/30/70," 23/2/AC 135, and "Operating Expenses for the NEROC Haystack Observatory, 7/1/69–6/30/70," 24/2/AC 135, MITA.
73.   Memorandum, Lilley to NEROC Board of Trustees, 21 November 1968, Box 1, UA V 630.159.10, PUHA; Documents in 64/1/AC 135, MITA.

of important problems in astronomy and astrophysics." The telescope was to be operated
as a national facility and located "primarily on the basis of scientific and technical criteria." The meeting resolved that the Smithsonian Institution should submit a proposal to
the appropriate federal agencies and carry general responsibility for the funding, design,
construction, and operation of the telescope. Finally, participants approved that the
"NEROC design for a 134-meter (440-ft) telescope in a radome is close in size and general specifications to a feasible optimum design," and endorsed it as the basis for the final
design of the Smithsonian telescope.[74]

The meeting was an unqualified success. James Bradley wrote to Ripley after the
meeting: "We have succeeded in gaining the support of thirty astronomers for our leg-
islative proposal to authorize the design and construction of a large-diameter, radio-radar
astronomical antenna."[75] The conference was only the first step in preparing to go directly to Congress. In the following weeks, the Smithsonian Institution and NEROC assembled
materials for the campaign. Among those materials was a publicity packet that included a
photograph of a model of the completed dish. Herb Weiss estimated the cost of the facility and compared the costs presented in the NSF proposal of June 1967 with projected
costs based on June 1969 and June 1970 starting dates.

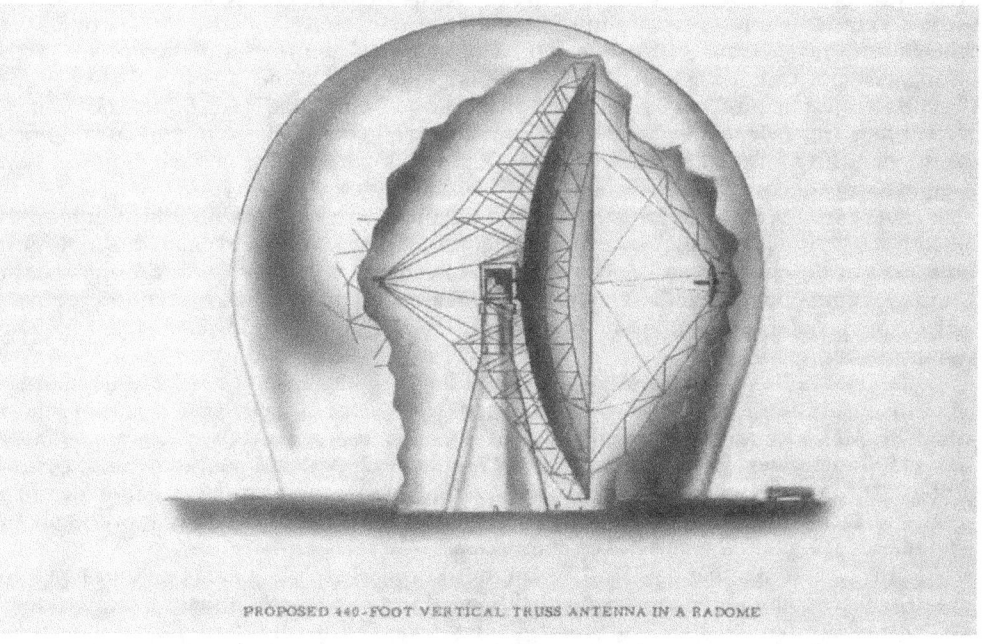

PROPOSED 440-FOOT VERTICAL TRUSS ANTENNA IN A RADOME

Figure 11
Artist's drawing of the proposed NEROC 440-foot (134-meter), radome-enclosed, fully-steerable antenna. This and other draw-
ings and models were prepared to raise funding for the radio-radar telescope. Its radar was to operate at 5 cm (6,000 MHz or
6 GHz), which was lower than Haystack Observatory's wavelength of 3.8 cm (7,750 MHz). (Courtesy of MIT Lincoln
Laboratory, Lexington, Massachusetts, photo no. 259646-1.)

---

74.   James C. Bradley, Charles A. Lundquist, and Lilley, draft letter to all regents, 20 November 1968, and
Memorandum, Lilley to NEROC Board of Trustees, 21 November 1968, Box 1, UA V 630.159.10, PUHA;
"Minutes, Radio and Radar Astronomers Meeting," pp. 1–5, 8-9, 11–12, 15–18 and 24, "List of attendees and
observers," Attachment 1, and "Conclusions and Recommendations," 61/137, SIAUSC, 1959-1972; J. W. Findlay,
"Summary of a meeting to consider a large filled-aperture radio-radar telescope," 1 December 1968, "SAO 1968,"
217, SIAOS, SIA.
75.   Memorandum, Bradley to Ripley, 16 December 1968, "SAO 1968," 217, SIAOS, SIA.

On 3 January 1969, STAG (Smithsonian Telescope Advisory Group), the radio astronomy advisory committee to Dillon Ripley, met at Lincoln Laboratory and reviewed detailed drawings of the design and the latest cost estimate. Meanwhile, the Smithsonian Institution Board of Regents approved requesting an initial $2 million for completing the NEROC design and authorized acquiring land for a site. The next step was to ask the Bureau of the Budget (BoB) for approval to include the $2 million request in the Smithsonian Institution budget for fiscal 1970. On 20 January 1969, Ripley submitted the proposed radio telescope legislation to the director of the BoB.[76]

Although the intention of approaching Congress directly was to circumvent the NSF review process, the Smithsonian Institution kept the Foundation informed. Meanwhile, in August 1968, NEROC submitted a proposal to the Foundation for expanded radio astronomy research at Haystack. The purpose of the proposal was to increase radio observing time to three shifts. It also included a three-year plan for shifting management and financial responsibility to NEROC, as well as a suggested management structure. The Haystack Scientific Advisory Committee, consisting of MIT and Harvard scientists, would assist the observatory director in approving experiments. Any qualified radio astronomer in the United States could request time.[77]

During what Haystack director Paul B. Sebring characterized as "the long, silent interval following the August 68 submission of the transfer plan" on 14 March 1969, Lincoln Laboratory, MIT, and NEROC concluded an interim agreement on the transfer of Haystack to NEROC and established the Haystack Observatory Office to evaluate and coordinate experiment proposals and to serve as a conduit for non-Lincoln Laboratory auxiliary funds for Haystack.[78]

The National Science Foundation turned the NEROC proposal over to the second Dicke Panel, which met in June 1969, nearly a year after NEROC submitted its proposal. The Panel recommended supporting Haystack radio astronomy. The blessing of the Dicke report turned into a one-year NSF grant effective 15 September 1969. The grant paid for wages, computer time, and other costs associated with adding two more shifts of observing time. Under the conditions of the grant, moreover, the Haystack telescope was opened to all qualified radio astronomers in the United States, subject to the approval of the Haystack Scientific Advisory Committee.[79]

The orderly transition of Haystack into a civilian radio observatory appeared on track, until a military auditor balked at the disparity between the Department of Defense and NSF shares of Haystack support. The NSF had bought two-thirds of the observing time for $200,000, while the Air Force paid about $1.3 million for only one-third. The

---

76.   Ripley to Haworth, 17 March 1969, 9/2/AC 135, documents in 10/2/AC 135, 12/2/AC 135, and 64/1/AC 135, MITA. Members of the Smithsonian Telescope Advisory Group, 18 February 1969: John W. Findlay, NRAO, Green Bank; Alan H. Barrett, MIT; Von R. Eshleman, Stanford; Richard M. Goldstein, JPL; Carl E. Heiles, UC Berkeley; John D. Krauss, Ohio State University; Frank J. Kerr, University of Maryland; A. Edward Lilley, Harvard; Alan T. Moffet, Caltech; Gordon H. Pettengill, Arecibo; Irwin I. Shapiro, MIT; Harold F. Weaver, UC Berkeley; and Gart Westerhout, University of Maryland. "NEROC Bd. of Trustees Minutes," Box 2, UA V 630.159.10, PUHA.
77.   NEROC, Proposal to the National Science Foundation, for Research Programs in Radio Astronomy Using the Haystack Facility, for the period 1 July 1969 to 30 June 1970, pp. 1–2 and 4–5, 11/64/AC 118, MITA. The proposal also can be found in "Research Proposals in Radio Astronomy Using the Haystack Facility, 7/1/69–6/30/70," 23/2/AC 135, and "Operating Expenses for the NEROC Haystack Observatory, 7/1/69–6/30/70," 24/2/AC 135, MITA. The scientific advisory committee consisted of Alan H. Barrett, William A. Dent, A. Edward Lilley, and Irwin I. Shapiro.
78.   Memorandum, Sebring to M. U. Clauser, 21 November 1969, 12/56/AC 118, and "Haystack Observatory Office, Agreement Establishing the H.O.O., 3/14/69," 31/2/AC 135, MITA.
79.   "Report of the Second Meeting of the Ad Hoc Advisory Panel for Large Radio Astronomy Facilities," 15 August 1969, p. 22, NSFL; Louis Levin to Wiesner, 12 September 1969, 18/2/AC 135, MITA; NEROC, Proposal to the National Science Foundation for Programs in Radio and Radar Astronomy at the Haystack Observatory, 8 May 1970, p. IV.3, LLLA.

arrangement conflicted with a Bureau of the Budget circular, and the auditor requested a written release from the Air Force before he would pass on the funding arrangement. Brig. Gen. R. A. Gilbert, Air Force Systems Command director of laboratories, refused to sign a written release; such a waiver, he judged, might commit the Air Force to underwriting Haystack through the end of fiscal 1970, a position he felt he could not take.[80]

The Mansfield Amendment cut this Gordian knot. Formally known as Section 203 of the Fiscal 1970 Military Procurement Authorization Act, the Mansfield Amendment compelled the Pentagon to demonstrate the mission relevance of basic research financed through its budget. Specifically, the Amendment stated: "None of the funds authorized to be appropriated by this Act may be used to carry out any research project or study unless such project or study has a direct or apparent relationship to a specific military function of operations." Sen. Mike Mansfield's goal had been to rechannel public funding for science through civilian rather than military agencies.[81]

The Air Force announced its intention to terminate operation of Haystack no later than 1 July 1970. The Mansfield Amendment was a key factor in that decision. Although the Air Force expressed its willingness to cooperate with the NSF in an orderly transfer, the decision brought chaos. With no Air Force money after 1 July 1970, Haystack was in a perilous financial situation. Sebring, as Haystack director, obtained NSF consent to reprogram its grant funds to defray the entire cost of Haystack radio astronomy operations. A small grant from the Cabot Solar Energy Research Fund supplemented the NSF money.[82]

The early withdrawal of the Air Force hastened agreements on Haystack ownership, management, and finances. The Air Force transferred Haystack to MIT, which already owned the land. Haystack personnel remained employees of MIT. The NEROC Board of Trustees appointed the observatory director, who reported to them through the board chair. NEROC took responsibility for Haystack research and financing.

To continue support of radio astronomy after 1 October 1970, NEROC submitted a new proposal to the NSF in May 1970. The proposal presented three alternative funding levels, but the NSF awarded less than that requested for a minimal program.[83] Subsequently, the NSF annually renewed its support of Haystack radio astronomy. The successful transition of Haystack from military to civilian funding and monitorship ultimately had an impact on the NEROC/SAO effort to fund the 134-meter (440-ft) telescope through Congress.

The Big Dish Bill

On 28 January 1969, Senators Clinton P. Anderson (D-N.M.), Hugh Scott (R-Pa.), and J. W. Fulbright (D-Ark.), all three regents of the Smithsonian Institution, introduced a bill in the Senate (S.705) "to authorize the Smithsonian Institution to acquire lands and to design a radio-radar astronomical telescope for the Smithsonian Astrophysical Observatory for the purpose of furthering scientific knowledge, and for other purposes."[84]

---

80.   Brunk, Memo to the Files, 18 December 1969, NHOB.

81.   James L. Penick, Jr., Carrol W. Pursell, Jr., Morgan B. Sherwood, and Donald C. Swain, eds., *The Politics of American Science 1939 to the Present*, rev. ed. (Cambridge: The MIT Press, 1972), pp. 338–349.

82.   W. D. McElroy to Grant Hansen, 5 May 1970, 18/2/AC 135, Hurlburt to Sebring, 20 May 1970, 18/2/AC 135, and Wiesner to Orlen, 16 July 1970, 16/2/AC 135, MITA; Hansen to Thomas O. Paine, 26 February 1970, NHOB.

83.   NEROC, Proposal to the National Science Foundation for Programs in Radio and Radar Astronomy at the Haystack Observatory, 8 May 1970, p. IV.1, LLLA; Wiesner to Wilbur W. Bolton, Jr., 15 October 1970, 18/2/AC 135, MITA.

84.   Documents in "Radio-Radar Telescope Legislation, 91st Congress, 7/1/69–12/31/69," 60, SIAOS, and "SAO 1968," 217, SIAOS, SIA; "Congress Gets 'Big Dish' Bill," Vol. 9, No. 4 *The SAO News* (March 1969): 1 and 4, 24/1/AC 135, MITA.

A STAG meeting of 2 April 1969 decided the site for the NEROC telescope. After set-tling upon a number of site criteria, STAG limited the site candidates to the continental United States, a decision, Fred Whipple pointed out, which led "almost inexorably to a final selection somewhere in the southern border states from western Texas through New Mexico and Arizona into California."[85]

The Smithsonian legislation, known popularly as the "Big Dish" bill,[86] requested $2 million for the fiscal year ending 30 June 1970. The bill was read twice, then referred to the Senate Committee on Rules and Administration. On 17 November 1969, Morris K. Udall (D-Arizona) introduced the legislation in the House (H. R. 14,837), where it was referred to the Committee on House Administration.

The Big Dish bill picked up approvals from NASA and the NSF. In February 1969, John Naugle, NASA associate administrator for Space Science and Applications, gave his blessing to the bill: "The addition of such a radio-radar telescope as a national facility would satisfy a need for the future of radio astronomy in the United States."[87] On 17 March 1969, Ripley asked Leland J. Haworth, director of the NSF, for his institution's sup-port of the Smithsonian legislation. The NSF's reply came in the form of an invitation. Robert Fleischer, head of the NSF Astronomy Section, wrote that the Dicke Panel would reconvene, on 9–11 June 1969, and invited NEROC to prepare a 30-minute presentation on the current status of its radome design.[88] If the Dicke Panel again deferred or reject-ed the NEROC design in favor of another project, passage of the Smithsonian Institution bill would be jeopardized.

Two years had passed since the first Dicke Panel met. "A need that was then urgent has now become critical," the second Dicke Report declared. "While our country has stood still, Great Britain, the Netherlands, Germany, and India have started new, large radio telescopes and several are essentially complete and ready for operation." The Panel reaffirmed the need to upgrade the Arecibo dish and supported the Owens Valley array and the Very Large Array. As for the NEROC antenna, the second Dicke Panel found it "clear that this instrument is not only feasible, but ready for final design and construc-tion." The Panel recommended that "the final design and construction...be started now...with the utmost dispatch." The Panel suffered amnesia, too; its report claimed that it had "highly recommended for continuation" of the NEROC design study two years ear-lier. In its conclusions, the second Dicke panel declared: "The urgent need for such a tele-scope is proven beyond doubt. The instrument is ready to go into the construction phase." Whether funded through the Smithsonian Institution or the National Science Foundation, "it is evident that this instrument should be operated as a national facility."[89]

The Dicke Panel report was released on 15 August 1969. Although Congress inter-preted the report as supporting the Big Dish, the Dicke Panel recommendations neither changed the playing field in Congress nor clarified the issues. After a two-hour hearing on 10 September 1969, Rep. Frank Thompson, Jr., (D-NJ), chairman of the Subcommittee on Library and Memorials, deferred the Big Dish legislation. He insisted on having reports

85.   Whipple to Bradley, 3 April 1969, "Miscellaneous Correspondence and Other Material," Box 1, UA V 630.159.10, PUHA.
86.   See, for instance, "Biggest Radio-Radar Scope Asked for U.S.," *Washington Evening Star*, 1 April 1969, p. A15, in "Radar Astronomy," NHO.
87.   John E. Naugle to Richard A. Buddeke, 18 February 1969, NHOB; "Radio-Radar Telescope Legislation, 91st Congress, 7/1/69–12/31/69," 60, SIAOS, SIA; "SAO 1968," 217, SIAOS, SIA; "Congress Gets 'Big Dish' Bill," pp. 1 and 4, 24/1/AC 135, MITA.
88.   Robert Fleischer to Wiesner, 20 May 1969, 18/2/AC 135, and Ripley to Haworth, 17 March 1969, 9/2/AC 135, MITA.
89.   "Report of the Second Meeting of the Ad Hoc Advisory Panel for Large Radio Astronomy Facilities," 15 August 1969, typed manuscript, pp. 1–3 and 15–17, NSFL. The membership of the second Dicke Panel was the same as the first, with the exception of Merle A. Tuve, Carnegie Institution of Washington, who was unable to attend.

from NASA, the NSF, and the Department of Defense before holding hearings. After the submission of the reports, hearings were set for 15 September 1969,[90] but the question was not settled before the end of the Congressional session.

House hearings took place on 29 July 1970, after Rep. Thompson reintroduced the legislation (H. R. 13,024) on 22 July 1970. The primary hurdle facing the bill was the tight budget, although money was available for the war in Vietnam. As Rep. Thompson quipped: "Maybe if we could get this [telescope] in the Defense budget it would be all right, but then I would be against it." In April 1971, Lilley and the Smithsonian Institution in fact did consider an amendment to the Big Dish bill that would include classified Navy research among its duties.[91]

During the 19 July 1970 hearings, astronomers argued that the telescope was need-ed because the United States was behind the rest of the world in radio astronomy. At no point, however, did anyone defend the telescope's radar research program. The bill went to the Subcommittee on Library and Memorials, which unanimously voted to report the bill to the Committee on House Administration with the recommendation that it be reported to the Congress for enactment into law.

The BoB torpedoed the Big Dish bill, however, citing the findings of a special NSF review committee, which had assigned higher priority to two other projects. The proposed expenditure, moreover, was not consistent with Nixon Administration efforts to limit fis-cal 1970 funding to items of the highest priority and to avoid commitments for fiscal 1971 and beyond. Among other issues, the BoB pointed out that the bill raised basic questions about the appropriate roles of the Smithsonian Institution and the NSF.[92]

The Big Dish bill returned to Congress in March 1971. On 31 March 1971, Rep. Thompson told Dillon Ripley that the bill would go through the House "with no trouble."[93] The Greenstein Panel, however, stopped the bill. Ripley wrote to Sen. Clinton Anderson on 23 June 1971 advising him to postpone action on the bill. The latest incarnation of the Dicke Panel, chaired by Jesse Greenstein, Caltech astronomy professor, was going to rec-ommend three facilities: the VLA, a large centimeter-wave antenna, and a large millimeter-wave antenna. It also was going to recommend that the VLA be started first. "In view of the priorities to be established by the Committee," Ripley wrote, "it does not seem wise to seek authorization now for the Smithsonian telescope. The three projects are all of great value to radio-radar astronomy and should not be put into a competition for limited Federal funds. If the array project is authorized on a reasonable time-scale, we look forward to a timely resumption of our efforts with you on the large Smithsonian telescope."[94]

The saga of the NEROC radio-radar telescope ended not in Congress, but within NEROC itself. Once Haystack was opened to radio astronomers from NEROC and other institutions, thanks to funding from the NSF, pressure to build the NEROC telescope eased. NEROC board members had come to realize, too, that the Big Dish bill was a lost

---

90.   "Radio-Radar Telescope Legislation, 91st Congress, 7/1/69–12/31/69," SIAOS, 60, SIA; "Statement by Herbert G. Weiss for Congressional Subcommittee Hearings, October 1969," 9/2/AC 135, MITA.

91.   Transcript of Congressional hearing of 29 July 1970, Subcommittee on Library and Memorials of the Committee on House Administration, pp. 381–382 and 393, "Miscellaneous Correspondence and Other Material," Box 1, and "Miscellaneous Correspondence and Other Material," Box 2, UA V 630.159.10, PUHA.

92.   Transcript of hearing, pp. 381–382 and 393, "Miscellaneous Correspondence and Other Material," Box 1; Memorandum for the record, James Bradley, 16 September 1969, and James M. Frey to Frank Thompson, Jr., 2 September 1969, "Miscellaneous Correspondence and Other Material," Box 2, UA V 630.159.10, PUHA; "SAO Radio-Radar Telescope, 1970," and Ripley to Lucien N. Nedzi, 2 April 1971, SIAOS, 61, SIA; Memorandum, Orlen to Wiesner, 4 February 1969, 9/2/AC 135, MITA.

93.   Ripley to Nedzi, 2 April 1971, "SAO Radio-Radar Telescope, 1971," 61, SIAOS, SIA.

94.   Ripley to Anderson, 23 June 1971, "440' Congress Suspension," Box 1, UA V 630.159.10, PUHA. The subpanel for radio telescopes included David S. Heeschen, NRAO; Geoffrey R. Burbidge, UC La Jolla; Bernard F. Burke, MIT; Frank Drake, Cornell; Gordon Pettengill, MIT; and Gart Westerhout, University of Maryland.

cause. In addition, radio astronomy was changing; millimeter frequencies were the newest frontier. So at an ad hoc meeting of 25 April 1972, Ed Lilley and the other NEROC members voted to terminate the Big Dish project. Instead, NEROC would concentrate on an NSF proposal to upgrade Haystack, so that it could operate at a wavelength of three millimeters.[95]

In retrospect, Herb Weiss, who voted at the ad hoc meeting, reflected on the demise of the NEROC project: "It's very difficult to judge the absolute priorities; it's a moving territory. I really felt that the country made the wrong decision not to pursue NEROC. Even though they might have dragged it out, they might have done something, but it's such small money and such a great step in the right direction, and not the ultimate. I mean you can go beyond that, but it'll take a long time; you've got to get new materials."[96]

For planetary radar astronomy, here was a lesson in Big Science. The need for the NEROC telescope, the decision to design and build it, and the entrepreneurial skills and energy to push the project all came from radio astronomers, not radar astronomers. Piggybacking onto a Big Science (radio astronomy) telescope helped to overcome many obstacles, but in the end, the loss of control that is inherent in piggybacking cost radar astronomy the telescope. Also, the episode illustrated that ultimately the instrument needs of radio and radar astronomers can be inharmonious.

Literally, they operate at different wavelengths. Whereas radio astronomers found a wavelength of three millimeters exciting, planetary radar astronomers could not operate at such short wavelengths. The generation of sufficient power to conduct radar experiments at millimeter wavelengths was, and remains, an insurmountable technological obstacle.

The Nadir of Radar

Three years after NEROC voted to terminate the Big Dish bill, all planetary radar stopped at Haystack; Lincoln Laboratory was out of the planetary radar business. The last Haystack planetary radar transmission traveled to Mercury on 22 March 1974.[97] The NSF supported radio astronomy at Haystack, but planetary radar depended on mission-oriented NASA grants. Topographical studies of the Moon and Mars supported the Apollo and Viking missions. In an exceptional move, when the hasty departure of the Air Force imperiled the telescope's finances, NASA patched together the required amount from the NASA Planetary Astronomy, Viking, and Manned Spacecraft Center program budgets.[98]

The obvious explanation for the end of planetary radar at Haystack is that the upgraded Arecibo telescope outclassed it. Yet reality was neither so obvious nor so simple. The upgraded Arecibo radar, in fact, was not operational until almost a year and a half after Haystack carried out its last planetary radar experiment. Although the upgraded Arecibo telescope was far more sensitive, it could look at a target for only two hours and forty minutes at best. With an ability to track targets for many more hours, Haystack could

95.   Memorandum, Lilley to Bradley, 1 May 1972, "SAO Radio-Radar Telescope, 1971," 61, SIAOS, SIA. Those attending the meeting included: Alan Barrett, MIT; Bernard Burke, MIT; Irwin Shapiro, MIT; Paul Sebring, Haystack and Lincoln Laboratory; Edward Purcell, Harvard; Herbert Weiss, Lincoln Laboratory; and Ed Lilley, Harvard and SAO.

A footnote to the NEROC story: a Haystack upgrade completed in January 1994 made it the premier United States radio observatory at 3 millimeters. An NSF review of Haystack carried out in the summer of 1994, only months after the NSF-funded upgrade, put funding for Haystack radio astronomy in jeopardy. Ramy A. Amaout, "NSF Review Puts Funding for Haystack in Jeopardy," *The Tech* vol. 114, no. 18 (5 April 1994): 1 and 9.

96.   Weiss 29 September 1993.

97.   Photocopy of Haystack logbook entry provided by Richard P. Ingalls and Alan E. E. Rogers.

98.   Memorandum, Henry J. Smith, 15 December 1969, and memorandum, Brunk to Distribution List, 10 June 1970, NHOB.

compensate for its lack of sensitivity by increasing signal integration time. Hardware alone was not the only reason for the end of planetary radar at Haystack.

Haystack radar use, heavy at first, did not stop suddenly in 1974, but declined gradually over the years. In 1970, radar accounted for about a third of observing time,[99] far more than at Arecibo or JPL. An optimistic NEROC proposal submitted to the NSF in 1971 stated: "It is believed that, for the next several years, the Planetary Radar instrumentation should continue to occupy the Haystack antenna for roughly 40 to 50 percent of the available time."[100] In fact, the actual total antenna time (exclusive of maintenance and improvements) for planetary radar observing fell from 17 percent in 1971 to 14 percent in 1972, then to 12 percent in 1973.[101]

Part of the problem was intense competition among radio astronomers for telescope time. The search for molecular spectral lines was frenetic and intensely competitive.

Figure 12
The Haystack Observatory planetary radar box. Technicians preparing the box for an experiment suggest the size of the box. A large forklift truck raised the box into position on the telescope. (Courtesy of MIT Lincoln Laboratory, Lexington, Massachusetts, photo no. P10.29-1785.)

99.   Sebring to Hurlburt, 27 March 1970, 18/2/AC 135, MITA. In March 1970, for example, of the 290 hours scheduled, 90 (31 percent) were spent on radar observations.
100.   "Plan for NEROC Operation of the Haystack Research Facility as a National Radio/Radar Observatory, 7/1/71–6/30/73," 26/2/AC 135, MITA.
101.   NEROC, *Semiannual Report of the Haystack Observatory*, 15 January 1972, p. 1; *ibid.*, 15 July 1972, p. 1; *ibid.*, 15 January 1973, p. 1; and *ibid.*, 15 August 1975, p. iii, MITA. For the 12 month period January through December 1973, out of 5,462.5 hours of total scientific use, planetary radar accounted for 658 hours, or about 12 percent. "Haystack Notes June 73–Dec 74," SEBRING.

Although Haystack installed radio astronomy equipment on the planetary radar box in early 1970 to increase observing time for radio astronomers, complaints about the box continued. Indeed, the planetary radar box could sit on the antenna for months at a time. In the second half of 1972, for example, planetary radar work kept the box on the antenna from 13 July to 24 September and from 9 October to 12 November.[102] As radar astronomer Gordon Pettengill reflected, "It wasn't convenient to make a change for a few hours from one box to another, and that's what really did it [Haystack] in I think."[103]

Another factor was NASA's decision to not fund research facilities. As the Air Force began withdrawing financial support from Haystack, NASA seemed to be a natural source of at least some operational funding. In his reply to the Air Force, NASA Deputy Administrator George M. Low explained that at NASA, "We consider, however, that within the present budgetary limitations and compared to other ongoing programs, the research programs that could be performed at the Haystack Facility have too low a priority to claim NASA support of the overall operational cost of the Facility." If another agency were to provide general operational support, NASA would be happy to underwrite specific, mission-oriented research, such as the topographic studies of Mars and the Moon.[104]

The Haystack radar transmitter klystron tubes, without which planetary radar could not be carried out, suffered from internal arcing on occasion. "At times," Haystack Associate Director Dick Ingalls explained, "it was hairy."[105] In 1973, Haystack asked NASA for a replacement klystron tube. NASA refused, accepting the risk that klystron failure meant the end of planetary radar research.[106] Of the two NASA missions for which Haystack conducted planetary radar research, Apollo and Viking, Apollo was over by 1973. Once Haystack radar data ceased serving the needs of the Viking mission, NASA no longer had any mission interest in Haystack planetary radar research.[107]

Thus, temperamental klystrons, complaints from radio astronomers, the end of NASA mission funds, and NASA's policy of not funding facility operations all contributed to bring Haystack planetary radar to its nadir and demise. Despite that demise and the fate of the NEROC telescope, planetary radar astronomers at Lincoln Laboratory and MIT were not without an instrument. The future was at the Arecibo Observatory.

102. Sebring to Hurlburt, 27 March 1970, 18/2/AC 135; NEROC, *Semiannual Report of the Haystack Observatory*, 15 August 1975, p. iii; and *ibid.*, 15 January 1972, p. 1, MITA. Also, see the references to complaints by radio astronomer William A. Dent in Memorandum, Sebring to Haystack Observatory Office Members, 2 February 1971, 44/2/AC 135, MITA.

103. Pettengill 28 September 1993.

104. George M. Low to Grant L. Hansen, 2 April 1970, NHOB.

105. Ingalls 5 May 1994.

106. Memorandum, Brunk to Joyce Cavallini, 25 July 1973, NHOB.

107. Haystack Observatory, *Final Progress Report: Radar Studies of the Planets* (Westford: NEROC, 29 August 1974). This was for NASA grant NGR-22-174-003.

Chapter Four

Little Science/Big Science

Lincoln Laboratory was not the only center where planetary radar took root. Cornell University had its Arecibo Observatory; JPL had its Goldstone facility. At each center, radar astronomy developed in the shadow of military, space, radio astronomy, and ionospheric Big Science. In fact, without those Big Science activities, planetary radar astronomy would not have had instruments to carry out research and, in short, would not have existed.

In 1961, when the first successful detections of Venus took place, virtually the sole funder of planetary radar astronomy in the United States was the military. The one exception was JPL's Goldstone facility, which NASA funded. Ten years later, the NSF took over the role of prime underwriter of the Arecibo Observatory from ARPA, and NASA agreed to support a major S-band upgrade of the facility's radar. As a result, NASA became the de facto patron of planetary radar astronomy at Arecibo, Goldstone, and Haystack. NASA supported planetary radar at those three centers through a variety of financial arrangements. Only at Arecibo, however, did NASA formally agree to support a planetary radar facility, as well as the research conducted with it. That agreement, moreover, was an obvious departure from its policy formulated in the wake of the Whitford Report.

The shift from military to civilian sponsorship at Arecibo, just as in the case of Haystack, was not in response to the Mansfield Amendment. Under the Kennedy Administration, the military, mainly the Office of Naval Research, already had started transferring research laboratories, especially nuclear physics facilities, to the NSF. The budgetary reforms introduced under Defense Secretary McNamara, whose first major reform was to make the DoD's budget reflect the military missions for which it was responsible, probably provided the initial impetus to those transfers.[1]

The emergence of NASA as the patron of planetary radar astronomy is obvious only in hindsight. Throughout the 1960s, NASA refused to fund radar construction, except for the Deep Space Network. The NSF was the prime underwriter of astronomy facilities, but did not support planetary radar research. Consequently, during the 1960s, planetary radar astronomers depended on the kindness of Big Science, whether the radio astronomers at Haystack, or the NASA space programs at Goldstone, or ARPA (the military sponsor of Arecibo), for its instruments.

But in 1971, NASA broke with its established policy and paid for S-band radar equipment and underwrote the research conducted with it at Arecibo. The result was not just a new NASA policy but also the creation of a permanent institutional and financial home for planetary radar astronomy that the field lacked elsewhere. This unique arrangement came about through the complex politics of science typical of Big Science facilities. Complicating relations between the Arecibo Observatory and its parent organization, as well as relations with its funding agency, were turf battles between competing Big Science fields (radio astronomy and ionospherics) and personality conflicts.

---

1.   Emilio Q. Daddario, "Needs for a National Policy," *Physics Today* 22 (1969): 33-38; James E. Hewes, Jr., *From Root to McNamara: Army Organization and Administration, 1900-1963* (Washington: U.S. Army Center of Military History, 1975), pp. 299-315.

The Arecibo Ionospheric Observatory

Cornell University's 1,000-ft (305-meter) Arecibo dish began as a UHF radar managed by a civilian institution, Cornell University, but funded by the military. The (Air Force) Rome Air Development Center largely funded Cornell ionospheric research, and the original purpose of the Arecibo telescope was to conduct ionospheric research. The Arecibo facility started in the mind of William E. Gordon, Cornell professor of electrical engineering, who had devised a new incoherent scatter technique for studying electrons in the upper levels of the ionosphere by bombarding them with powerful radar waves. He worked on the technique with Cornell electrical engineering colleagues Henry Booker, his dissertation advisor, and Ben Nichols, both of whom shared Gordon's interest in ionospheric research. A Cambridge graduate, Booker had worked in the radio section of the Cavendish Laboratory, and during World War II he led the theoretical division of the radar effort at the Telecommunications Research Establishment.[2]

In order to apply his technique, Gordon realized he needed a powerful radar, which he proposed to build with state-of-the-art components. Gordon also realized that the instrument would be costly, too costly to have a single purpose. He proposed, therefore, that it also do radar astronomy experiments. Henry Booker added radio astronomy, a field that interested him.

Funding for the initial radar design studies, completed by Gordon, Booker, and Nichols in December 1958, came from the military: the Office of Naval Research, the Rome Air Development Center, and the Aerial Reconnaissance Laboratory, Wright Air Development Center, Wright-Patterson Air Force Base, Ohio. The studies outlined the radar parameters: a pulse radar with 2.5 megawatts of peak power and 150 kilowatts average power (essentially the Millstone radar transmitter), a low noise temperature, and an operating frequency around 400 MHz (430 MHz; 70 cm in the final design). The availability of components and antenna technical limits largely determined the operating frequency. The antenna itself was to be a parabolic dish 305 meters (1000 ft) in diameter fixed in a zenith-pointing position and fed from a horn on a 152-meter (500 ft) tower.[3]

Concurrent with these design studies, Bill Gordon sought funding. The budget of the NSF, the agency of choice for basic research, was not large enough for the project, and NASA was interested in building spacecraft. The National Bureau of Standards already had built ionospheric radars and was building a dipole array radar in Jicamarca, outside Lima, Peru, that incorporated Gordon's incoherent scatter technique. Its director, Ken Bowles, a Cornell graduate, had demonstrated the feasibility of Gordon's technique with a Bureau of Standards meteor radar in Illinois.

Gordon first presented his project to ARPA in the summer of 1958. ARPA was an entirely new agency. Although Gordon was not aware of it at the time, ARPA's Defender Program was an effort to research, develop, and build a state-of-the-art defense against

---

2.   Gordon 28 November 1994; Benjamin Nichols, telephone conversation, 14 December 1993; "Cornell University Center for Radiophysics and Space Research," typed manuscript, 12 August 1959, Office of the Administrative Director, NAIC; Gordon, "Incoherent Scattering of Radio Waves by Free Electrons with Applications to Space Exploration by Radar," *Proceedings of the IRE* 46 (1958): 1824–1829; George Peter, *Evolution of Receivers and Feed Systems for the Arecibo Observatory* (Ithaca: NAIC, 1993), pp. 4–5; *SCEL Journal* Vol. S-1, no. 32 (6 August 1953): 2, "Signal Corps Engineering Laboratory Journal/R&D Summary," HAUSACEC; Gillmor, "Federal Funding," p. 126.

3.   Gordon 28 November 1994; Benjamin Nichols, telephone conversation, 14 December 1993; Gordon, Booker, and Nichols, *Design Study of a Radar to Explore the Earth's Ionosphere and Surrounding Space,* Research Report EE 395 (Ithaca: Cornell School of Electrical Engineering, 1 December 1958), pp. 1 and 10–11; Gordon, *Antenna Beam Swinging and the Spherical Reflector,* Research Report EE 435 (Ithaca: Cornell School of Electrical Engineering, 1 August 1959), pp. 1 and 8; CRSR, *Construction of the Department of Defense Ionospheric Research Facility - Final,* Research Report RS 55 (Ithaca: CRSR, 30 November 1963), p. 2; Gillmor, "Federal Funding," p. 127.

Soviet missiles. Though some ARPA scientists saw the scientific value of Arecibo, ARPA's main interest in the project was as part of the Defender Program to track the ion trails created by missile exhaust.[4]

Gordon campaigned in Washington for two years. The Sugar Grove dish was a barrier to gaining approval; reviewers wanted to know why he needed to build the 305-meter (1000-ft) dish, when the Navy had a fully-steerable antenna under construction. Finally, Gordon met Ward Low of the Institute for Defense Analysis and an ARPA adviser, and ARPA agreed to finance the engineering and construction of the dish. The Air Force Office of Scientific Research (AFOSR), through the Electronics Research Directorate, Air Force Cambridge Research Laboratories (AFCRL), Bedford, Massachusetts, monitored the contract. The AFCRL now influenced the design of the telescope. Low introduced Bill Gordon to the AFCRL antenna group, which had been studying spherical reflectors for over a decade. They redesigned the fixed, zenith-looking paraboloid into a spherical reflector with a movable antenna feed mounted on a suspended platform.[5]

The antenna was larger than any other attempted for radar or radio astronomy, larger even than the Sugar Grove dish. The size required an unprecedented support structure. Cornell civil engineering professors proposed placing the dish in a natural bowl in the earth. The proposal was practical from an engineering perspective and cut costs, according to preliminary studies by William McGuire and George Winter, Cornell School of Civil Engineering.

Topographical, political, and scientific factors influenced the choice of a site. In the tropics, the planets would pass nearly overhead and into the antenna's cone of view. After considering sites in Hawaii, central Mexico, Cuba, Puerto Rico, and some smaller Caribbean islands, the search narrowed to the Island of Kauai, the Matanzas area of Cuba, and northern Puerto Rico. Political and import problems eliminated Cuba; Hawaii was too far and too remote. Puerto Rico had a favorable location, political stability, and minimum distance, as well as a karst topography full of sinkholes in which to locate the giant reflector. After looking at locations in Puerto Rico, Cornell chose a natural bowl in the mountains south of the city of Arecibo.[6]

With feasibility and location established, ARPA and Cornell signed a contract on 6 November 1959 in which the University agreed to perform three tasks: 1) conduct design studies on a vertically-directed ionospheric radar probe; 2) consider ionospheric and other scientific uses for the instrument, then propose a priority list of the first experiments; and 3) lay out structures and buildings needed for the initial facility.[7]

Meanwhile, also in 1959, Henry Booker launched the Center for Radiophysics and Space Research (CRSR), an umbrella organization for mainly astronomy and electrical engineering faculty research, as well as management of the Arecibo facility. Booker shared its administration with fellow Cambridge graduate Thomas Gold. Like Booker, Gold had

---

4.   Gordon 28 November 1994; Nichols, telephone conversation, 14 December 1993; Jack P. Ruina, "Arecibo," *Electronics* 7 April 1961, n.p., article in publicity folder, Office of the Administrative Director, NAIC; CRSR, *Ionospheric Research Facility*, p. 2; Herbert F. York, *Making Weapons, Talking Peace: A Physicist's Odyssey from Hiroshima to Geneva* (New York: Basic Books, 1987), pp. 142–143; Gillmor, "Federal Funding," p. 126.

5.   Gordon 28 November 1994; Philip Blacksmith, "DODIRF 1000-foot Spherical Reflector Antenna," and Alan F. Kay, *A Line Source Feed for a Spherical Reflector*, Technical Report 529 (Hanscom AFB: AFCRL, 29 May 1961), Phillips Laboratory; Roy C. Spencer, Carlyle J. Sletten, and John E. Walsh, "Correction of Spherical Aberration by a Phased Line Source," *Proceedings of the National Electronics Conference* 5 (1949): 320–333; Gillmor, "Federal Funding," p. 127.

6.   Gordon 28 November 1994; Gordon, "Arecibo Ionospheric Observatory," *Science* 146 (2 October 1964): 26; Gordon, "Arecibo Ionospheric Observatory," p. 26; Gordon, Booker, and Nichols, pp. 12–13; Donald J. Belcher, "Site Locations for a Proposed Radio Telescope," Appendix C in ibidem; R. E. Mason and W. McGuire, "The Fixed Antenna for a Large Radio Telescope: Feasibility Study and Preliminary Cost Estimate," Appendix B in ibidem.

7.   CRSR, *Design Studies for the Arecibo Radio Observatory*, Research Report RS 9 (Ithaca: CRSR, 30 June 1960), NAIC, p. 1.

worked on radar during World War II, but at the Admiralty Research Establishment. After the war, Cambridge, the Cavendish Laboratory, and the Royal Observatory in Greenwich, Gold arrived in the United States in 1957 and taught astronomy at Harvard. Booker thought Gold ideal for running the CRSR.[8]

The CRSR staff, professors from the astronomy, electrical engineering, and physics departments, drew up a research program for the Arecibo telescope. Following ARPA guidance, they listed 20 experiments arranged in order of priority. The first three explored the ionosphere. Then came proposals for planetary, lunar, solar, and other radar work, followed by three more ionospheric experiments. The last 10 were all radio astronomy experiments. The first 10, the CRSR staff concluded, were "clearly within the scope of the ARPA missions," but the "relation of experiments 11 through 20 [in radio astronomy] to the ARPA mission is not so clear." ARPA did not appear interested in radio astronomy. Well before the telescope's inauguration, however, radio astronomy had been assigned a major role in its scientific mission.[9]

Cornell next began building the Department of Defense Ionospheric Research Facility, as the telescope was named originally. Construction of the structure, antenna, concrete towers, and electronics were let out to over a half dozen commercial subcontractors, while the Army Corps of Engineers supervised the construction and civil engineering. The raising of the 300-ton feed platform from the bottom of the bowl, where it had been assembled, to its approximate final position 152 meters (500 ft) overhead, was an awe-inspiring sight. As Bob Price recalled, the raising of the pylons was also "Very impressive....They had all these very strong Puerto Ricans pulling at cables. It was like some 1930s Mexican mural painting. Labor at its best. All coordinated pulling at these cables, and pouring cement at the same time, and getting the right tension on everything."[10]

8.   Gold 14 December 1993; Nichols, telephone conversation, 14 December 1993; "Center for Radiophysics;" *Annual Summary Report, Center for Radiophysics and Space Research, July 1, 1965—June 30, 1966,* 30 June 1966, p. 10; *Arecibo Observatory Program Plan, October 1, 1970—September 30, 1971,* May 1971, pp. 62–63, AOL.

9.   CRSR, *Scientific Experiments for the Arecibo Radio Observatory,* Research Report RS 5 (Ithaca: CRSR, 31 March 1960), pp. vii and 31–33; AIO, *Research in Ionospheric Physics,* Research Report RS 41 (Ithaca: CRSR, 30 June 1962), p. 7.

10.   Price 27 September 1993; CRSR, *Construction of the Department of Defense Ionospheric Research Facility,* Research Report RS 22 (Ithaca: CRSR, 30 June 1961), pp. 1–2; *ibid.,* Research Report RS 34 (Ithaca: CRSR, 31 December 1961); *ibid.,* Research Report RS 40 (Ithaca: CRSR, 30 June 1962), pp. 12–15; *ibid.,* Research Report RS 45 (Ithaca: CRSR, 31 December 1962), pp. 1 and 11–12; various items in publicity binder, Office of the Administrative Director, NAIC; Thomas C. Kavanagh and David H. H. Tung, "Arecibo Radar-Radio Telescope Design and Construction," *Journal of the Construction Division, Proceedings of the American Society of Civil Engineers* 91 (May 1965): 69–98.

Figure 13

Aerial view of the Arecibo Observatory showing its location in a natural sinkhole in the hills of north central Puerto Rico. The antenna is so large that its can only be seen in its entirety from above. (Courtesy of National Astronomy and Ionosphere Center, which is operated by Cornell University under contract with the National Science Foundation.)

After its inauguration on 1 November 1963, the Arecibo Ionospheric Observatory (AIO) was not just a Cornell-ARPA facility; it also became part of an international agreement for the exchange of faculty and graduate students between Cornell and the University of Sydney, signed in September 1964. The University of Sydney was a major, worldwide center for radio astronomy. The agreement gave Americans access to some of the most advanced radio astronomy instruments in the world, as well as some of the most renowned researchers.[11]

Bill Gordon directed the observatory at Arecibo. After meeting Gordon Pettengill at Millstone, Thomas Gold "twisted his arm" to get Pettengill to take the position of associate director. At Lincoln Laboratory, Pettengill had carried out radar astronomy experiments, but more as a hobby. When he arrived at Arecibo in July 1963, "A totally new world opened up down there. This was a university-operated facility....And there was no direct military work!" Pettengill devoted his entire time to planetary radar and achieved recognition in the field.[12]

What made the Arecibo world so different, apart from the lack of "military work" that was the bread and butter of Lincoln Laboratory, was the fact that planetary radar astronomy was an integral part of the scientific agenda. Arecibo's university connection would supply graduate student researchers. Moreover, as associate director, Pettengill could hire people to do planetary radar. Thus, the earliest Arecibo planetary radar

11.   *CRSR Summary Report, July 1, 1964—June 30, 1965,* 1 July 1965, CRSR, p. 5; *Cornell-Sydney University Astronomy Center,* 1965, p. 4, AOL; Gold and Harry Messel, "A New Joint American-Australian Astronomy Center," *Nature* 204 (1964): 18–20.
12.   Pettengill 28 September 1993; Gold 14 December 1993.

astronomer was not trained in the traditional way, as a graduate student in an academic setting, but was hired to do planetary radar. The first such hire was Rolf B. Dyce.

Pettengill first met Dyce years earlier, when Dyce was with the Rome Air Development Center, Griffiss Air Force Base, in Rome, New York. Dyce had a B.A. in Physics and a Ph.D. from Cornell, where he did radar studies of auroras. Dyce eventually landed a job with the Stanford Research Institute (SRI) at Menlo Park, California, where he worked on classified ionospheric and radar research, including auroral, meteor, and lunar studies. Dyce and Pettengill also toured Europe together and visited key radar research centers, including Jodrell Bank, the Dutch facility at Dwingeloo, the Chalmers Institute in Gothenburg, Sweden, and the Norwegian Defense Research Establishment. Pettengill hired Dyce in January 1964, just weeks after the Arecibo dedication in November 1963.[13]

Arecibo was different from Lincoln Laboratory and Haystack in many other ways, too, because of the relationships between Arecibo and Cornell and between Arecibo and Lincoln Laboratory. While MIT did not train radar astronomers to work at Lincoln Laboratory, Cornell sent graduate students to Arecibo to work on doctoral dissertations in radar astronomy. MIT students also carried out radar astronomy dissertation research at Arecibo. As a result, Arecibo became a training ground for future radar astronomers.

Some of the earliest graduate student radar research was done on the Sun and Moon, not the planets. Vahi Petrosian, a Cornell graduate student working on a masters thesis, attempted some solar radar work in July and August 1964. After later attempts by two other graduate students, solar echo experiments were abandoned; the results were neither as good nor as productive as those achieved by the El Campo solar radar.[14]

On the other hand, starting in 1965, Arecibo undertook a far more vigorous and productive program of lunar radar research with supplementary funding from NASA, which hoped to use the results to help select Apollo landing sites.[15] Carrying out the lunar work in collaboration with Dyce and, occasionally, Pettengill was Cornell graduate student Thomas W. Thompson. The research formed the basis of his 1966 doctoral dissertation. Thompson worked briefly at Haystack, then again at Arecibo, before he found a position at JPL. He returned to Arecibo occasionally to make lunar radar observations.[16]

---

13.   Dyce 22 November 1994; Pettengill 28 September 1993.

14.   AIO, *Research in Ionospheric Physics*, Research Report RS 61 (Ithaca: CRSR, 31 December 1964), pp. 46–48; *ibid.*, Research Report RS 72 (Ithaca: CRSR, 31 January 1968), p. 127; Vahi Petrosian, *Two Possible Methods of Detecting UHF Echoes from the Sun*, Research Report RS 54 (Ithaca: CRSR, 30 September 1963), which was his masters thesis. His doctoral thesis, completed in June 1967, however, was on "Photoneutrino and Other Neutrino Processes in Astrophysics." Petrosian later went to Stanford. CRSR, "Proposal to National Science Foundation for Research Ionospheric Physics, Radar-Radio Astronomy, October 1, 1969 through September 30, 1971," April 1969, pp. 138–140, Office of the Administrative Director, NAIC.

Donald B. Campbell obtained solar continuous-wave echoes at 40 MHz (7.5 meters) during the summer of 1966. AIO, *Research in Ionospheric Physics*, Research Report RS 70 (Ithaca: CRSR, 31 January 1967), p. 75. Alan D. Parrish, a NASA Trainee, and Campbell made more solar observations in 1967. ibid., Research Report RS 71 (Ithaca: CRSR, 31 July 1967), pp. 78–79; Campbell 8 December 1993.

15.   Thompson and Dyce, "Mapping of Lunar Radar Reflectivity at 70 Cm," *Journal of Geophysical Research* 71 (1966): 4843-4853; Thompson, "Radar Studies of the Lunar Surface Emphasizing Factors Related to Selection of Landing Sites," Research Report RS 73 (Ithaca: CRSR, April 1968); Gold, *CRSR Summary Report, July 1, 1964— June 30, 1965*, 1 July 1965, CRSR, p. 4; *Annual Summary Report, Center for Radiophysics and Space Research, July 1, 1966—June 30, 1967*, 30 June 1967, p. 10; *Annual Summary Report, Center for Radiophysics and Space Research, July 1, 1968—June 30, 1969*, 30 June 1969, p. 4; AIO, *Research in Ionospheric Physics*, Research Report RS 61 (Ithaca: CRSR, 31 December 1964), pp. 39–41.

16.   Thompson 29 November 1994; NAIC QR Q2/1970, n.p.; Thompson, "The Study of Radar-Scattering Behavior of Lunar Craters at 70 Cm," Ph.D. diss., Cornell, February 1966; Thompson, "Radar Studies of the Lunar Surface Emphasizing Factors Related to Selection of Landing Sites," Research Report RS 73 (Ithaca: CRSR, April 1968). The lunar radar measurements were made at 40 MHz (7.5 meters) and 430 MHz (70 cm) at the AIO.

The next graduate student was Raymond F. Jurgens, whose 1969 dissertation used Arecibo radar data to form some of the first range-Doppler images of Venus. Then came Donald B. Campbell, originally from Australia. Using the radar interferometric method developed at Haystack, and working under both Dyce and Arecibo director Frank Drake, Campbell began a lifelong career devoted to the radar imaging of Venus.[17] Both he and Jurgens later were key figures in planetary radar astronomy.

While its relationship with Cornell turned Arecibo into a breeding ground of radar astronomers, its relationship with Lincoln Laboratory and Haystack, forged through the presence at Arecibo of Gordon Pettengill, provided entree to the software, techniques, and ephemerides developed by Lincoln Laboratory. Pettengill was a vital factor not only as associate director from 1963 to 1965, but also as Arecibo director from 1968 to 1970.

At the heart of that relationship was the business of creating radar ephemerides. The standard planetary ephemerides issued by the U.S. Naval Observatory were simply not accurate enough for radar work, so special ephemerides computer programs had to be developed. In order for them to be as accurate as possible, these radar ephemerides had to draw on a data base of radar observations. At Lincoln Laboratory, Irwin Shapiro started such a radar ephemerides computer program. Haystack provided a large amount of the ephemerides data, and so did Arecibo at the instigation of Gordon Pettengill, with a modest grant from NASA. Pettengill recalled the speed with which radar observational data arrived at Lincoln Laboratory: "I remember we used to send it back by special delivery mail. We would mail it by six in the evening at Arecibo, and it would be delivered in Lexington, Massachusetts, at nine the next morning. Very efficient. Then it would be put into the Lincoln Laboratory ephemeris program."[18] In addition to the ephemerides, Lincoln Laboratory supplied Arecibo with software and techniques. As mentioned earlier, Don Campbell adopted the Haystack radar interferometry technique at Arecibo, and the special fast Fourier transform software created for the Haystack interferometer also migrated to Arecibo.[19]

When Pettengill left Arecibo in 1970, he returned not to Lincoln Laboratory, but to MIT, where he became professor of planetary physics in the Department of Earth and Planetary Sciences. The change from Lincoln Laboratory to MIT was as stimulating to Pettengill as the original move to Arecibo. He continued planetary radar research, using both Haystack and Arecibo. He was not alone; both Tommy Thompson and Don Campbell used both telescopes.[20] Moreover, Pettengill, who already had guided the radar astronomy dissertations researched at Arecibo, began offering a course in radar astronomy at MIT and sending MIT graduate students to Arecibo to do their doctoral research.

The fruit of this cross-fertilization between Arecibo and MIT and Lincoln Laboratory was that Arecibo evolved into a common research facility for both Cornell and MIT, so that by the time planetary radar astronomy research ended at Haystack, Arecibo already was in position to continue the research programs underway at Haystack. That did not mean, though, that the Arecibo telescope provided the same amount of observing time as Haystack.

At Haystack, planetary radar astronomy accounted for a greater percentage of observing time than at Arecibo. Although planetary, lunar, and solar radar experiments occupied roughly 9 percent of Arecibo antenna time for the period December 1965 through September 1969, only 2.4 percent of total observing time was given over to radar

---

17.   CRSR, "Proposal to National Science Foundation for Research Ionospheric Physics, Radar-Radio Astronomy, October 1, 1969 through September 30, 1971," April 1969, Office of the Administrative Director, NAIC, pp. 138–140; Jurgens, "A Study of the Average and Anomalous Radar Scattering from the Surface of Venus at 70 Cm Wavelength," Ph.D. diss., Cornell, June 1968.

18.   Pettengill 28 September 1993.

19.   Rogers 5 May 1994; Hine 12 March 1993. For a discussion of radar interferometry at Haystack and Arecibo, see Chapter Five.

20.   Pettengill 28 September 1993.

astronomy in 1970, while radar accounted for about a third of Haystack antenna time in the same year.[21] Moreover, as radar astronomy use of Haystack declined from 17 percent in 1971 to 12 percent in 1973, radar use of the Arecibo telescope increased, but not proportionally, and peaked in 1972 at 9.5 percent, somewhat lower than the lowest use at Haystack. The combined absolute number of total observing hours on the two telescopes suggests that planetary radar astronomy activity in the early 1970s was not increasing or even remaining stable, but was declining. It was Little Science becoming smaller.

From ARPA to the NSF

In November 1974, eleven years after the dedication of the Arecibo Ionospheric Observatory (AIO), a second dedication ceremony took place to denote the instrument's upgrading to S-band. The upgrade was not achieved by simply adding higher-frequency equipment. The reflector surface had to be refinished, the suspended platform accommodated to the new equipment, a new power supply provided, and the S-band transmitter and maser receiver designed, built, and installed. Each component of the instrument had to be adapted in order that the whole might function in the higher frequency range. For planetary radar astronomy, the upgrade essentially created a new instrument with entirely different and expanded capabilities. Nonetheless, however critical the upgrade was for radar astronomy, both radio astronomy and ionospheric research benefited significantly from the resurfacing and equipment improvements, too.

The conversion of the AIO into an S-band radar telescope was a long, indirect, and difficult process, even if considered only as a technological feat. The conversion paralleled and was inextricably enmeshed in the transformation of the AIO into a National Science Foundation National Research Center. That transformation was set in motion by cutbacks in the ARPA budget, not the Mansfield Amendment.

The realization that the S-band upgrade was possible is said to have been born in August 1966, during Hurricane Inez. The 100-kilometer-per-hour (62-mile-per-hour) winds moved the telescope less than a half inch (1.27 cm), instead of the foot (30 cm) it was feared. A subsequent study of the telescope structure showed that it was sufficiently stable to operate at wavelengths of the order of 10 cm (3,000 MHz). Optimistically, Frank Drake, successor to Bill Gordon as observatory director, thought that the dish could be resurfaced in less than two years for under $3 million.[22]

But funds were not readily available. Moreover, the annual budget allotted by ARPA started to shrink, from over $2 million initially to $1.8 million in the period 1965 through 1969. Although ARPA was cutting back all research in order to support the Vietnam War,[23]

21.   Sebring to Hurlburt, 27 March 1970, 18/2/AC 135, MITA; AIO, *Research in Ionospheric Physics*, Research Report RS 69 (Ithaca: CRSR, 30 June 1966), p. 87; *ibid.*, Research Report RS 70 (Ithaca: CRSR, 31 January 1967), pp. 124–125; *ibid.*, Research Report RS 71 (Ithaca: CRSR, 31 July 1967), pp. 113–124; *ibid.*, Research Report RS 72 (Ithaca: CRSR, 31 January 1968), pp. 125-134; *ibid.*, Research Report RS 74 (Ithaca: CRSR, 31 July 1968), pp. 137–145; *ibid.*, Research Report RS 75 (Ithaca: CRSR, 31 March 1969), p. 51; *ibid.*, Research Report RS 76 (Ithaca: CRSR, 30 September 1969), p. 44; NAIC QR Q1-Q4/1970, passim. The Arecibo Observatory quarterly reports for the years 1971 to 1975 indicate the fraction of radar astronomy use of the antenna: 2.9 percent in 1971; 9.5 percent in 1972; 6.9 percent in 1973; 1.9 percent in 1974; and 7.2 percent in 1975. At Haystack, in March 1970, for example, of the 290 hours scheduled, 90 (31 percent) were spent on radar observations. See Chapter 3 for Haystack radar use.
22.   Peter, p. 12; AIO, *Research in Ionospheric Physics*, Research Report RS 70 (Ithaca: CRSR, 31 January 1967), p. 1.
23.   John Lannan, "An Example of Scientific Research under Scrutiny," *The Sunday [Washington] Star*, 30 March 1969, p. F-3. For the AIO budget, see *CRSR Summary Report, July 1, 1964—June 30, 1965*, 1 July 1965, CRSR, p. 1; *Annual Summary Report, Center for Radiophysics and Space Research, July 1, 1965—June 30, 1966*, 30 June 1966, pp. 2 and 7; *ibid., July 1, 1966—June 30, 1967*, 30 June 1967, pp. 8, 10 and 12; *ibid., July 1, 1967—June 30, 1968*, 30 June 1968, pp. 1, 11 and 13; *ibid., July 1, 1968—June 30, 1969*, 30 June 1969, pp. 1 and 15. AFOSR contract F44-620-67-C0066 allocated $5,210,200 for the term 1 February 1967 through 30 September 1969.

the Arecibo budget suffered because ARPA felt that the telescope performed below expectations.

The antenna feed operated at only 21 percent efficiency; the dish received less than half the power it should have received. That was a huge dollar loss, too; the cost of building a dish half the area would have been much less. Nonetheless, it was still an extremely sensitive telescope. The inefficiency of the antenna feed became a source of friction between Thomas Gold and Bill Gordon, who insisted that the feed could be improved, and between AIO management and ARPA.[24]

Figure 14

Linear antenna feeds attached to the suspended platform of the Arecibo Observatory. (Courtesy of National Astronomy and Ionosphere Center, which is operated by Cornell University under contract with the National Science Foundation.)

24.   Gordon 28 November 1994; Gold 14 December 1993; Campbell 7 December 1993; L. Merle Lalonde and Daniel E. Harris, "A High-Performance Line Source Feed for the AIO Spherical Reflector," *IEEE Transactions on Antennas and Propagation* AP-18 (January 1970): 41.

The line feeds were an ongoing serious problem. After a three-day visit to Arecibo in October 1967, Bart J. Bok, director of the Steward Observatory, Tucson, observed that the line feed problem "seems to be the most critical one facing the Arecibo-Cornell group." A number of Ithaca researchers attempted to improve the feeds. One Cornell graduate student considered the use of Gregorian optics, an option also studied by the AFCRL's Antenna Laboratory. However, not until 1988 was the first Gregorian feed tested and installed at Arecibo.

Arecibo had three feed research programs going on at the same time. Only one, for a high-powered, 430-MHz radar feed operating at both circular polarizations, was vital to its radar functions. Of two competing radar feed designs, the AIO selected that of Alan Love of the Autonetics Corporation, a subsidiary of North American Rockwell. Love worked with Cornell's L. Merle Lalonde to construct an appropriate feed, which was installed on the antenna in early 1972. The new radar feed was a success.[25]

ARPA's funding of the AIO dropped to a great extent because of the inefficient feed. Too, radio astronomy at the AIO was expanding rapidly in the wake of the discovery of pulsars (the AIO had tremendous advantages for investigating them), and ARPA felt more and more that it should support just the facility's ionospheric work, which was the only research relevant to Department of Defense interests. The AIO, though, hoped that ARPA would pay for the resurfacing and a new radar feed.

Although the ARPA contract did not allow the AIO to seek funding from other agencies, ARPA was now receptive to the idea of sharing the AIO budget with the NSF. So with ARPA's blessing, Thomas Gold and Frank Drake approached the National Science Foundation about civilian operational money for the AIO. The AIO also submitted a proposal to the NSF in early 1967 for detailed engineering studies and a cost estimate to resurface the reflector.[26] The search for both resurfacing and operational funds thus proceeded concurrently and was boosted by the report of the Dicke Panel.

Thomas Gold, Frank Drake, and Rolf Dyce pitched the Arecibo resurfacing project before the Dicke Panel. The Panel gave the project highest priority. As a result, Cornell obtained an NSF grant for a study and cost estimate of the reflector resurfacing. The AIO selected the Rohr Corporation, which also built JPL's Mars Station, to conduct the study. Rohr planned to install light aluminum panels for the reflector surface at a total cost of $3.5 million.[27]

The NSF, however, did not ask Congress to underwrite the resurfacing of the Arecibo reflector. The feed problem stood in the way. At its meeting of 16–17 October 1967, the NSF Astronomy Advisory Panel resolved:[28]

> *The NSF Advisory Panel will be hesitant to favor the improvements of the surface of the Arecibo dish or the undertaking of substantial operating expenses for Arecibo until a successful radio astronomy feed has been constructed and made operational at frequencies low enough that the surface is not critical.*

25.   Bok to George B. Field, "Arecibo," NSFHF; Kay, *A Line Source Feed*; J. Pierluissi, *A Theoretical Study of Gregorian Radio Telescopes with Applications to the Arecibo Ionospheric Observatory*, Research Report RS 57 (Ithaca: CRSR, 1 April 1964), NAIC; Peter, p. 18; Campbell 7 December 1993.

26.   Diary note, Hurlburt, 15 December 1967, and Long to Haworth, 27 July 1967, "Arecibo," NSFHF; *Annual Summary Report, Center for Radiophysics and Space Research, July 1, 1965—June 30, 1966*, 30 June 1966, pp. 12 and 18; *ibid., July 1, 1966—June 30, 1967*, 30 June 1967, p. 8; AIO, *Research in Ionospheric Physics*, Research Report RS 71 (Ithaca: CRSR, 31 July 1967), p. 1; Gold 14 December 1993.

27.   National Science Board, Approved Minutes of the Open Sessions, pp. 113:14–113:15, National Science Board; "Report of the Ad Hoc Advisory Panel for Large Radio Astronomy Facilities," 14 August 1967, typed manuscript, pp. 2–3, NSFL; Lalonde and Harris, p. 42; AIO, *Research in Ionospheric Physics*, Research Report RS 72 (Ithaca: CRSR, 31 January 1968), p. 4; AIO, *Ibid.*, Research Report RS 74 (Ithaca: CRSR, 31 July 1968), pp. 8–9.

28.   Haworth to John Foster, 9 November 1967; Memorandum, Gerard Mulders to Haworth, Randal M. Robertson, and William E. Wright, 25 August 1967; and Memorandum, Mulders to Robertson, 3 January 1968, "Arecibo," NSFHF.

In short, if an adequate feed design were not feasible, investing in an expensive resurfacing of the reflector for operation at higher frequencies made no sense. The feed problem held up the resurfacing and by implication the entire S-band upgrade. Consequently, Cornell undertook an in-house effort to design a 327-MHz feed at its own expense.

Although the reflector resurfacing project came to a temporary halt, the drive to secure NSF operational support succeeded in the wake of the Dicke Panel report. In July 1967, as the Dicke Panel was meeting in Washington, Cornell Vice President for Research and Advanced Studies Franklin A. Long asked Leland Haworth, director of the NSF, for a meeting about the possibility of jointly funding the operation of the AIO with ARPA. The NSF and ARPA soon entered into discussions and, by late August 1967, the NSF was agreeable to replacing the AFOSR as the government agency monitoring the Arecibo contract.[29] This was the first step in converting the AIO into a civilian observatory.

ARPA was prepared to underwrite the full AIO budget to the end of September 1968. Beginning 1 October 1968, for fiscal years 1969 through 1972, ARPA would pay for a third of the AIO budget, representing the portion of telescope time spent on ionospheric work. "It is very much hoped," the ARPA negotiator expressed, "that the entire facility will be identified as a National Science Foundation Observatory with ARPA as one of several users."[30]

In December 1967, well before passage of the Mansfield Amendment, Cornell and ARPA came to an agreement on the AIO contract. Cornell, NSF, and ARPA would negotiate a one-year contract for AIO operation from 1 October 1968 through 30 September 1969. The ARPA-NSF Memorandum of Understanding, signed in late April 1969, left the AIO under ARPA and the AFOSR until 1 October 1969, when the NSF took over, thereby anticipating the effect of the Mansfield Amendment. For the fiscal year starting 1 October 1968, each agency agreed to pay half the facility's annual budget. For the two years beginning 1 October 1969, ARPA agreed to transfer to NSF a third of the annual budget to support just ionospheric research. ARPA did not commit any funding after 1 October 1971, but left the door open to the possibility.

The Memorandum of Understanding defined ARPA's step-by-step divestment of Arecibo. Although ARPA initially had funded Arecibo for Project Defender, the telescope was never engaged in classified military research. Moreover, one clause in the Memorandum of Understanding specifically forbade the participation of the AIO in secret work: "The Observatory shall not be used to make measurements which are themselves classified nor be used as a repository for classified information."[31] The AIO was on the rocky road to civilian supervision and funding.

What's In a Name?

The transformation of the AIO into an NSF National Research Center involved two interconnected issues, the observatory's management structure and the status of ionospheric research, both of which were complicated by personality conflicts and turf fights between Big Science fields. Implicit in being a National Research Center was free access to the telescope for all qualified scientists. The AIO always maintained that it operated as

29.   Hurlburt diary note; Long to Peter Franken, 23 August 1967, and Long to Leland Haworth, 27 July 1967, "Arecibo," NSFHF.

30.   Franken to William Wright, 23 August 1967, "Arecibo," NSFHF.

31.   S. J. Lukasik to Long, 12 December 1967; Memorandum of Understanding, AIO, attached to letter, Haworth to John Foster, 30 April 1969; and Memorandum of Understanding, AIO, attached to letter, S. E. Clements to Haworth, 12 May 1969, signed by Haworth and Foster, "Arecibo", NSFHF.

a national center, and the Cornell-Sydney agreement opened the observatory to foreign scientists. The real problem was that radio astronomy use of the telescope had skyrocketed, especially in contrast to ionospheric research. From December 1965 through September 1969, for example, ionospheric research accounted for 30 percent, while radio astronomy took up 50 percent of antenna time.[32]

Ionospheric research had been the reason for creating the AIO in the first place, and it was more interesting to the electrical engineering than to the astronomy department. The name of the facility changed to the Arecibo Observatory, discarding the "ionospheric" of the original name. To some individuals, the name change did not reflect the facility's multiple research agenda, which was the intent of the change, but instead signified lack of interest in ionospheric work. As Gordon Pettengill explained: "We settled on that name early, because it encompassed the radio astronomy, radar astronomy, and ionospheric research. There was quite a group that wanted to call it the Arecibo Ionospheric Observatory, which was the original name under Bill Gordon."[33] Many accused Thomas Gold, who had fostered the expansion of radio astronomy, of thwarting ionospheric work, but Gold insisted that no ionospheric researchers ever were turned down.

Perceptions outside Arecibo and Cornell confused the presumed reduction of ionospheric studies with the rift between astronomy and electrical engineering within the CRSR, and colored everything with the friction between Bill Gordon and Thomas Gold. Gold found Gordon "a little difficult, because he really wanted to cut himself off from Cornell, from everything completely, and I realized that if he did so, then the telescope would never be used for radio astronomy and radar, and it would become merely an ionospheric instrument, and that I was very opposed to, being nominally in charge of building such a huge wonderful instrument and then finding it's not used for what it's capable of."[34] Bill Gordon, for his part, stated, "If you ask me, I was mad at the time, and whatever I tell you has some personal bias built in." In short, he explained, "I thought I was removed from a job that I deserved to have."[35]

Frank Drake, radio astronomer and one-time Arecibo director, explained the conflict rather precisely. "I had picked up enough innuendo in Gold's tone and Gordon's words to realize that the two of them were engaged in a bitter battle for the Arecibo turf," he wrote. Gold "wanted the Arecibo telescope freed to do more research in radio astronomy. He was lobbying the university administration to put it under his jurisdiction." Gordon "could not bear to relinquish control of it." He left, however, after Gold pointed out to the university administration that Gordon had been off-campus far longer than the university bylaws allowed. "It was a fact people might have been willing to overlook, but once Gold seized on it, Gordon was forced to make a choice."[36]

Feelings about the friction between Gold and Gordon, as well as the perceived neglect of ionospheric work, also shaped how the NSF handled the AIO. The chief personality at the NSF was Tom Jones, director of the Division of Environmental Sciences. He explained the situation to the NSF director in 1968:[37]

32.   Maintenance and equipment improvements were 11 percent and radar astronomy 9 percent of antenna time. AIO, *Research in Ionospheric Physics*, Research Report RS 69 (Ithaca: CRSR, 30 June 1966), p. 87; *Ibid.*, Research Report RS 70 (Ithaca: CRSR, 31 January 1967), pp. 124–125; *Ibid.*, Research Report RS 71 (Ithaca: CRSR, 31 July 1967), pp. 113–124; *Ibid.*, Research Report RS 72 (Ithaca: CRSR, 31 January 1968), pp. 125–134; *Ibid.*, Research Report RS 74 (Ithaca: CRSR, 31 July 1968), pp. 137–145; *Ibid.*, Research Report RS 75 (Ithaca: CRSR, 31 March 1969), p. 51; and *Ibid.*, Research Report RS 76 (Ithaca: CRSR, 30 September 1969), p. 44.
33.   Pettengill 28 September 1993.
34.   Gold 14 December 1993.
35.   Gordon 28 November 1994.
36.   Frank Drake and Dava Sobel, *Is Anyone Out There?* (New York: Delacorte Press, 1992), pp. 77 and 79.
37.   Jones to Haworth, 8 February 1968, "Arecibo," NSFHF.

The operation of AIO has been tainted by a great deal of political infighting on the Cornell campus. Results of these confrontations included the departure from Cornell of Drs. W. Gordon and H. Booker, both aeronomers, who were the originators of the backscatter concept for probing the ionosphere and who saw the Arecibo venture through from the proposal stage right on up to its final construction and initial operation. There are indications that, aside from accepting opportunities for professional growth, they left Cornell because the administrative control of AIO was removed from the director of the Observatory and placed in the hands of another individual on the Cornell campus. We do know, from conversations with aeronomers, that they do not want to give up the use of the Arecibo instrument.

Jones maintained a vigil on the AIO case, as he moved from the Division of Environmental Sciences to the Office of National Centers, which directly oversaw the Arecibo Observatory. Thomas Gold found that Jones "kept expressing a sort of paranoia about ionospheric work, but constantly. I mean, I couldn't talk to him without getting a lecture that far too little ionospheric work was being done, and he couldn't support any funding for Arecibo if this were done, even though at the time it was doing very good work in radio and radar astronomy, but not enough ionosphericists wanted to go there. I couldn't help it!"

According to Gold, Jones told him that he could not support funding for Arecibo if the reduction of ionospheric research continued. As for his relations with Bill Gordon, Thomas Gold insisted that it had nothing to do with ionospheric research. He and Gordon disagreed over the management of the observatory. According to Gold, Gordon wanted to operate it "in a way independent of Cornell," and he did not want to return to Cornell. Bill Gordon "wanted to make all the decisions as to who gets what time and all that," and Gold objected.[38]

Control of the observatory was the key issue dividing Gordon and Gold. The issue of where management of the AIO should rest, at Arecibo or at Ithaca, was precisely the concern of the NSF, too. The issue was clouded by both personality conflicts and the status of ionospheric research. On 27–28 February 1968, the NSF Advisory Panel for Atmospheric Sciences, which included Bill Gordon, issued a formal statement on the future of ionospheric research at the AIO: "As the NSF assumes increasing operational responsibility, the Panel strongly recommends that any management changes be made in such a way as to insure the availability of the AIO for experimental research in aeronomy and solar-terrestrial physics." Moreover, "The Panel considers it important to establish a management structure for the AIO whereby scientists from institutions throughout the United States may use the Observatory. To accomplish this, it is suggested that the scheduling and operating policy be established by the scientific community and implemented by the resident director. An appropriate way to assure representation of the scientific community would be to place the management of the AIO in the hands of a consortium of interested universities."[39]

The Advisory Panel was not alone in suggesting management by a university consortium along the lines of NEROC or the NRAO.[40] However, Cornell and Gold wanted to retain control of the Arecibo Observatory (AO). Harry Messel, head of the University of Sydney School of Physics and joint director, with Gold, of the Cornell-Sydney University

38.   Gold 14 December 1993. Bill Gordon declined comment on the whole affair. Gordon 28 November 1994.

39.   Statement of the National Science Foundation Advisory Panel for Atmospheric Sciences to the Director of the National Science Foundation, 21 March 1968, "Arecibo," NSFHF.

40.   See, for instance, Haworth to Long, 23 January 1968, "Arecibo," NSFHF.

Astronomy Center, protested to Donald F. Hornig, the special presidential assistant for science and technology, that any change in the AO management structure would affect the Cornell-Sydney arrangement, too. Despite Hornig's assurances to the contrary, the evolving AO management structure led to the termination of the Cornell-Sydney agreement.[41]

However, the crux of the management structure question—all personality and turf conflicts aside—was separation of observatory administration from all academic departments, like the CRSR. The NSF did not want to fund National Research Centers that were prisoners of an astronomy department or of any other academic unit. It was clear, though, that if the AO were to become a National Research Center, with a secured budget from the NSF, Cornell would have to draft a new management structure; otherwise, a university consortium might take over Cornell's managerial role.

In March 1969, as the NSF looked toward assuming full responsibility for the AO on 1 October 1969, the Foundation asked Cornell to prepare a proposal for the operation of the AO for the two-year period beginning 1 October 1969. The proposal was to discuss the AO management structure, "bearing in mind our opinion that a director of a National Center should report to a level of management significantly above that of a department or similar unit."[42] The April 1969 proposal outlined a management structure drafted the previous summer. The director of the AO reported to a policy committee, which consisted of only the university provost, the director of the CRSR (Gold), and the vice president for research and advanced studies.[43]

A special National Science Foundation AIO Group reviewed the proposal. Their major objection was the management plan: "It does not show much change from the existing management structure at Cornell and does not appear to be suitable for a National Center. No member of the AIO group finds it acceptable." Specifically, the problem was the three-man policy committee. "This Committee seems clearly intended by Cornell to be the group which runs the show. It is proposed that it be made up exclusively of Cornell employees resident in Ithaca. The suggestion that such a group should be considered 'national management' has reduced the undersigned [Fregeau] to a conviction that his education in the art of strong language is grossly inadequate."

The AIO Group felt that a more appropriate structure would have the observatory director report directly to the vice president of research, a single individual, and not a committee; otherwise, "the implication [is] that the *committee* is the AIO director's boss." In the judgement of the AIO Group, "The Cornell proposal is not, in its present form, suitable for review by the scientific community. If it were to be sent out in this form, the community reaction would probably poison the beginnings of what we expect to be a fruitful venture for NSF."[44]

On 1 October 1969, when monitorship of the Arecibo contract passed to the NSF, Cornell reorganized the AO's management structure to conform more closely to the Foundation's guidance. The observatory was removed from CRSR supervision and placed under an Arecibo Project Office headed by Assistant Vice President for Research (Arecibo Affairs) Thomas Gold.[45]

    41.   Messel to Donald Hornig, 12 June 1968, and Hornig to Messel, 9 July 1968, "Arecibo," NSFHF; Gold 14 December 1993.
    42.   Randal N. Robertson to Long, 17 March 1969, "Arecibo," NSFHF.
    43.   CRSR, "Proposal to National Science Foundation for Research Ionospheric Physics, Radar-Radio Astronomy, October 1, 1969 through September 30, 1971," April 1969, Office of the Administrative Director, NAIC; advanced draft, "The Management of the AIO as a National center," July, 1968, "Arecibo," NSFHF.
    44.   Memorandum, J. H. Fregeau to Associate Director (Research), NSF, 28 April 1969, "Arecibo," NSFHF.
    45.   *Annual Summary Report, Center for Radiophysics and Space Research, July 1, 1969—June 30, 1970,* 30 June 1970, p. 9.

In the following months, Cornell and the NSF continued to consider the observatory's management structure. The result was a new organizational structure effective 1 July 1971 that brought it more in line with other National Research Centers, and a new name, the "National Astronomy and Ionospheric Center" (NAIC). The name and acronym were intended to emulate the NRAO as a model and gave assurances of the importance of ionospheric research.

In the new management structure, the title of Assistant Vice President of Research (Arecibo Affairs) was discontinued. Gold had quit. Those duties were given to the observatory director, who was responsible to Cornell, through the vice president for research, for the overall management and operation of Arecibo. He prepared the annual budget, annual program plan, and long-range plans for the AO. The observatory director was to be located primarily in Ithaca and was also the director of the NAIC. The director of observatory operations, who answered to the director, had responsibility for the operation, maintenance, administration, and improvement of the facility, oversaw personnel and time allocations, and helped prepare the budget. He was required to be located in Arecibo.[46]

For the new director of observatory operations, the NAIC hired Tor Hagfors in 1971. His selection reassured those who worried about the status of ionospheric research. Hagfors had an impressive background in ionospheric (and radar) research and administration at the Norwegian Defense Research Establishment, Stanford University, the Jicamarca Radio Observatory (where he was director, 1967–1969), and Lincoln Laboratory's Millstone Hill radar.[47]

The NSF and NASA Agreement

As the new management structure emerged, and as the National Science Foundation took over the Arecibo contract, the search to fund the S-band upgrade continued. In May 1969, the Subcommittee on Science, Research, and Development of the Committee on Science and Astronautics, headed by Emilio Q. Daddario (D-Conn.), recommended deferring the NSF request for resurfacing money. Gordon Pettengill, speaking as Arecibo director, pointed out that "many throughout the radio astronomy community were seriously disappointed at the failure of the Congress to authorize funds for the resurfacing of the AO reflector." When the Dicke Panel reconvened in June 1969 and reaffirmed the need for the resurfacing, they too expressed disappointment that the new reflector surface had not yet been started.[48]

A major breakthrough occurred when NASA took an interest in the project. Throughout the 1960s, NASA had funded only mission-oriented radar research, but not radar telescope construction. In January 1969, Harry H. Hess, chair of the Space Science Board, wrote to John Naugle, associate administrator of Space Science and Applications at NASA, urging NASA to fund the Arecibo radar upgrade. The cost, estimated to be $5 million for the resurfacing plus $2 or $3 million more for the radar equipment, was "small in comparison with the construction of a new radar facility but would make it possible to

46.   Campbell 9 December 1993; *Arecibo Observatory Program Plan, October 1, 1970—September 30, 1971,* May 1971, pp. 35–39, Office of the Administrative Director, NAIC.

47.   Pettengill 28 September 1993; Campbell 7 December 1993; *Arecibo Observatory Program Plan, October 1, 1971—September 30, 1972,* January 1972, NAIC, pp. 25–31; *Arecibo Observatory Program Plan, October 1, 1970—September 30, 1971,* May 1971, p. 63, AOL; NAIC QR Q3/1971, 9.

48.   AIO, *Research in Ionospheric Physics,* Research Report RS 76 (Ithaca: CRSR, 30 September 1969), p. 1; "Report of the Second Meeting of the Ad Hoc Advisory Panel for Large Radio Astronomy Facilities," 15 August 1969, typed manuscript, p. 3, NSFL.

map the surface of Venus with a resolution of a few kilometers. Such a map would obviously be a tremendous step forward in our knowledge of the planet. The NASA contribution to the total cost of improving the Arecibo facility would be very small compared to the cost of obtaining the same information from some future orbiter."[49]

Hess's argument closely resembled that of Werner von Braun in an anecdote related by Don Campbell: "I don't know if it's apocryphal or not, but there is the story that Werner von Braun said, if you can get a two-kilometer resolution on Venus for $3 million, which is roughly what we were talking about, that it was an immense bargain, and that NASA should take it straight away."[50]

Indeed, NASA became interested in funding the upgrade for one major reason. The S-band equipment could make radar maps of the Venusian surface with a resolution of two to five kilometers. The space agency was interested in a one-megawatt radar operating at 10 cm (3,000 MHz). And as NASA chief of planetary astronomy William Brunk came to realize, the total cost of the upgrade was a fraction of the initial cost of the facility.[51]

The country was discovering that it could not afford both guns and butter, the Vietnam War and the Great Society. NASA and NSF were under serious pressure to cut their budgets, and in December 1969, the new Republican President shut down the NASA Electronics Research Center in Cambridge. Budgetary austerity perhaps led NASA to support the Arecibo S-band upgrade, at a cost of a few million dollars, over the NEROC telescope, with an estimated price tag of $30 million dollars. Furthermore, given the superior transmitter power and receiver sensitivity of the upgraded Arecibo S-band radar over the proposed NEROC radar, NASA would be getting a better investment for its dollars.

Budgetary belt tightening also induced NASA to realize that every planetary mission had to do something that could not be done from the ground, and missions would have to rely on ground-based results more than ever. The radar images obtainable from the upgraded radar would be invaluable to the exploration of the planets. Thus, the mission-oriented logic of NASA, combined with budgetary restraint, led to its adopting the Arecibo upgrade project.[52] NASA now approached the NSF.

When NASA and NSF representatives met on 2 December 1969, the NASA budget for fiscal 1971 was among the topics of discussion. The space agency was going to ask for an extra $1 million to build a major planetary research facility as part of its Planetary Astronomy Program. Three candidate projects were under consideration: a 60-inch (1.5-meter) planetary telescope at Cerro Tololo, Chile; a large-aperture infrared telescope; and the Arecibo upgrade. The final choice pivoted on the NSF budget submission to the Bureau of the Budget and Congress.

The NASA strategy was to pay for the resurfacing, if the NSF failed to win funds from Congress, and to worry about the rest of the S-band upgrade later. Brunk knew that NASA had to be prepared to pay for the radar equipment. Because radar equipment was "not a high priority item for general radio astronomy," he reasoned, "the development of a high power radar transmitter at a wavelength of 10 centimeters will be a low priority for NSF funding and must therefore be included in the NASA Planetary Astronomy budget."[53]

Soon after the NASA-NSF meeting, in February 1970, Cornell submitted a funding proposal for the S-band upgrade to both NASA and the NSF. The proposal asked for $3 million over three years, with work to begin February 1971. Both the NSF and NASA fiscal 1971 budget requests contained money for Arecibo. The NSF proposed to underwrite

49.   Hess to Naugle, 27 January 1969, NHOB.
50.   Campbell 7 December 1993.
51.   Brunk, Planetary Astronomy New Starts, FY 1971, n.d., NHOB.
52.   Tatarewicz, p. 98. For the creation and demise of the NASA Electronics Research Center, see Ken Hechler, *Toward the Endless Frontier: History of the Committee on Science and Technology, 1959-1979* (Washington: USGO, 1980), pp. 219–231.
53.   Henry J. Smith, Memo to the files, 11 December 1969, NHOB.

the reflector resurfacing, while NASA budgeted for the radar equipment and its installation. Congress approved both the NASA and NSF Arecibo S-band expenditures. The NSF funds were frozen, however, until a new cost estimate became available. The estimate, completed in November 1970, was $5.6 million.[54]

The upgrade brought together the NSF and NASA into a special relationship that started with joint discussions in December 1970 between William Brunk, chief of the NASA Planetary Astronomy Program, and Daniel Hunt, head of the NSF Office of National Centers and Facilities Operations. As discussions progressed, on 6 March 1971, NASA formally expressed its intent to enter into an agreement with the NSF for the addition of the S-band equipment. The two agencies entered into negotiations and, on 24 June 1971, signed a Memorandum of Agreement, which went into effect 1 July 1971.[55]

Under the agreement, the NSF funded the resurfacing and NASA the addition of a one megawatt S-band radar transmitter, receivers, and associated changes to the antenna to provide radar capability at a wavelength of ten centimeters (3,000 MHz). The project was to be managed under the existing NSF-Cornell contract. The NSF would serve as the monitoring agency, and NASA would transfer its portion of the funds to the NSF. The agreement deferred the issue of S-band operational costs until later, although the two agencies intended to share those costs proportionally.[56]

In this way, NASA came to fund radar instrument construction and committed itself to supporting the research performed with the instrument. This deviation from earlier policy was motivated by an interest in mission-oriented research, namely, to obtain radar images in support of space missions to the planets, particularly to Venus. The consequence was a permanent institutional and funding arrangement for planetary radar astronomy at Arecibo, as well as a unique instrument.

Arecibo Joins the S-Band

The NASA-NSF agreement, backed by Congressionally-approved funds, provided the legal, financial, and managerial framework for the actual upgrade work to take place. The upgrade began with a search for a contractor to undertake the reflector resurfacing. Two firms bid, the Rohr Corporation and LTV Electrosystems of Dallas, and in November 1971 Cornell awarded the contract to LTV Electrosystems, which shortly afterward changed its name to E-Systems. The original spherical reflector consisted of 1/2-inch (1.3 cm) steel wire mesh (chicken wire) supported by heavy steel cables. Over 38,000 thin aluminum panels, fabricated on-site by E-Systems, replaced the chicken wire. From the beginning, inefficiencies and equipment failure plagued panel installation, but they were overcome, and the last panel was installed in November 1973.[57]

---

54. AIO, *Proposal to NSF and NASA for Major Additions and Modifications to the Suspended Antenna Structure and Equipment of the Arecibo Observatory, February 1971 through February 1973*, February 1970, and Daniel Hunt to Brunk, 11 December 1970, NHOB; National Science Board, Minutes of the Open Meetings, 132:6 and 133:7–8, National Science Board; NAIC QR Q3/1971, p. 13.

55. Hunt to Brunk, 11 December 1970; NASA Deputy Associate Administrator for Space Science and Applications to Assistant Administrator, Office of DoD and Interagency Affairs, 6 March 1971; Memorandum, Director of Planetary Programs, Office of Space Science and Applications, NASA, to Associate Administrator for Office of Tracking and Data Acquisition, 20 May 1971; and Memorandum of Agreement between NASA and the NSF for the Addition of a High-Power S-Band Radar Capability and Associated Additions and Modifications to the Suspended Antenna Structure of the NAIC at Arecibo, 24 June 1971, NHOB.

56. Memorandum of Agreement between NASA and the NSF for the Addition of a High-Power S-Band Radar Capability and Associated Additions and Modifications to the Suspended Antenna Structure of the NAIC at Arecibo, 24 June 1971, NHOB; "High Power Transmitter to Boost Arecibo Radar Capability," NSF press release, 17 August 1971, "Radar Astronomy," NHO.

57. NAIC QR Q1/1971, p. 5; Q2/1971, p. 6; Q3/1971, p. 6; Q4/1971, p. 7; Q1/1972, p. 10; Q2/1972, p. 13; Q3/1972, p. 11; and Q1/1973, p. 13; National Science Board, Minutes of the Open Meetings, 144:4-5, National Science Board.

Concurrently, the NAIC oversaw the design and construction of the S-band radar transmitter, receivers, and associated equipment; the necessary modifications to the suspended feed platform; and construction of a new carriage house to hold the S-band equipment. Ammann & Whitney, a well-known structural engineering consulting firm, reported on the suspended structure and reflector cable anchorages as well as on the feasibility of upgrading the suspended structure. They found no basic deficiencies in the structure that would make upgrading impractical or inadvisable.[58]

In order to develop specific transmitter characteristics that met scientific goals, yet represented realistic state-of-the-art feasibility, NAIC staff discussed its design with experienced radar astronomers and with experts from Varian Associates, Raytheon, and Continental Electronics. The operating frequency, 2380 MHz (12.6 cm), appeared to be the optimum choice for both radar and radio astronomy and was close to the JPL planetary radar frequency (2388 MHz; 12.6 cm).[59]

Originally, the transmitter was to produce 800 kilowatts using two klystrons. The NAIC based the decision on the experience of JPL Goldstone, where a single klystron produced 400 kilowatts of average continuous-wave power at 2388 MHz. Although Varian was developing a one megawatt continuous-wave klystron, the advantages of proven reliability and ready availability of spares militated against using a single, experimental one-megawatt klystron. Finally, after extensive discussion with representatives of NASA and the NSF, the NAIC reduced the transmitter power requirement to 450 kilowatts, thereby lowering costs "without impacting upon scientific goals of the program." In contrast, the original UHF radar transmitter produced only 150 kilowatts of average power.[60]

The experience of JPL in operating at S-band proved invaluable to the Arecibo radar upgrade. In addition to providing expert advice to the NAIC staff, a former JPL employee reviewed technical matters for the NASA technical monitor. The maser receivers, moreover, were excess Deep Space Network equipment. The agreement between the NAIC and JPL for the transfer of the masers noted that JPL was "a pioneer in the development of maser systems," and that "no commercial firms have the required capability, experience and expertise to produce an S-band maser system that would be operational at 2.38 GHz."[61]

The renovated reflector was dedicated on 15–16 November 1974. After delivery of the keynote speech, Rep. John W. Davis (D-Ga.) gave the signal for the transmission of The Arecibo Message, 1974, an attempt to communicate with extraterrestrial civilizations. The radar upgrade, however, was not yet completed and would not be entirely ready until the following year.[62]

Yet, shortly after the resurfacing dedication, Gordon Pettengill (Arecibo) and Richard Goldstein (JPL) used the S-band transmitter to bounce signals off the rings of Saturn. Because the Arecibo maser receiver was not yet installed, Arecibo sent and JPL's Mars Station received. The bistatic experiment worked, despite line feed and turbine generator problems at Arecibo.

58.   AIO, *Proposal to NSF and NASA for Major Additions and Modifications to the Suspended Antenna Structure and Equipment of the Arecibo Observatory, February 1971 through February 1973*, February 1970, pp. 7–8, 11–15, NHOB; NAIC QR Q3/1972, pp. 11–12.
59.   NAIC QR Q4/1971, p. 8; and Q2/1972, p. 13.
60.   Campbell 8 December 1993; NAIC QR Q2/1970, p. 4, Q4/1970, p. 9, Q1/1972, p. 12, Q3/1972, p. 12, and Q1/1973, p. 14; AIO, *Proposal to NSF and NASA for Major Additions and Modifications to the Suspended Antenna Structure and Equipment of the Arecibo Observatory, February 1971 through February 1973*, February 1970, pp. 14–15, NHOB; AIO dedication brochure, no page numbers, Cornell, 1974, NHOB.
61.   Brunk to Claude Kellett, 18 April 1973, NHOB; Jack W. Lowe to W. E. Porter, 29 March 1977, Office of the Administrative Director, NAIC; NAIC QR Q1/1972, p. 11; Peter, p. 13. Documents relating to the transfer can be found in the Office of the Administrative Director, NAIC.
62.   Drake and Sobel, pp. 180-185; Campbell 8 December 1993; Dedication publication, Cornell University, 1974, NHOB; National Science Board, Minutes of the Open Meetings, 168:2, National Science Board; NAIC QR Q3/1974, p. 10, Q2/1975, p. 4, and Q3/1975, p. 4.

As Don Campbell recalled: "The initial feeds that we used with the transmitter needed cooling; we were having trouble attaching the cooling lines. People would line up late
at night in front of the control room during this experiment. We would turn on the transmitter, and there would be this sort of flash of light, as things burned up up there and
everybody went 'Ah!' It was a bit like fireworks. When the problem finally got solved, I
think everyone was rather disappointed that there wasn't any flash of light up there!"[63]

The Arecibo S-band upgrade literally created a new instrument with which to do
planetary radar astronomy, a field whose outer limits of capability still leaned strongly on
the availability of new hardware. Although the Arecibo telescope made S-band radar
observations of Mars beginning August 1975 for NASA's Viking mission, the arrangement
with NASA freed Arecibo also to do radar research that was not mission related. So, late
the following month, on 28 September 1975, the radar detected Callisto, followed two
nights later by Ganymede, the first detections of Jupiter's Galilean moons.[64]

The NASA agreement guaranteed planetary radar astronomy an instrument and a
research budget. Nowhere else did planetary radar astronomy operate with such extensive
institutional and financial support. These unique advantages, combined with its relations
with Cornell and MIT, have sustained Arecibo as the focal center of planetary radar
astronomy to the present day.

The JPL Mars Station

In sharp contrast to Arecibo, JPL did not formally recognize planetary radar astronomy as a scientific activity. Planetary radar had neither a budget line nor a program at JPL;
it was invisible. Its role was to test the performance of the Deep Space Network (DSN). Eb
Rechtin, the architect of the DSN, deliberately avoided creating a radar astronomy program. He saw no reason, other than for science, why NASA ought to fund it. Instead, planetary radar became, in the words of JPL radar astronomer Richard Goldstein, the "cow-
catcher on the DSN locomotive," financed at the "budgetary margin" of the DSN. In
Goldstein's own words, "I was the cow-catcher and still am."[65]

Radio astronomy, on the other hand, held a more privileged position. Nick Renzetti,
the DSN manager responsible for links between the Network and its users (NASA space
missions), forged an agreement with NASA Headquarters that permitted qualified radio
astronomers to perform experiments on Goldstone antennas at no cost, provided the
experiments did not conflict with the antennas' prime mission, spacecraft communications and data acquisition.[66]

Not only was JPL planetary radar astronomy invisible, but relations between JPL and
its oversight institution, the California Institute of Technology, were about as distant as
those between Lincoln Laboratory and MIT. JPL employees, like their peers at Lincoln
Laboratory, could not have graduate students, unless they held a joint appointment at
Caltech. Although Dick Goldstein taught a radar astronomy course at Caltech during the
1960s, his students did not become radar astronomers, but went into other fields. An
unusual case was Lawrence A. Soderblom, who took Goldstein's course in the fall of 1967.
Soderblom later joined the U.S. Geological Survey, where he interpreted planetary radar
data.[67]

63.   Campbell 8 December 1993; Campbell 7 December 1993; NAIC QR Q1/1975, p. 4.
64.   NAIC QR Q3/1976, pp. 4–5.
65.   Goldstein 14 September 1993; Goldstein 7 April 1993; Rechtin, telephone conversation,
13 September 1993.
66.   Renzetti 16 April 1992; Renzetti 17 April 1992.
67.   Soderblom 27 June 1994.

Unlike Lincoln Laboratory or Arecibo, JPL did not hire people to do radar astronomy, because JPL officially did not have a radar astronomy program. Goldstein and Dewey Muhleman had participated in the 1961 Venus radar experiment, not because they were Caltech graduate students, but because they were JPL employees in Walt Victor's group. Roland Carpenter, another JPL planetary radar astronomer of the 1960s, also worked under Walt Victor.[68] Once Roland Carpenter and Dewey Muhlemen left JPL, Dick Goldstein remained the sole JPL radar astronomer for several years.

Planetary radar astronomy subsisted at JPL during the 1960s on money earmarked for various space missions and on the budget of the DSN. The NASA budget then was more generous. NASA paid the cost of operating and maintaining the Goldstone radar as part of the DSN, so that the costs of the radar instrument were paid. When Goldstein needed a piece of hardware designed and built, he assigned the job to one of the employees he supervised as manager of Section 331. The Advanced Systems Development budget of the DSN paid for hardware design and construction.[69]

In order to obtain time on the Goldstone radar, Goldstein went from mission to mission and explained why the mission ought to support his radar experiments. With approval from a mission, Goldstein could then request antenna time from the committee in charge of allocating antenna use. Mariner missions supported many of the radar observations, while Viking and Voyager supported experiments on Mars, Saturn's rings, and the Galilean satellites of Jupiter. Officially, the experiments were done for neither radar improvement nor the science, but for "better communications" with spacecraft.[70]

If the NASA Headquarters Planetary Science Program, then headed by William Brunk, approved of a particular set of radar experiments, getting antenna time was much easier. As Goldstein explained, Brunk "would support me a little and I took that as a positive thing, and I guess later on he turned that off, but it wasn't very big in the first place....It was a kind of a way to get legitimacy. If he funds you a little, that means it's important. If he doesn't fund you at all, that means it's not important....I would go to great lengths to get antenna time and a little funding from Brunk was helpful."[71]

Planetary radar astronomy at JPL thus came into existence and continued to function because of the Laboratory's Big Science space missions and Deep Space Network activities. In particular, it was the idea, put forth by the DSN's chief architect Eb Rechtin, that radar astronomy would have the dual function of testing the DSN's ability to support interplanetary missions and developing new hardware for the DSN (that is, the justification for having Advanced Systems Development underwrite radar astronomy hardware). It was specifically for developing and testing new DSN hardware that, shortly after the 1961 Venus radar experiment, Rechtin arranged with NASA to set aside a Goldstone radar for that purpose. On that instrument, Goldstein and Carpenter made Venus radar observations during the 1962 and 1964 conjunctions. Planetary radar research benefitted from the developmental work, which increased the continuous-wave radar's average power output from 10 to 13 kilowatts in 1962 and then to 100 kilowatts for the 1964 Venus experiment.[72]

When Goldstein made observations during the 1967 Venus conjunction, however, he used a new, more powerful 64-meter-diameter (210-ft-diameter) S-band antenna, the Mars Station. The need to handle missions at ever increasing distances from Earth furnished

68.   Carpenter, telephone conversation, 14 September 1993.
69.   Jurgens 23 May 1994; Goldstein 14 September 1993; Downs 4 October 1994.
70.   Memorandum, Carl W. Johnson to Murray, 31 October 1977, 62/3/89-13, JPLA; Goldstein 14 September 1993.
71.   Goldstein 14 September 1993.
72.   Victor, "General System Description," p. 3 in Goldstein, Stevens, and Victor, eds., Goldstone Observatory Report for October-December 1962, Technical Report 32-396 (Pasadena: JPL, 1 March 1965); Waff, ch. 6, pp. 17 & 19; Goldstein and Carpenter, "Rotation of Venus," pp. 910-911; Carpenter, "Study of Venus by CW Radar," p. 142.

the raison d'être for JPL's entry into the Big Dish arena, and incidentally supplied its future radar astronomers with an ideal instrument for imaging and other planetary radar work. With the Mars Station, Goldstein and his colleagues discovered three rugged sections of Venus; the largest received the name Beta. The needs of NASA space missions, not radar astronomy, dictated the design of the Mars Station.

The Mars Station represented the DSN's commitment to the S-band and its need for large antennas capable of communicating with probes at great distances from Earth. Starting in 1964, all new space missions were to use the higher S-band. Despite the commitment to S-band, NASA still had active missions operating at lower frequencies. The switch to S-band throughout the Deep Space Network therefore required a hybrid technology capable of handling missions operating at either the higher or lower frequency bands. A JPL design team devised the equipment, which was installed throughout the DSN.

The hybrid equipment, however, was only a transitional phase before the construction of more powerful and more sensitive antennas specifically intended to handle unmanned missions to the planets. In order to determine the essential characteristics and optimal size for those antennas, JPL initiated a series of studies in 1959 that culminated in the Advanced Antenna System. NASA's Office of Tracking and Data Acquisition, which oversaw the DSN, sponsored a pioneering conference on large antennas on 6 November 1959. Speakers reported on three kinds of antennas: steerable parabolic dishes, fixed antennas with movable feeds (e.g., Arecibo), and arrays, the same antenna types considered later by NSF panels.

NASA and JPL decided to stay with the proven design of steerable dishes. The next decision was antenna size. JPL engineering studies showed that antenna diameters between 55 and 75 meters (165 and 225 ft) were near optimal and the most cost-effective. The final choice, 64 meters (210 ft), was the same size as the recently-completed Australian radio telescope at Parkes. This was no coincidence. JPL engineers had received a lot of help from the Australian designers. Their studies of the Parkes telescope provided JPL engineers with a wealth of data and ideas to use in the design of their 64-meter (210-ft) dish.

JPL also commissioned private firms to carry out feasibility and preliminary design studies for the Advanced Antenna System beginning in September 1960, before awarding a construction contract to the Rohr Corporation in June 1963. Construction proceeded after JPL analyzed and approved the Rohr design in January 1964. Rohr completed the antenna in May 1966, following the formal dedication on 29 April 1966.

Figure 15
JPL Goldstone Mars Station (DSS-14) upon completion in 1966. (Courtesy of Jet Propulsion Laboratory, photo no.
333-5967BC.)

  The dish was dubbed the Mars Station, because its mission was to support Mariner
on its journey to Mars in 1964, long before the antenna was operational. Nonetheless, on
16 March 1966, the big dish received its first signals from Mariner 4 and provided operational support for Pioneer 7, launched in August 1966. The Mars Station subsequently
supported several other missions, including the first Surveyor flights, and made possible
live Apollo television pictures from the Moon, not to mention planetary radar images and
topographical maps. In order to systematize its growing number of antennas around the
world, the DSN instituted a numbering system, so that each Deep Space Station (DSS)
would bear a unique number. The original Echo antenna became DSS-12, while the antenna used in the Venus radar experiments became DSS-13. The Mars Station was DSS-14.[73]
  The Mars Station, as part of the Deep Space Network, underwent a major upgrade
in the 1970s in order to accommodate the needs of the Viking and Mariner Jupiter-Saturn
spacecraft (later known as Voyager). For the Viking mission, each DSN station would have
to handle six simultaneous data streams from the two Viking Orbiters and the one Lander.

  73. Corliss, *Deep Space Network*, pp. 37-38, 50, 60-61, 82, 84, 87, 129 and 131; Renzetti, *A History*,
pp. 25–26, 32, 52 and 54; Robertson, pp. 255–261; *The NASA/JPL 64-Meter-Diameter Antenna at Goldstone,
California: Project Report*, Technical Memorandum 33–671 (Pasadena: JPL, 15 July 1974), pp. 7–17; Rechtin, Bruce
Rule, and Stevens, *Large Ground Antennas*, Technical Report 32–213 (Pasadena: JPL, 20 March 1962), pp. 7–10.

Viking, in fact, was a dual-frequency craft; it used both S-band and X-band frequencies. For the Mariner flight to Jupiter and Saturn, the telemetry rates were the same as those for Mariner 10, but the data were coded and transmitted at X-band from distances up to nine astronomical units. Operating in the higher X-band range gave the increased sensitivity needed to remain in contact with Mariner, as it flew by Jupiter and Saturn. Construction of the 400-kilowatt, X-band (8495 MHz; 3.5 cm) transmitter for the Mars Station was completed by Advanced Systems Development, and the DSS-14 began operating at X-band in 1975.[74]

During the 1970s, the population of JPL planetary radar astronomers grew. Jurgens had an undergraduate and graduate degree in electrical engineering from Ohio University and had taught electrical engineering at Clarkson College (Ohio), before pursuing a doctoral degree at Cornell. Sometime after he finished researching his dissertation, a study of the radar scattering properties of Venus, at Arecibo, JPL hired Jurgens in 1972 to serve on the technical staff of the Telecommunications Research Section, not to do planetary radar astronomy.[75]

Also working in Goldstein's section was George Downs, who had studied radio astronomy at Stanford University under Ronald Bracewell. Goldstein had Downs analyze Mars radar data and make observations at Goldstone to assist in the selection of the Viking landing site, a project funded by the Viking Project Office. The planetary radar work, however, was in addition to his regular JPL duties, which involved studying newly discovered radio sources as potential timing sources for the Deep Space Network.[76]

During the heyday of the Viking Mars radar observations, Goldstein called upon other JPL employees, such as Howard C. Rumsey, Jr., who had a strong background in physics and mathematics, and the hardware experts George A. Morris and Richard R. Green. Jurgens described the atmosphere at JPL: "We all knew each other's talents. It was very efficient. Nobody ever felt like we were working terribly hard. It was just like a big playpen. Everybody came here, and we sort of did our thing and thought about what we wanted to do. We'd talk to each other, and we'd go out to lunch. It was the period of the long lunches sometimes. We had the Gourmet Society. The Gourmet Society was really headed by Howard Rumsey, who really liked good food. He would read the Sunday gourmet page and the Thursday gourmet page in the *L. A. Times*, and pick out interesting restaurants. At least one day a week, we went trudging off-lab to eat decent food at some interesting place that Howard had selected. These things often involved bicycle trips as far as Long Beach."[77]

Once Viking project funding ended in 1976, JPL radar astronomy hit hard times. Getting time on the DSN become more difficult. It was easy to get time in the early and middle 1960s, when the DSN was tracking few spacecraft. As Dick Goldstein explained: "Back in the sixties I thought of myself as director of the Goldstone Observatory. I got to choose what we could do, if I could get support for it."[78] During the 1960s, the JPL radar experiments conducted on Venus involved hundreds of hours of runs; for example, the 1961 Venus experiment involved 238 hours of data collected over two months. But by the end of the decade, the amount of time available had declined. The JPL 1969 Venus observations were not made daily for a period of months during inferior conjunctions, but only "on 17 days spaced from 11 March to 16 May 1969."[79]

---

74.   Rob Hartop and Dan A. Bathker, "The High-Power X-Band Planetary Radar at Goldstone: Design, Development, and Early Results," *IEEE Transactions on Microwave Theory and Techniques* MIT-24 (December 1976): 958-963; JPL Annual Report, 1974-1975, p. 22, JPLA.

75.   Jurgens 23 May 1994.

76.   Downs 4 October 1994.

77.   Jurgens 23 May 1994.

78.   Goldstein 14 September 1993.

79.   Golomb, "Introduction," in Victor, Stevens, and Golomb, p. 4; Goldstein and Howard C. Rumsey, Jr., "A Radar Snapshot of Venus," *Science* 169 (1969): 975.

The reduction in available antenna time was in direct proportion to the increasing number of spacecraft with which the Deep Space Network communicated. By 1977, the DSN was in communication with a record 14 spacecraft. In addition to the three Viking craft (two orbiters and one lander), the DSN communicated with Helios 1 and 2, Pioneer 11 (Saturn), Pioneer 10 (which was leaving the solar system), Voyagers 1 and 2, and Pioneers 6, 7, 8, and 9. That number grew to 19, a new record, the following year, when the DSN also handled communications with Pioneer Venus, which was an orbiter and four probes.[80]

Then the Deep Space Network stopped funding radar astronomy hardware. The ability to carry out radar astronomy without official recognition was maintained thanks to the presence at high levels of JPL management of Eb Rechtin and Walt Victor, who watched over planetary radar activities. But Rechtin left JPL, and Victor transferred in December 1978 from the DSN to the Office of Planning and Review.[81] Without their guardianship, JPL radar astronomy was vulnerable.

As Goldstein explained: "From a chauvinistic point of view, it was a disaster, because the rest of the world passed us by....We went from being a couple years ahead to being a couple years behind."[82] Without funding for hardware, the radar system was at risk. Moreover, the Goldstone Mars Station was in desperate need of repairs, and the equipment was becoming harder and harder to maintain. In 1976, the antenna already was ten years old, and the electronic equipment transferred to the Mars Station from the Venus Station (DSS-13) was even older.[83]

The termination of Deep Space Network funding of planetary radar astronomy grew out of two concerns, one within JPL and the other within the Deep Space Network. One of Bruce Murray's chief concerns after taking over as laboratory director was the state and status of science and scientists at JPL. The basic criticism was that JPL lacked a commitment to scientists. But the problem had a cultural side; technologically-centered teamwork dominated laboratory culture. Also, many of those doing science were like Dick Goldstein and Ray Jurgens; trained and hired as electrical engineers, they carried out radar astronomy science experiments. Murray made the problem the topic of mini-retreats, meetings, and seminars and, as a first step in elevating the status of science at JPL, appointed in October 1977 the first JPL chief scientist, Caltech physics professor Rochus E. Vogt, who had authored a report on relations between Caltech and JPL, another topic of great concern.[84]

Despite, or rather because of, Murray's concerns for science at JPL, planetary radar astronomy did not fair well under his reign as laboratory director. Goldstein was transferred out of the section where he had guided and supported the JPL planetary radar effort. JPL management decided that it was not proper to do science under the guise of improving the DSN. Radar astronomy should compete with other JPL science activities, and the Office of Space Science and Applications (OSSA; now the OSSI, Office of Space Science and Instruments) at NASA Headquarters should fund it, they ruled.[85]

At the same time, the DSN budget was suffering from monetary and manpower limitations.[86] To make matters worse, a routine review of the Deep Space Network, chaired

---

80.   JPL Annual Report, 1976–1977, p. 22, and *ibid.*, 1978, p. 20, JPLA.

81.   Murray to Allen M. Lovelace, 30 November 1978, 75/5/89-13, JPLA.

82.   Goldstein 14 September 1993.

83.   Jurgens 23 May 1994.

84.   Agenda, Director's Mini-Retreat, "How Does Science Fit In at JPL?," 22 March 1977, 55/3/89-13; Director's Letter, no. 22, 30 September 1977, 61/3/89-13; Roger Noll to Murray, 23 November 1977, 63/3/89-13; and typed manuscript, First Annual "State of the Lab" Talk by Murray to Management Personnel, 1 April 1977, 55/3/89-13, JPLA.

85.   Goldstein 14 September 1993.

86.   Notes from a discussion of TDA problems discussed during a mini-retreat held 8 November 1977, 63/3/89-13, JPLA; Jurgens 23 May 1994; and Stevens 14 September 1993.

by Eb Rechtin, declared that radar astronomy was no longer the "cow-catcher" of the DSN, meaning that the role of radar astronomy in creating new hardware to help drive forward the Deep Space Network had come to an end. It was, therefore, time to pull the plug on planetary radar astronomy, after previous reviews had lauded it.

From time to time, acting under instruction from NASA Headquarters, the Deep Space Network called into existence the TDA (Tracking and Data Acquisition) Advisory Panel to review DSN long-term plans. Planetary radar was held high as an integral part of DSN development activities by Ed Posner, a DSN manager. Among the hardware contributions of radar astronomy he listed were microwave components, signal processing techniques, and station control concepts, all of which were tested in a "realistic environment." Planetary radar fell from that favorable position during the 1978 review. The head of the review panel, now called the DSN Advisory Group, was none other than Eb Rechtin, the architect of the Deep Space Network and the one responsible for making planetary radar a testbed of DSN technology. DSN management asked the panel to consider, among many other questions, radar astronomy. In the opinion of the Advisory Group, which Eb Rechtin wrote, "Another DSN technology which may have had its day as a foundation for DSN technology is DSN radar astronomy. Radar astronomy served the DSN very well for many years. The Advisory Group wonders what the next area might be."[87] Radar astronomy no longer produced the cutting edge hardware that justified the support of Advanced Systems Development. As a result, planetary radar astronomy at Goldstone went begging for money.

A good part of the problem was the perception of planetary radar as *just* a testbed for DSN technology. The value of the *science* was simply not recognized by either DSN management or NASA Headquarters. After all, in accordance with Eb Rechtin's plan, planetary radar was not to occupy a budget line nor to have program status; it was simply a DSN activity to assist in the development and testing of new technology.

The lack of money to even maintain the Goldstone radar, whose age and one-of-a-kind design engineering made it all that much harder to maintain, began to frustrate the performance of experiments. By 1980, the Goldstone radar was in such bad shape that planetary radar astronomy experiments were no longer carried out on a regular basis. The radar was resurrected for attempts at asteroid 4 Vesta on 28 May 1982 and comets IRAS-Araki-Alcock and Sugano-Saigusa-Fujikawa in 1983, but only Comet IRAS-Araki-Alcock was detected successfully. As Ray Jurgens reflected on the situation: "Basically, it looked like it was the end of the radar."[88]

Goldstack

In discussing radar systems available for planetary research, an instrument that one must not overlook is the bistatic Goldstack radar, which used Haystack as the transmitting antenna and the JPL Goldstone DSS-14 radar as the receiving antenna. In the past, planetary radar astronomers seldom used bistatic radars, let alone radars requiring the coordination of two unrelated institutions. Bistatic radars require a daunting amount of coordination on both the technical and institutional level. Nonetheless, transmitting power and antenna receiver sensitivity can combine to create a radar capable of doing more than either facility operating monostatically. In theory, Goldstack could outperform either Haystack or DSS-14 separately and achieve a nearly tenfold increase in overall radar performance.

---

87.   "TDA Advisory Panel, 1971–1972," and "TDA Advisory Council, 1978–1981," JPLPLC.
88.   Jurgens 23 May 1994; Jurgens, "Comet Iras," pp. 222 and 224.

When radar astronomers Irwin Shapiro and Gordon Pettengill pitched Goldstack to NASA in 1968, they outlined an ambitious program of research: 1) observations of the Galilean satellites Ganymede and Callisto; 2) maps of the surfaces of Mercury and Venus; 3) a Moon-Earth-Moon triple-bounce experiment to study the Earth's radar-reflecting properties; 4) topographical studies of Mars at a resolution of 150 meters; and 5) a radar test of General Relativity.[89] Haystack and JPL engineers worked out the technical details of those experiments, and by May 1970 JPL had installed an X-band maser tunable to the Haystack frequency. The demands of the space program on the Mars Station, however, forced postponement of the experiments. As Shapiro recalled: "DSN always had scheduling problems. Scheduling was the biggest pain in the neck. From the point of view of science, I never felt the best things were done with scheduling; the engineering and mission pressures were too enormous. It always seemed to be as impossible as possible to schedule ground-based science experiments, but good science, in fact, was done." Goldstack eventually searched for Ganymede and Callisto in late May and early June 1970.[90]

Jodrell Bank

Several years before planetary radar astronomy ended at Haystack and declined at JPL, radar research at Jodrell Bank came to an end, too. In contrast to JPL, Jodrell Bank officially recognized and funded its radar astronomy program, and Sir Bernard Lovell proudly and, in the face of adversity, stubbornly maintained radar research. The Jodrell Bank facility was an example of British Big Science; private and civilian governmental funding underwrote the building of the large dish. While the U.S. military funded some meteor radar research, Jodrell Bank radar astronomy was not, in any sense, an extension of American Big Science. The demise of planetary radar astronomy at Jodrell Bank was a lesson in the dangers inherent in Little Science, not Big Science.

Thanks to NASA and the American military, Jodrell Bank did not lack for radar equipment. The still secret agreement between Lovell and an unidentified Air Force officer had as its immediate objective the sending of commands to the Pioneer 5 spacecraft. The U.S. Air Force funded Space Technology Laboratories (STL), a Los Angeles-based wholly-owned subsidiary of Ramo-Wooldrige (later TRW), to install a continuous-wave 410.25-MHz (73-cm) radar transmitter and other equipment on the Jodrell Bank telescope in order to track lunar rocket launches. Although the STL transmitter had only a few kilowatts of power, it was stable, reliable, and free of the problems that plagued the pulse radar apparatus pieced together by John Evans. Ownership of the STL transmitter passed to NASA, which provided operational funds between 1959 and 1964 to track rocket launches, not to perform radar experiments. NASA left the equipment on the Jodrell Bank antenna "on an indefinite loan basis," so that the University of Manchester might use it for scientific research.[91]

---

89. Memorandum, NEROC Project Office to Wiesner, 19 September 1968, regarding "Proposed Contact with Newell Regarding Possible Partial Support of Haystack by NASA," 8/2/AC 135, MITA; NEROC, Proposal to the National Science Foundation for Programs in Radio and Radar Astronomy at the Haystack Observatory, 8 May 1970, pp. III.8-III.10, LLLA; Brunk to Distribution List, 4 October 1968, NHOB.

90. Shapiro 4 May 1994; Shapiro 1 October 1993; "Funding Proposal, 'Plan for NEROC Operation of the Haystack Research Facility as a National Radio/Radar Observatory,' NSF, 7/1/71-6/30/73," 26/2/AC 135, and Sebring to Hurlburt, 27 March 1970, 18/2/AC 135, MITA; NEROC, Proposal to the National Science Foundation for Programs in Radio and Radar Astronomy at the Haystack Observatory, 8 May 1970, pp. III.8-III.10, LLLA; JPL 1970 Annual Report, p. 14, JPLA.

91. Lovell, 11 January 1994; Evans 9 September 1993; Ponsonby 11 January 1994; Lovell, "Astronomer by Chance," pp. 322–325 and 328–329; Edmond Buckley to R. G. Lascelles, 8 November 1961, and related documents in 2/53, Accounts; Able, Thor, and Pioneer 5 materials in 4/16, Jodrell Bank Miscellaneous; materials in 1/4, Correspondence Series 2; 2/53, 2/52, 2/55, 7/55, 8/55, 1/59, and 3/59, Accounts; and 4/16, Jodrell Bank Miscellaneous, JBA.

In 1962, the Jodrell Bank radar group consisted of only John Thomson and his graduate student John E. B. Ponsonby. Evans had sought his fortune at Lincoln Laboratory. As Ponsonby characterized the radar group, "We were always two men and a boy [K. S. Imrie]." When Ponsonby arrived at Jodrell Bank in 1960, he was shocked to discover that he was the only one in the radar group with a flare for electronics; Thomson, according to Ponsonby, was happy doing computations. Jodrell Bank radar astronomy was small not only in terms of staff, but also in observing time, which varied between 1 and 10 percent.[92]

Thomson and Ponsonby abandoned much of the equipment Evans had been using; they used a simpler approach with less technical risk. The old apparatus used vacuum tubes; the new was all solid-state digital electronics. A grant from the DSIR underwrote the cost of these modifications, as well as the purchase of a parametric amplifier and spare klystron tubes. The 1962 and 1964 Jodrell Bank Venus radar experiments were carried out with this digital continuous-wave equipment.[93]

The focus of Jodrell Bank's Venus radar research after 1962 was a bistatic experiment with the Soviet Long-Distance Space Communication Center located near Yevpatoriya in the Crimea. The experiment was possible only because Lovell had succeeded in thawing Cold War relations. The opportunity came in March 1961, when Soviet space trackers lost contact with a Venus probe launched the previous month. The Soviet Academy of Sciences approached Lovell to use the Jodrell Bank telescope to search for signals. As the months passed, and Jodrell Bank attempted to make contact with the probe, communications between the British and Soviets increased. The collaboration led to the establishment of a telex link between Jodrell Bank and the Yevpatoriya radar station, as well as an invitation for Lovell to visit the Soviet Union two years later.

The idea of doing the bistatic experiment came to Lovell during his visit to Yevpatoriya, when he discovered the extremely powerful Soviet transmitter. Vladimir Kotelnikov, who headed the Soviet planetary radar effort, joined Lovell as the other moving spirit behind the bistatic project. An Iron Curtain of secrecy hindered the project, however. In order to set up the bistatic radar, Jodrell Bank had to know the frequency and precise coordinates of the Yevpatoriya radar. The Soviets were loathe to disclose their frequency, transmitter size, location, or even antenna dimensions, but the British established those parameters step by step. Nonetheless, the experiment did not work initially, because Jodrell Bank lacked the correct Doppler shift. After testing the bistatic arrangement on the Moon, the Yevpatoriya facility began to transmit radar signals to Venus, and Jodrell Bank received them from January through March 1966. Data tapes were delivered to Kotelnikov by way of the British Embassy. As a long-distance bistatic radar experiment, the effort was a first. However, it was an opportunity lost.[94]

Ponsonby set forth a cogent analysis of the bistatic Venus experiment in his dissertation: "Planetary radar has proved to be a field in which new results are only obtained by the groups which have the most sensitive systems and the data processing capacity to make the best use of the data acquired. In both respects the group at Jodrell Bank has never been in a leading position."[95]

Lovell disagreed; the bistatic experiment was "just too late." JPL and Lincoln Laboratory already had determined the rate and direction of rotation of Venus. "But if

92. Ponsonby 11 January 1994; summaries of telescope use in 1/2, Correspondence Series, JBA.

93. Ponsonby 11 January 1994; 2/51, Accounts, JBA; Ponsonby, Thomson, and Imrie, "Radar Observations of Venus," pp. 1–17; Ponsonby, Thomson, and Imrie, "Rotation Rate of Venus Measured by Radar Observations, 1964," *Nature* 204 (1964): 63–64.

94. Lovell, 11 January 1994; Ponsonby 11 January 1994; Lovell, "Astronomer by Chance," pp. 370–372; Lovell, *Out of the Zenith*, pp. 186–188 and 201–204; Ponsonby and Thomson, "U.S.S.R.-U.K. Planetary Radar Experiment," pp. 661–671 in R. W. Beatty, J. Herbstreit, G. M. Brown, and F. Horner, eds., *Progress in Radio Science, 1963–1966* (Berkeley: URSI, 1967); Ponsonby, "Planetary Radar," pp. 6.11–6.22.

95. Ponsonby, "Planetary Radar," p. 6.21.

only my 1963 conversations and agreement with the Soviet Union could have been facilitated without trouble at this end and without trouble at the transmitter," Lovell argued,
"we would have been first on that."[96]

Lovell's analysis, as well as that of Ponsonby, raises the vital question of the ability of
the Jodrell Bank radar group to effectively compete against American radar astronomers.
The STL transmitter operated in the UHF band (410.25 MHz; 73 cm). Although the
Arecibo Observatory operated in the same band, the trend in planetary radar astronomy
was toward higher frequency ranges, the S-band at JPL (and later at Arecibo) and the X-
band at Haystack. The higher frequencies allowed the radar to do much more radar
astronomy science than was possible at UHF.

Ponsonby raised another point in his dissertation: "If a true state-of-the-art transmitter were acquired it would cost an appreciable fraction of the cost of the telescope, and
clearly to justify investment on that scale it would have to be used much more extensively
than would be compatible with the predominantly passive radio astronomical programs at
Jodrell Bank. Passive radio-astronomy may appropriately be done as a secondary line of
research at a primarily radar installation, but experience has shown that the two activities
do not combine well the other way. Appreciating this, the research reported in this thesis
is not, at least for the present, being pursued further." Indeed, Ponsonby continued, "The
limited computing facilities available in the University and the lack at the time of on-line
computers at Jodrell Bank in effect prevented a thorough analysis of the data that was
acquired, and this took away much of the value of the observations."[97]

The acquisition of new radar and computer equipment certainly would have constituted a significant expenditure, but Lovell probably could have raised the necessary
money. Could Jodrell Bank have kept up with the development of planetary range-
Doppler mapping in the United States? Thomson was working on an aperture synthesis
technique for making lunar radar maps. The mathematical process for constructing the
image was analogous to that now used for tomographic brain scanners and differed entirely from that used in the United States to construct range-Doppler maps. The technique
was not very practical, however; it required computer capacity not then available at Jodrell
Bank and ultimately could not be generalized to the planets.[98]

In the end, the small scale of planetary radar astronomy at Jodrell Bank did it in.
Thomson and Ponsonby grew tired of the Soviet bistatic Venus experiment. They carried
the main load of the work at irregular hours of the day and night. Finally, on 18 March
1966, Thomson and Ponsonby could take no more. They handed Lovell a list of ten good
reasons for ending the experiment. Kotelnikov agreed to "an interval" in the observations,
which never resumed.[99] Ponsonby remained rather cynical about the venture, which he
has characterized as a political exercise. "The signals were recorded on magnetic tape and
sent off to Russia, and I never heard from them again!"[100]

Ponsonby already was tired of the bistatic experiments, when the death of John
Thomson from an inoperable brain tumor in August 1969 devastated the Jodrell Bank
planetary radar program. Without Thomson, and certainly without Ponsonby's interest,
Jodrell Bank had no radar group. Through sheer stubbornness, however, Lovell tried to
keep the radar program going. In October 1969, he and his Jodrell Bank colleagues drew
up a scientific program for a proposed 122-meter (400-ft) telescope, the Mark V. The program included a series of planetary radar experiments outlined by Ponsonby. Was this the

96.   Lovell 11 January 1994.
97.   Ponsonby, "Planetary Radar," pp. 6.21–6.22.
98.   Ponsonby 11 January 1994; Lovell, "Astronomer by Chance," pp. 373–375; various documents in
2/51, Accounts, JBA.
99.   Lovell, Out of the Zenith, pp. 207–208.
100. Ponsonby 11 January 1994.

telescope that could have revived Jodrell Bank radar research? Like its American cousin the NEROC telescope, the Mark V was never built. In retrospect, Lovell realized that "It was now out of the question for us to continue....I saw the passing of radar as inevitable, but with regret."[101]

The sixties was the era of the Big Dish; large antenna projects came and went, and so did planetary radars. In 1965, four antennas supported planetary radar experiments: Arecibo, Haystack, Jodrell Bank, and the Goldstone Mars Station. A fifth dish, the NEROC telescope, was on the drawing board. But ten years later, the NEROC telescope had not been built; Haystack and Jodrell Bank no longer performed planetary radar experiments. By 1980, Goldstone had joined their number. Only Arecibo remained. Planetary radar astronomy appeared to be a collapsing field.

At Arecibo, nonetheless, radar astronomy had found a patron in NASA. Planetary radar there also had a recognized and guaranteed budget, as well as a world-class research instrument, and both Cornell and MIT fed graduate students to the Arecibo facility. Given the financial, institutional, technological, and other resources available at Arecibo for planetary radar astronomy, one would have expected the field to have occupied an increasing amount of antenna time from 1974, when Haystack ceased radar astronomy, to 1980, when JPL activity virtually ended. Instead, antenna use remained relatively stable, averaging about six percent between 1971 and 1980 and passing seven percent concurrently with the inferior conjunctions of Venus.[102]

In terms of personnel, one could count the field of planetary radar astronomy as consisting of nine individuals. At MIT was Gordon Pettengill; at JPL, Dick Goldstein, Ray Jurgens, and George Downs. The Arecibo Observatory supported four radar practitioners: Don Campbell, associate director at the Arecibo Observatory since 1979; John Harmon, AO research associate since 1978; Steven J. Ostro, Cornell assistant professor of astronomy since 1979; and Barbara Ann Burns, a graduate student of Don Campbell.

In 1980, planetary radar astronomy was indeed a small field in terms of available instrumentation and active practitioners. It was an example of Little Science, but one which depended on Big Science for its very existence. Moreover, although that Big Science had been as diverse as military, space, ionospheric, and radio astronomy research at the emergence of radar astronomy, by 1980 Big Science had come to mean one thing: NASA. The financial and institutional arrangements with NASA influenced the kind of science done. In order to understand how that science was influenced, we must first look at the evolution of planetary radar astronomy as a science.

101. Lovell 11 January 1994; Ponsonby 11 January 1994; Lovell, *Out of the Zenith*, p. 203; Lovell, *The Jodrell Bank Telescope*, Chapters 5–6 and 9–10, especially pp. 55–56 and 257. In analyzing the demise of radar astronomy at Jodrell Bank, though the smallness of the active radar astronomy staff, technical and technological factors, and the American lead had a more determinant role, to be sure, one must not overlook the lure of radio astronomy.

102. These figures are based on the NAIC quarterly reports for the years 1971–1980. The percentage of radar use annually was 2.9 percent in 1971; 9.5 percent in 1972; 6.9 percent in 1973; 1.9 percent in 1974; 7.2 percent in 1975; 5.8 percent in 1976; 7.3 percent in 1977; 4.7 percent in 1978; 5.0 percent in 1979; and 7.8 percent in 1980. The average percentage for the period 1971-1980 was 5.9, while the average for 1971–1975 was 5.68 percent and for 1976–1980 6.12 percent.

Chapter Five

Normal Science

Starting with the initial detections of Venus in 1961, planetary radar astronomy grew rapidly by discovering the rate and direction of Venus's rotation, by refining the value of the astronomical unit, and by rectifying the rotational period of Mercury. Data gathered from radar observations made at Haystack, Arecibo, and Goldstone formed the basis for precise planetary ephemerides at JPL and Lincoln Laboratory. In sum, the results of planetary radar astronomers served the needs of the planetary astronomy community. In addition, radar also served to test Albert Einstein's General Theory of Relativity.

Planetary radar astronomy concerned itself with two different but related sets of problems. One set of problems related to planetary dynamics and ephemerides, for instance, orbits, rotational and spin rates, and the astronomical unit. A second set related to the radar characteristics, or what is called the radar signature, of the planets, such as surface scattering mechanisms, dielectric constants, and radar albedos. The latter problems are epistemological; that is, they deal with how radar astronomers know what they know.

What defines this second set of epistemological problems is the fact that planetary radar astronomy is based on the use of techniques particular to radar. These problems have remained unchanged over time. In contrast, the first set of problems, those dealing with planetary and dynamics ephemerides, have changed over time. The nature of that change has been additive; at each stage of change, new problems are added to the old problems, which remain part of the set of problems radar astronomers seek to solve.

Both the epistemological and scientific sets of problems are interrelated. For example, planetary radar astronomers derive the ability to solve astronomical problems out of the resolution of epistemological questions. The development of range-Doppler mapping, for example, led to the solution of a set of problems entirely different from ephemerides problems, yet the solution of ephemerides problems was sine qua non to the creation of range-Doppler maps. Conversely, the attempts to solve certain scientific questions required reconsideration of the radar techniques themselves.

The philosopher of science Thomas S. Kuhn has attempted to explain the conduct of scientific activity.[1] Although Kuhn has used the term "paradigm" differently over time, initially it had a limited meaning. Stated simply, a paradigm, as used by Kuhn, is a core of consensus within a group of practitioners. The essence of the paradigm consensus is a set of problems and their solutions. Planetary radar astronomy quickly achieved and maintained a paradigmatic consensus on which problems to solve.

Moreover, the field often achieved scientific success by solving problems left unsolved or unsatisfactorily solved by optical means. Just as radar astronomy had resolved earlier that meteors were part of the solar system, so the determination of the rotational rates of Venus and Mercury and the refinement of the astronomical unit were astronomical problems inadequately solved by optical methods, but resolved through the analysis of radar data.

---

1. The works of Kuhn, which span over thirty years, have been summarized, explained, and analyzed in Paul Hoyningen-Huene, *Reconstructing Scientific Revolutions: Thomas S. Kuhn's Philosophy of Science*, trans. Alexander T. Levine (Chicago: University of Chicago Press, 1993). Especially relevant to the discussion here are pp. 134–135, 143–154, 169, 188–190 and 193–194.

For Kuhn, "normal science" was a specific phase of scientific development distinguished by universal consensus within a given scientific community over the problems to be solved and the ways of solving those problems. In other words, normal science was paradigm science. Preceding its evolution into normal science, according to Kuhn, a scientific activity passes through a developmental phase in which the problem-solving consensus that characterizes normal science does not yet exist. In this "preconsensus" or "pre-paradigm" phase, and immediately before a phase of normal science, groups of investigators addressing roughly the same problems but from different, mutually incompatible standpoints compete with each other. As a consensus emerges, members of the competing schools join the group whose achievements are better, as measured by scientific values.

Planetary radar astronomy did not pass through Kuhn's "preconsensus" phase, however. Complementary, not competing, groups marked the emergence of the field. The "bistatic radar" approach of Von Eshleman at Stanford University complemented the efforts of ground-based planetary radar astronomers, and that complementarity had been Eshleman's intention.[2] Ground-based planetary radar astronomers distinguished themselves from the Stanford approach. In a review article on planetary radar astronomy published in 1973, Tor Hagfors and Donald B. Campbell, both at the Arecibo Observatory, explained, "We have, however, chosen to omit this work [space-based radar] here since it is our opinion that it properly belongs to the realm of space exploration rather than to astronomy."[3] Space exploration versus astronomy, then, was how planetary radar astronomers established turf lines.

Planetary radar astronomy was, above all else, a set of techniques used with large-scale ground-based radar systems. As a result, planetary radar was an algorithm in search of a problem, a data set in search of a question. Hence, the success of planetary radar inexorably depended on its ability to link its techniques and results to the problem-solving of a scientific discipline. Initially, those problems came from planetary astronomy, but as the types of techniques accumulated, radar came to solve new problems posed by planetary geology. Furthermore, the solving of those problems tied planetary radar astronomy to NASA's space missions.

Despite its mercurial nature, planetary radar astronomy did exhibit an essential characteristic of Kuhn's normal science, a paradigm. The paradigm consisted of a consensus on a particular set of problems (e.g., orbital parameters) and agreement on a particular way of solving those problems (the analysis of range, Doppler, and other radar data obtained with ground-based radars from solar system objects). The detections of Venus, Mercury, and Mars between 1961 and 1963 opened the field, but rotational rates, as well as the refinement of the astronomical unit, established the field. With the successful application of range-Doppler mapping to Venus, the paradigm began to shift in a new direction.

Around the Sun in 88 Days

The first radar detection of Mercury was announced by the Soviet scientists working under Vladimir A. Kotelnikov and associated with the Institute of Radio Engineering and Electronics (IREE) of the Soviet Academy of Sciences and the Long-Distance Space Communication Center near Yevpatoriya, in the Crimea. Kotelnikov's group made 53 radar observations of Mercury during the inferior conjunction with that planet in June 1962. At that time, the distance from Earth to Mercury was between 83 and 88 million

2.   Eshleman 9 May 1994.
3.   Hagfors and Campbell, "Mapping of Planetary Surfaces by Radar," *Proceedings of the IEEE* 61 (September 1973): 1219–1225, esp. 1224.

kilometers, twice the distance to Venus during inferior conjunction. Although the weakness of the return echoes prevented their use as a reliable indicator of the astronomical unit, Kotelnikov and his colleagues claimed a technical tour de force and a first in planetary radar astronomy.[4]

Richard Goldstein and Roland Carpenter at JPL took up the Soviet challenge and bounced radar waves off Mercury the following year in May 1963 using the Goldstone experimental radar. The experiment established a distance record that overshadowed the Soviet claim. Mercury was then farther from Earth, over 97 million kilometers away. In addition, the JPL experiment confirmed what astronomers already knew about Mercury, that its period of rotation was 88 days. Goldstein had no reason to believe it was otherwise.[5]

However, when Gordon Pettengill and Rolf Dyce observed Mercury in April 1965 with the new Arecibo telescope, they reported a rotational rate of 59 ± 5 days. This discovery, one of the earliest major achievements of planetary radar astronomy, astounded astronomers, who sought to explain the new, correct rotational rate. As Pettengill and Dyce concluded, "The finding of a value for the rotational period of Mercury which differs from the orbital period is unexpected and has interesting theoretical implications. It indicates either that the planet has not been in its present orbit for the full period of geological time or that the tidal forces acting to slow the initial rotation have not been correctly treated previously."[6]

Pettengill, Dyce, and Irwin Shapiro next published a lengthier discussion of their radar determination of Mercury's 59-day rotational period based on additional observations made in August 1965.[7] Working with Giuseppe "Bepi" Colombo, an astronomer from the University of Padova visiting the Smithsonian Astrophysical Observatory, Shapiro began to develop an explanation for the new rotational period. Colombo, Shapiro recalled, "realized almost immediately that 58.65 days was exactly two-thirds of 88 days. Mercury probably was locked into a spin such that it went around on its axis one-and-a-half times for every once around the planet. The same face did not always face the Sun. That meant that near Mercury's perihelion, that is, when its orbit is closest to the Sun, Mercury tends to follow the Sun around in its orbit. Near perihelion, then, the orbital motion and spin rotation of Mercury were very closely balanced, so that Mercury almost presented the same face to the Sun during this period."[8]

In a joint paper, Colombo and Shapiro analyzed Mercury radar data, as well as optical observations from the past, and presented a preliminary model.[9] In a seminal paper, Peter Goldreich and Stanton J. Peale pointed out the need to consider the capture of Mercury into the resonant rotation as a probabilistic event. If initial conditions during the

4.   Kotelnikov, G. Ya. Guskov, Dubrovin, Dubinskii, Kislik, Korenberg, Minashin, Morozov, Nikitskiy, Petrov, G. A. Podoprigora, Rzhiga, A. V. Frantsesson, and Shakhovskoy, "Radar Observations of the Planet Mercury," *Soviet Physics—Doklady* 7 (1963): 1070–1072. Given the stated weakness of the Mercury echoes, as well as their difficulty in obtaining accurate and verifiable Venus results, the Soviet announcement of a detection of Mercury, a much farther radar target than Venus, raised doubts in the United States about the validity of the Soviet claims.

5.   Carpenter and Goldstein, "Radar Observations of Mercury," *Science* 142 (1963): 381.

6.   Pettengill and Dyce, "A Radar Determination of the Rotation of the Planet Mercury," *Nature* 206 (19 June 1965): 1240.

7.   Dyce, Pettengill, and Shapiro, "Radar Determination of the Rotations of Venus and Mercury," *The Astronomical Journal* 72 (1967): 351–359.

8.   Shapiro 30 September 1993; Giuseppe Colombo, "Rotational Period of the Planet Mercury," *Nature* 208 (1965): 575.

9.   Colombo and Shapiro, "The Rotation of the Planet Mercury," *The Astrophysical Journal* 145 (1966): 296–307. Earlier, it had appeared as an internal SAO publication: Colombo and Shapiro, *The Rotation of the Planet Mercury*, SAO special report no. 188 (Cambridge: SAO, 13 October 1965).

formation of the solar system had been slightly different, the capture may not have taken place.[10]

Irwin Shapiro's graduate student, Charles C. Counselman III, then did his doctoral thesis on the rotation of Mercury. Counselman developed a theory of capture, escape, recapture, and escape, as the eccentricity of Mercury's orbit changed, in a two-dimensional statistical model of the capture problem. Later, Norman Brenner, a graduate student working with both Shapiro and Counselman, expanded the analysis into a three-dimensional model in his 1975 doctoral dissertation. Meanwhile, Stanton Peale published his own three-dimensional analysis.[11]

The Outer Limits

Although Venus became the prime target of planetary radar astronomers, other planets drew their attention from the earliest opportunity to detect echoes from that planet. Richard Goldstein made the first radar detection of Mars during the opposition of February 1963, when the distance to Mars from Earth was over 100 million kilometers. Goldstein found Mars "a very difficult radar target because of its great distance from Earth and rapid rate of rotation."[12]

Mars defined the farthest limits of planetary radar detections until after the addition of the S-band radar to the Arecibo telescope and the X-band upgrade of the Goldstone Mars Station. Farther out, neither American nor Soviet efforts ever resulted in an unambiguous radar detection of Jupiter. Certainly no echoes returned from any solid surface features. Nonetheless, US and Soviet investigators claimed detections. The case of Jupiter demonstrates the difficulty of obtaining radar echoes from a "soft" target, that is, one that is not a solid body, especially at such an extreme distance.

Soviet investigators working with Vladimir Kotelnikov at the Yevpatoriya radar center claimed to have detected radar echoes from Jupiter as early as September 1963 in the 29 December 1963 issue of *Pravda*. The planet was in opposition at a distance of about 600 million kilometers, six times farther than Mars at opposition in 1963. Not surprisingly, Kotelnikov and his colleagues reported that the echoes were weak.[13]

Between 17 October and 23 November 1963, during the same opposition of Jupiter, Dick Goldstein attempted observations of the planet with the Goldstone experimental radar. He found few if any echoes. Occasionally, though, a single run did indicate a "statistically significant" return. Goldstein noticed that the time interval between these "significant" returns were most often a multiple of the rotation period of Jupiter, about 10 hours. It seemed that a single localized area on Jupiter, which did not coincide with the celebrated red spot, was both a good and a smooth reflector of radar waves.

---

10.   Shapiro 30 September 1993; Peter Goldreich, "Tidal De-spin of Planets and Satellites," *Nature* 208 (1965): 375–376; Goldreich and Stanton Peale, "Resonant Spin States in the Solar System," *Nature* 209 (1966): 1078–1079; Goldreich, "Final Spin States of Planets and Satellites," *The Astronomical Journal* 71 (1966): 1–7; Goldreich and Peale, "Spin-Orbit Coupling in the Solar System," *The Astronomical Journal* 71 (1966): 425–438. Also, in a joint paper, Peale and Gold attempted to explain the rotational period of Mercury in terms of a solar tidal torque effect. Peale and Gold, "Rotation of the Planet Mercury," *Nature* 206 (1965): 1241–1242.

11.   Shapiro 30 September 1993; Counselman, "Spin-Orbit Resonance of Mercury," Ph.D. diss., MIT, February 1969. See also Counselman, "The Rotation of the Planet Mercury," Chapter 14, pp. 89–93 in R. G. Stern, ed., *Review of NASA Sponsored Research at the Experimental Astronomy Laboratory* (Cambridge: MIT, 1967).

12.   Goldstein and Willard F. Gillmore, "Radar Observations of Mars," *Science* 141 (1963): 1171–1172.

13.   Memorandum, O. Koksharova to I. Newlan, 9 January 1964, translation of *Pravda* article, microfilm 22–314, JPL Central Files. The article later appeared as Kotelnikov, Apraksin, Dubrovin, Kislik, Kuznetsov, Petrov, Rzhiga, Frantsesson, and Shakhovskoi, "Radar Observations of the Planet Jupiter," *Soviet Physics-Doklady* 9 (1964): 250–251.

To investigate further, Goldstein divided Jupiter into eight "time zones" and averaged all the runs which illuminated a single "time zone." The zone centered about the Jovian longitude 32° gave a response that Goldstein characterized as "statistically significant," although, he admitted, "this detection cannot be considered absolutely conclusive." The amount of return was simply too high to be believable. Goldstein later attempted to obtain echoes from Jupiter, using a Goldstone radar that was "a hundred times better," but he did not find any echoes. "We never were able to repeat it," he confessed.[14]

During the next oppositions of Jupiter, in November 1964, December 1965, and February 1966, Gordon Pettengill, Rolf Dyce, and Andy Sánchez, from the University of Puerto Rico at Rio Piedras, bounced radar waves off Jupiter using the 430-MHz Arecibo telescope. They designed their experiments to duplicate both the Soviet and JPL approaches; however, they failed to validate either the Soviet or JPL claims.

The Arecibo investigators obtained results that were many times smaller than those reported by Goldstein. As for the Soviet results, which were close to the noise level, the Arecibo investigators concluded: "The results reported in the U.S.S.R., which exceed the associated system noise by only 1.3 standard deviations of the fluctuations in that noise, should probably not be taken seriously." The Arecibo investigators suggested that the echoing mechanism was located in the upper levels of Jupiter's atmosphere "and that echoes might be returned only in exceptional circumstances." They concluded: "Many more observations of Jupiter spanning a long period of time and carried out at many widely separated frequencies must be made before the behavior of Jupiter as a radar target can begin to be understood."[15]

Those observations never took place. Jupiter remained a misunderstood and disregarded radar target. The outer reaches of planetary radar astronomy remained confined to the terrestrial planets. Jupiter and Saturn had to await the Arecibo S-band and the Goldstone X-band upgrades. Even then, however, planetary radar astronomers focused on solid targets, Jupiter's Galilean moons and Saturn's rings.

Icarus Dicarus Dock

In contrast to the attempts on Jupiter, the radar detection of Icarus was unambiguous. Icarus is an Earth-crossing asteroid, meaning its orbit around the Sun crosses that of Earth. On occasion, Icarus comes within 6.4 million kilometers of Earth, as it did in June 1968. Nonetheless, Icarus was a difficult radar target, because of its small size. Its radar detectability was extremely small, one thousandth that of Mercury at its closest approach and only 10^{-12} (one trillionth) that of the Moon.[16]

Only Haystack and the Goldstone Mars Station succeeded in detecting the asteroid. Although Icarus was within the declination coverage of Arecibo, attempts on 15 and 16 June 1968 yielded ambiguous results. "A successful search would have been more likely," Rolf Dyce reported, "if the full performance of the line feed had been available."[17]

Investigators at Haystack Observatory leaped over imposing hurdles to make the first radar detection of Icarus. Irwin Shapiro and his Lincoln Laboratory colleagues prepared an ephemeris based on 71 optical observations of Icarus between 1949 and 1967. Radar observation began in earnest at Haystack on the morning of 12 June 1968. Late that

---

14.   Goldstein 14 September 1993; Goldstein, "Radar Observations of Jupiter," *Science* 144 (1964): 842–843.

15.   Dyce, Pettengill, and Sánchez, "Radar Observations of Mars and Jupiter at 70 cm," *The Astronomical Journal* 72 (1967): 771–777.

16.   Goldstein, "Radar Observations of Icarus," *Science* 162 (1968): 903.

17.   Dyce, "Attempted Detection of the Asteroid Icarus," in AIO, *Research in Ionospheric Physics*, Research Report RS 74 (Ithaca: CRSR, 31 July 1968), pp. 90–91.

evening, the Haystack observers received a new set of optical positions from astronomer Elizabeth Roemer at the University of Arizona. Michael Ash, of Lincoln Laboratory Group 63, immediately integrated the optical data into the radar ephemeris, and by midnight Haystack was observing with the new ephemeris.

Despite these heroic efforts to organize an improved ephemeris, rain, which severely attenuates X-band radar signals, bedeviled the observations. As a result, Haystack did not obtain a reasonably firm indication of an echo from Icarus until the afternoon of 13 June. Another particularly successful run that evening confirmed the presence of an echo, and by the morning of 14 June success was certain. Haystack terminated observations the morning of 15 June. To achieve its results, the Haystack radar had operated non-stop for 20 hours. Analysis of the data suggested that the radius of Icarus was between 0.8 and 1.6 km.

The effort to detect Icarus in spite of the rain and the difficult nature of the asteroid as a radar target inspired Louis P. Rainville, a Lincoln Laboratory technician who participated in the observations, to compose the following poem:[18]

> *Anode to Icarus*
> *Icarus Dicarus Dock*
> *We worked around the clock*
> *For three straight days*
> *We aimed our rays*
> *And an echo showed on the plot.*
>
> *But as always, there's a woe*
> *The rain made a better show*
> *As bleary our eyes*
> *Stared at the skies*
> *We hoped that the clouds would go.*
>
> *Oh for the roar and yell*
> *And the glory for old double "L"*
> *If on that crucial day*
> *When it came and went away*
> *We'd had one more decibel!*
>
> *Now as Icarus speeds from our sphere*
> *These words are for all men to hear*
> *T'was a good show men!*
> *Let's try again* ——
> *In another nineteen years!*
>
> *And so this was to be our lot*
> *We hoped for more than we got*
> *But we beat the worst;*
> *We did it* **first!**
> *Icarus Dicarus Dock*
>
> *EURICARUS!*

---

18.   "Weekly Reports, 5/13/68–8/11/69," 36/2/AC 135, MITA. The results appeared as Pettengill, Shapiro, Michael E. Ash, Ingalls, Louis P. Rainville, Smith, and Melvin L. Stone, "Radar Observations of Icarus," *Icarus* 10 (1969): 432–435.

Although Rainville's verse implied a contest to detect Icarus, no such competition existed; notwithstanding the rain, the spin direction of the Earth would assure Haystack the first look at the asteroid. At JPL, Dick Goldstein also successfully detected Icarus on 14–16 June 1968. Goldstein used a bistatic radar; the Mars Station received signals from a newly-developed 450-kilowatt transmitter installed on a nearby 26-meter (85-ft) dish. Although the Goldstone transmitter had nearly twice the power of Haystack Observatory, it still received only weak echoes.[19]

Using optical methods, asteroid astronomers Tom Gehrels, Elizabeth Roemer, and others calculated values for the period and the direction of the spin axis of Icarus and found that it appeared to be a rough stony-iron body, nearly spherical, with nonuniform reflectivity over the surface and with a spin period of 2 hours and 16 minutes. Its radius, they calculated, was at least 750 meters, which was close to the low end of the Haystack estimate. Armed with these results, Goldstein then reinterpreted his radar data and concluded that the surface of the asteroid was rocky and varied in roughness.[20]

The detection of Icarus was an important achievement of planetary radar, the first detection of an asteroid. Icarus also served to bring together radar and optical planetary astronomers in a special symposium on Icarus organized by Gordon Pettengill and chaired by Arvydas Kliore. Held in Austin, Texas, on 10 December 1968, the symposium was part of the pre-inaugural meeting of the Division for Planetary Science (DPS) of the American Astronomical Society (AAS). Appropriately, the symposium papers appeared in the journal of planetary science *Icarus*.[21]

The Icarus symposium was a pivotal moment for both planetary radar astronomy specifically and planetary astronomy in general. Previously, no organization dedicated exclusively to planetary astronomy existed. The AAS had approved the formation of the DPS only a few months earlier in August 1968. In 1973, the DPS opened its ranks to planetary scientists other than AAS members, such as chemists, geologists, and geophysicists, and the DPS endorsed *Icarus* as the primary publication for planetary research. Under the editorial direction of Carl Sagan, a champion of radar astronomy, *Icarus* began to solicit more articles in planetary astrophysics, as opposed to the earlier focus on celestial mechanics.[22]

The Icarus symposium typified the normal science paradigm of planetary radar astronomy in the 1960s. Activity centered on detecting a solar system object with a ground-based radar instrument and analyzing range and Doppler data to obtain information on orbital parameters and radii and related questions. Radar astronomers then presented these results to asteroid astronomers, echoing the fruitful joining of radar observers and astronomers that led to the discovery of the origin of meteors.

The Planetary Ephemeris Program

Starting in the 1960s, the raw data for the improvement of planetary ephemerides was provided by the accumulation of radar range and other data. Traditional observations of planetary positions involved only angular determinations, which provide a position in a two-dimensional plane (the sky). Radar added new dimensions with range and Doppler shift data and included the astronomical unit and the radii and masses of Mercury, Mars,

19.   Goldstein, "Icarus," pp. 903–904.
20.   T. Gehrels, E. Roemer, R. C. Taylor, and B. H. Zellner, "Minor Planets and Related Objects: 4. Asteroid (1566) Icarus," *The Astronomical Journal* 75 (1970): 186–195; J. Veverka and W. Liller, "Observations of Icarus: 1968," *Icarus* 10 (1969): 441–444; Goldstein, "Radar Observations of Icarus," *Icarus* 10 (1969): 430–431.
21.   "Editor's Introduction to: A Symposium on Icarus," *Icarus* 10 (1969): 429.
22.   Tatarewicz, pp. 122–123.

and Venus. JPL and Lincoln Laboratory undertook separate radar ephemerides programs.

The Lincoln Laboratory radar ephemerides program, known as the Planetary Ephemeris Program or PEP, had its roots in the anti-ICBM early warning systems. As a member of a Lincoln Laboratory task force charged with the early detection of incoming enemy ICBMs with radar, Irwin Shapiro became expert in the mathematics of deducing ballistic missile trajectories from radar observations. He wrote up his results in a Lincoln Laboratory report in early 1957. After the launch of Sputnik, the New York publishing house McGraw-Hill released Shapiro's report as a book in April 1958, because his ballistic missile techniques were applicable (with some modification) to satellite tracking. That book then became the basis for the JPL ephemeris program.[23]

Shapiro and the radar group at Arecibo worked very closely on gathering data for the Lincoln Laboratory planetary radar ephemerides. As Don Campbell explained, "He has always been our ephemerides person, and we provide him with input."[24] The close connection between the Arecibo and MIT Lincoln Laboratory groups resulted from the appointment of Gordon Pettengill of Lincoln Laboratory as the first associate director of the Arecibo Ionospheric Observatory. Pettengill set up the program so that the Arecibo radar ephemerides would always come from the PEP group.

Acquiring input for the PEP required extensive data taking that involved long hours of observations, often late at night. Don Campbell and Ray Jurgens, both graduate students at the time, did a lot of the work on Venus, Mars, and Mercury, under the supervision of Rolf Dyce and Gordon Pettengill. Campbell remembered the Mars observations in particular:[25]

> *This involved a lot of late nights, unfortunately, because the Mars opposition was around midnight. Every time the radar system was used, you had to go up to the suspended platform and actually change the receiver over. Then you had to go up after you finished to change them over again. Since I was very much at the lowest end of the totem pole at the time, it was my job to get on the cable car, go up to the structure, dabble with the thing late at night, change the receivers, come back, then when we finished, go back up and change them again. I suppose in retrospect you think of it as painful, although at the time I don't remember being particularly worried about it. I probably thought it was fun initially, although there were a lot fewer fences and safety devices on the platform then than there are now. It was quite possible to fall right through the thing.*

The initial PEP calculations performed with the planetary radar data served to refine the astronomical unit. Shapiro, however, also saw the need to refine the planetary ephemerides and the planetary masses. "It was also clear to me," he explained, "that we should not do it the way astronomers did, that is, with analytical series expanded out to huge numbers of terms. It seemed to me that with computers, even with those available at that time, we should be able to do this numerically, integrating the equations of the motions of the planets, integrating the partial derivatives, and doing everything digitally."[26]

The PEP required a large computer as well as an immense computer program. Today, the program has well over 100,000 Fortran statements. Computer programming,

23.   Shapiro, *Prediction of Ballistic Missile Trajectories from Radar Observations* (New York: McGraw-Hill, 1958); Shapiro 4 May 1994.
24.   Campbell 9 September 1993.
25.   Campbell 7 December 1993.
26.   Shapiro 30 September 1993.

however, was not Shapiro's forte. "I am pretty much a computer ignoramus," he confessed. So he hired a summer student, Michael E. Ash, who was a Princeton graduate student in mathematics. After graduating from Princeton, Ash worked at Lincoln Laboratory for about twelve years before taking a position at MIT's Draper Laboratory. Ash was the chief architect of the PEP computer program. John F. Chandler, a graduate student of Shapiro, took over the PEP from Ash and worked on it for over twenty years. Chandler expanded its applications so that now, in the words of Shapiro, "it does everything but slice bread."

Originally, PEP also analyzed optical observations of the Sun, Moon, and planets, including optical data from the U.S. Naval Observatory back to 1850. "I spent more time than I care to admit," Shapiro confessed, "transferring to machine-readable form all the optical observations recorded in history since 1750 of the Sun, Moon, and planets. In the end, I didn't think it was worth it. I never published our results, to Michael Ash's chagrin. We had this manuscript about so high [nearly seven and a half centimeters or three inches], but I could never find enough time to polish it to my satisfaction. History passed us by. That was the biggest unfinished task of my life. Michael Ash put in a lot of work on that, though not as much as I did. But the ball was in my court to finish it off, and I did not do it. So this is a guilt session."[27]

Today the PEP is a very complicated program that analyzes a variety of observations, including lunar laser ranging data. When he moved to Draper Laboratory, Michael Ash modified it for satellite and lunar work. It is still used in planetary radar and by astronomers at the Harvard-Smithsonian Center for Astrophysics. It can process pulsar as well as Very Long Baseline Interferometry (VLBI) observations. For a while, most of the pulsar observers in the world used the PEP; however, they shifted to the JPL ephemeris program in recent years. A lack of funding has left the PEP just able to keep up with the Arecibo ephemeris work.

In contrast, JPL has had the manpower and funding to support it. JPL developed its radar planetary ephemerides to support NASA spacecraft missions. Today, the JPL planetary ephemeris program, under the direction of E. Myles Standish, Jr., employs about a half dozen people who work on planetary, lunar, cometary, asteroidal, and satellite ephemerides. JPL initially called their ephemeris programs DE followed by the version number, with DE standing for "Development Ephemeris." In the late seventies, JPL sent over fifty copies of its ephemeris DE-96 to observatories, space agencies, and astronomical research groups around the world.

Next came the DE-200 series, which used a new equator and equinox. All major national almanac offices, including the U.S. Naval Observatory, and the French, British, German, Japanese, and Russian almanac offices, now use the JPL DE-200 ephemeris program, as do many universities, the European Space Agency, and radio astronomers. Moreover, the DE-200 program, formerly available on magnetic tape, now is distributed through the Internet as an FTP file.[28]

The Lincoln Laboratory and JPL planetary ephemeris programs were uses of planetary radar data that did not necessarily lead to publications. Moreover, the vast amount of data routinely collected by radar astronomers and stored in the data bases of those ephemeris programs did not result from experiments designed to achieve a special purpose. Many planetary radar experiments quickly became routine operations. A glance at the extant Haystack radar log books indicates that radar astronomers rarely ran experiments themselves; expert technicians, like Haines Danforth and Lou Rainville, operated

27. Shapiro 30 September 1993.
28. Shapiro 30 September 1993; Shapiro 4 May 1994; E. Myles Standish, Jr., telephone conversation, 20 May 1994; Paul Reichley, telephone conversation, 19 May 1994; Memorandum, Standish to R. Green, 10 May 1979, Jurgens materials.

the radar equipment, and the software consisted of "cookbook programs."[29] This routinization of experimentation is one aspect of Kuhnian "normal" or paradigm science.

Testing Albert Einstein

According to Shapiro and his colleagues at Lincoln Laboratory, the main purpose in gathering radar data for the planetary ephemeris program was "to test Einstein's theory of General Relativity."[30] The Shapiro test of the gravitational time delay predicted by General Relativity is interesting for its contribution to theoretical physics and astrophysics, as well as a major early achievement of planetary radar. Its development underscores the close and necessary connection between the capabilities of radar instruments and the kinds of scientific problems that one can solve with radar. It also illustrates the emotional intensity with which scientists struggle to assert their claims of discovery and priority of publication.

After announcing his theory of Special Relativity in 1905, Albert Einstein spent another ten years developing the theory of General Relativity.[31] The theory of General Relativity traditionally has found support in three principal experimental areas. The first came from its accounting for the precession of Mercury's perihelion, the point at which Mercury is closest to the Sun. Traditional theoretical physics had been incapable of explaining the precession of Mercury's perihelion without leaving certain discrepancies unexplained. The ellipse of Mercury's orbit was turning faster than traditional physics said it ought to by an amount of some 43 seconds of arc per century. Einstein found that his equations gave just that amount of deviation from the measure predicted by traditional physics. The perihelion motion came out not only with the right numerical value but also in the correct direction.

Einstein's theory of General Relativity predicted that a gravitational field would bend or deflect the path of light rays. For a light ray glancing the Sun, the theory of General Relativity predicted a deflection of 1.7 seconds of arc, about 1/1,000 of the angular width of the Sun as seen from the Earth. The theory of General Relativity also predicted that the gravitational field would cause the speed of a light wave to slow.

Three types of experimental tests conducted over several decades confirmed the precession of Mercury's perihelion, the deflection of light rays in a strong gravitational field, and the red shift. Consequently, Irwin Shapiro called his the Fourth Test of General Relativity. Initially, Shapiro was interested in using radar to confirm the precession of Mercury's perihelion. He hit upon that idea in 1959, but Shapiro was not sure whether a check on a widely accepted physical theory would be a worthwhile experiment. So in April 1960, he asked a visiting French physicist, Cyrano de Dominicis, about the experiment.

29.   Hine 12 March 1993; Log books, Haystack Planetary Radar, HR-70-1, 9 December 1970 to 11 August 1971; HR-71-1, 16 August 1971 to 14 April 1972; HR-73-1, 27 June 1973 to 26 November 1973; and HR-73-2, 9 December 1970 to 11 August 1971, SEBRING. There is a lacuna in the log book records; observations made after 14 April 1972 and before 27 June 1973 are not represented.
30.   Ash, Shapiro, and Smith, "Astronomical Constants and Planetary Ephemerides Deduced from Radar and Optical Observations," *The Astronomical Journal* 72 (1967): 338.
31.   The section on Einstein's general theory of relatively draws loosely from Banesh Hoffmann, *Relativity and its Roots* (New York: W. H. Freeman and Company, 1983); Peter G. Bergmann, *The Riddle of Gravitation*, revised and updated (New York: Charles Scribner's Sons, 1987); and Mendel Sachs, *Relativity in Our Time: From Physics to Human Relations* (Bristol, PA: Taylor and Francis, 1993). See also Klaus Hentschel, "Einstein's Attitude towards Experiments: Testing Relativity Theory, 1907–1927," *Studies in History and Philosophy of Science* 23 (1992): 593–624.

Dominicis told Shapiro he thought the experiment worth doing, because scientists had so few tests of relativity.[32]

No radar at the time, however, had the necessary sensitivity to carry out the precession experiment. At any rate, the Fourth Test was not to measure the precession of Mercury's perihelion, but the slowing down of light waves caused by solar gravity. The new idea came to Shapiro in the spring of 1961, as he was attending a briefing for the military on some of the research conducted at Lincoln Laboratory and MIT with Department of Defense funds. After his lecture on measuring the speed of light, George Stroke, in a conversation with Shapiro, mentioned that the speed of light is not the same everywhere, but depends on the gravitational field through which it is passing. Shapiro was surprised. He refreshed his memory on General Relativity and realized that there was a misunderstanding: according to General Relativity, a (freely) falling observer would measure at any location the same speed of light, independent of the (local) gravitational field. However, Shapiro reasoned, the effect of the gravitational field on the speed of light would be cumulative over a round-trip path (unlike the red shift) and that a radar experiment, therefore, ought to be able to detect this gravitational time delay.

Shapiro now had the idea of testing the gravitational time delay predicted by General Relativity, but he realized that extant radars could not measure this small relativistic effect. Moreover, Shapiro did not write up the idea at that time. "I just kept it in the back of my mind," he explained.[33]

The inauguration of the Arecibo Ionospheric Observatory in November 1963 revived Shapiro's interest in testing General Relativity. In July 1964, Shapiro and his wife, pregnant with their first child, travelled to Arecibo at Gordon Pettengill's invitation to spend the summer working at the AIO. When Charles Townes, then MIT Provost, visited Arecibo that summer, Shapiro briefed him on his proposed relativity test and told him that Arecibo could not perform the test. "We would never be able to see this effect," Shapiro explained. "The plasma effect of the solar corona would be of the same general type, and the variations would be much larger than the relativistic effect we were looking for. We would never be able to pick it out."[34]

Shapiro then returned home and learned that Haystack was to be dedicated in October 1964. Suddenly it occurred to Shapiro that Haystack might have enough capability to do the experiment. He did some quick back-of-the-envelope calculations and concluded that Haystack might be able to do the experiment. Shapiro sent his manuscript to *Physical Review Letters*, and the journal received it on 13 November 1964.[35]

From his realization that Haystack could do the experiment to his submission of the paper took only one week. After doing the calculations more accurately, Shapiro realized that the sensitivity of the Haystack radar was not good enough to detect the relativistic effect. He and his Group Leader then requested an upgrade of the Haystack radar from the head of Lincoln Laboratory, Bill Radford, who subsequently obtained a funding

---

32.   Shapiro's recounting of the conception of the Fourth Test is included here, because the three-decade-long feud that resulted from it has become a part of the lore of radar astronomy. The sources for Shapiro's and Muhleman's versions of the story are oral histories conducted specifically for this history, namely Shapiro 1 October 1993 and Muhleman 19 May 1994. Paul Reichley, in a telephone conversation of 19 May 1994, refused any other comment than to state that he agreed with whatever Muhleman said.

33.   Shapiro 1 October 1993.

34.   Shapiro 1 October 1993.

35.   Shapiro, "Fourth Test of General Relativity," *Physical Review Letters* 13 (28 December 1964): 789.

Although little noted at the time, Shapiro in his 1964 paper also pointed out that a possible change with atomic time of Newton's universal gravitational constant could be tested with radar observations of Mercury. Such a change was predicted by Paul Adrien Maurice Dirac in 1937 in his "large numbers hypothesis." Evidence for such a change is being actively sought still from monitoring orbits, as Shapiro suggested, because any such change would have profound effects on the evolution of the universe and the formation of structure within it.

commitment from the Rome Air Development Center for the upgrade. The upgrade consisted of design and construction of a new electronic plug-in unit, boosting the continuous-wave transmitter from 100 to 500 kilowatts, and replacement of the cooled parametric amplifier with a lower noise maser.[36]

In January 1965, as the design and construction of the radar upgrade was underway, a colleague showed Shapiro a JPL internal publication dated 31 October 1964 in which an article by Duane Muhleman discussed using radar to measure the general relativistic effect.[37] Shapiro was upset. He recalled vividly that in January 1964, he was walking near Harvard Square with Muhleman. When Muhleman asked him why he was still interested in radar astronomy, Shapiro told him about his idea to test this new effect predicted by General Relativity. Yet Muhleman did not acknowledge that conversation in his JPL report. Furthermore, that report only discussed the test being done near the inferior conjunction of Venus, where such a test was, and remains, infeasible. Shapiro noted that several years later, he approached Muhleman's co-author of the JPL report, Paul Reichley, and asked him how he got involved in that project. To Shapiro's amazement, Reichley responded directly that Muhleman had said to him, "Shapiro says there's an effect here, let's look into it."

The Muhleman and Shapiro relativity experiments both involved using radar and finding the relativistic time delay, but the design of their experiments differed widely. The Shapiro Test sent radar waves from Earth to graze past the Sun and bounce from Mercury (or Venus) at superior conjunction, that is, as the planet was just going behind the Sun (or emerging from behind the Sun) when seen from Earth.

The radar waves then returned from Mercury (or Venus) and again passed near the Sun on their return trip to Earth. The Sun's gravitational field would slow down or delay the radar waves. General Relativity predicted that the cumulative time delay due to the direct effect of the Sun's gravitational field might be somewhat more than 200 microseconds. On the other hand, this time delay for radar waves bounced from, say, Venus at its inferior conjunction amounted to only about 10 microseconds.[38]

Muhleman's experiment grew out of his theoretical work at JPL on communications with spacecraft flying near the Sun. Spacecraft navigation was at that time essentially a matter of measuring Doppler shift to a high degree of accuracy. Because JPL also was considering ranging systems, Muhleman was studying the effects of the solar corona on both Doppler and range signals. "While working on that problem," he explained, "I realized that the main effect of the solar corona on the radio signal was that the signal was bent as it went around the Sun." Muhleman considered the solar gravitational field as though it were a lens with an index of refraction, an idea he later discovered in various relativity books. On a practical level, the Muhleman and Shapiro relativity studies differed widely. Whereas Shapiro intended to bounce radar waves off Mercury (or Venus) at superior conjunction, Muhleman proposed measuring at inferior conjunction, when the relativistic effect would not be detectable.[39]

---

36.   C. Robert Wieser to Gen. B. A. Schriever, 31 May 1966, 13/56/AC 118, MITA.

37.   See Shapiro, "Fourth Test," pp. 789–791; Muhleman and Reichley, "Effects of General Relativity on Planetary Radar Distance Measurements," in *Supporting Research and Advanced Development*, Space Programs Summary 37–29 (Pasadena: JPL, 31 October 1964), pp. 239–241. Although Muhleman's note had an earlier publication date, it was in an internal report with a tightly limited distribution, whereas Shapiro published in a widely distributed scientific journal. Paul Reichley, Muhleman's co-author, was a young college graduate recently hired at JPL and worked with Muhleman on occultation studies of radio signals. Reichley, telephone conversation, 19 May 1994; and Muhleman 19 May 1994.

38.   Shapiro, *Effects of General Relativity on Interplanetary Time-Delay Measurements*, Technical Report 368 (Lexington: Lincoln Laboratory, 18 December 1964), pp. 1–2; and Shapiro, "Testing General Relativity with Radar," *Physical Review* 145 (1966): 1005–1010.

39.   Muhleman 19 May 1994; Shapiro 1 October 1993; Shapiro, *Effects of General Relativity*, p. 2.

The judgement of general texts is that Irwin Shapiro originated the Fourth Test.[40] Muhleman, for a number of reasons, dropped out of radar astronomy for over twenty years. Shapiro and his Lincoln Laboratory coworkers eventually did perform the Fourth Test at Haystack during the superior conjunction of Mercury in November 1966. Haystack made subsequent measurements during the superior conjunctions of 18 January, 11 May, and 24 August 1967. The results confirmed General Relativity to an accuracy of about ten percent.[41]

Additional observations of Mercury and Venus made at both Haystack and Arecibo during several superior conjunctions helped to refine the Fourth Test results. Subsequent experiments carried out on spacecraft further improved the accuracy of the test. The best accuracy yet achieved was from a combined MIT and JPL experiment on the Viking mission to Mars; it confirmed Einstein's theory of General Relativity to a tenth of a percent. The accuracy of the measurement of the relativistic effect had improved by an impressive factor of 100, or two orders of magnitude, in 10 years.[42]

A Shifting Paradigm

The planetary radar research discussed up to this point shared a consensus on problem-solving activities in a way typical of a Kuhnian paradigmatic science. Among the forces driving the evolution of planetary radar astronomy was the interaction between the two kinds of problems radar astronomers attempted to solve. One set related to the larger theoretical framework which the results of radar observations and analysis attempt to address; the other related to epistemological questions and included radar techniques. Because the two problem sets are necessarily linked to one another, the invention or adaptation of new radar techniques impacted on the kinds of scientific problems addressed by radar astronomy and, as a result, expanded the paradigm without altering the original problem-solving activities and techniques.

One of the most powerful new radar techniques was planetary range-Doppler mapping. It added a whole gamut of answers that radar astronomy previously could not provide. The successful application of the new technique depended on the availability of a generation of highly sensitive radars, Haystack, Arecibo, and DSS-14. Technology continued to drive radar astronomy. Because the kinds of problems range-Doppler mapping solved were related more to geology than to astronomy, planetary radar grew close to the theoretical framework of planetary geology. This shift of the paradigm (without alteration of the original astronomy-oriented paradigm) also reflected the evolving social context of planetary radar, which in 1970 found itself a patron in NASA and its missions of planetary exploration. Thus, changing problem sets and theoretical frameworks on the one hand and the evolution of financial and institutional patronage on the other became inextricably linked.

Planetary Range-Doppler Mapping

Both range and Doppler were standard radar measurements long before they united to provide range-Doppler maps of planetary surfaces. Range or time-delay

---

40.   See, for example, Peter G. Bergmann, *The Riddle of Gravitation*, rev. ed. (New York: Charles Scribner's Sons, 1987), p. 158.

41.   Shapiro 1 October 1993; Shapiro, Pettengill, Ash, Stone, Smith, Ingalls, and Brockelman, "Fourth Test of General Relativity: Preliminary Results," *Physical Review Letters* 20 (1968): 1265–1269.

42.   Shapiro 1 October 1993.

measurements determine how far away a target is by the amount of time the echo takes to return to the radar receiver. The greater the distance to the target, the longer the echo takes to appear in the receiver. Conversely, the shorter the distance to the target, the less time the echo takes to appear in the receiver. Knowing that radar waves travel at the speed of light, one can calculate the distance traveled by a radar signal from the amount of time between transmission of a radar signal and reception of its echo.

If one assumes that a planetary target is a perfect sphere, then when a transmitter directs radar waves at it, the waves arrive first at a circular area at the center of the planet as viewed from Earth. The point on the planet's surface that has the radar at its zenith, and is thus closest to the observer, is called the subradar point. Thus, the radar waves first hit a circular area on the planet surrounding the subradar point and form what is called a range ring. Within each range ring, the distance from Earth to the planet's surface, that is, the range or delay in time, is the same. The longest delays (and therefore ranges) generally correspond to echoes from near the planetary limbs.

When a radar transmits, it sends a signal that contains only a very narrow band of frequencies and appears almost line-like. Such would be the case, too, for the echo received back at the radar were there no difference in the relative motion between the radar and its target. In reality, when looking from the Earth at a planetary target, this relative motion is always a factor. The combined motions of the Earth as it spins on its axis and orbits around the Sun, and of the planetary target as it also spins on its axis and orbits about the Sun, cause what is known as the Doppler effect or Doppler shift, which is the difference between the frequencies of the radar transmission and the radar echo. The differences in the relative motions of the radar and the target broaden the frequency of the returning signal. Instead of a (nearly) single frequency, the returning signal exhibits a spectrum of frequencies "shifted" or set off from the transmitted frequency.

In order to remove the Doppler shift caused by the relative motion of the observer and the target, planetary radar astronomers generally use a radar ephemeris program. The program automatically adjusts the incoming signal for the expected Doppler shift, which itself changes over time because of the changes in relative motion of the observer and the target. Thus, the predicted Doppler shift must be accurate enough to avoid smearing out the echo in frequency. This requirement places stringent demands on the quality of the observing ephemeris. Thus, the Lincoln Laboratory PEP and the JPL Development Ephemeris series were of vital importance to the successful execution of planetary range-Doppler mapping.

A given portion of the echo frequency spectrum corresponds to a slice or strip on the planet's surface. Each slice is parallel to the plane containing the line from the observer to the planet and the spin axis of the planet, and each slice has the same Doppler shift value, because each portion of that slice of the planet's surface has the same motion relative to the observer. When Doppler shift and range data are combined, the slices of equal Doppler shift intersect the range rings to form "cells." In general, each range-frequency cell corresponds to two particular areas on the planet's surface. The amount of surface area corresponding to a particular range-frequency cell represents the resolution of the radar image on the planet's surface and varies over the planetary surface (see Technical Essay, Figure 42).

The amount of power returned from the target for each range-frequency cell can be converted into a two-dimensional image of the planetary surface through a series of complex mathematical manipulations. Each spectrum has the attributes of power, bandwidth (the maximum spread of line-of-sight velocities), shape, minor features, and a weak broadband component. The total power received depends on such instrument factors as transmitter power, antenna gain, and pointing accuracy, as well as on the reflecting (backscattering) properties of the target and the so-called radar equation. The radar equation states that the amount of power in an echo is inversely proportional to the fourth power of the distance. That means that the echo power received by a radar decreases sharply with increasing distance of the target. For instance, a radar echo is 1/16th as strong if the distance to the target is doubled, all other conditions being equal.

In general, rough planetary surfaces backscatter more power from near the planet's limbs than a corresponding smooth sphere. Thus, a rough surface leads to a broadening of the frequency spectrum of the planetary echo. A smooth planetary surface (relative to the size of the wavelength of the radar waves) would broaden the return signal to a far lesser degree. The amount of power returned to the receiver in each resolution cell therefore corresponds to the planet's surface characteristics.

The idea of combining range and Doppler data to form a radar image came together at Lincoln Laboratory in the late 1950s. There, Paul Green began to consider the calculation of a planet's spin velocity from a simultaneous measure of range and Doppler spread. Another member of the Lincoln Laboratory radar group, Roger Manasse, pointed out that when you look at a spinning object, the planes of equal Doppler shift are parallel to the plane containing the line of sight and the rotation vector.[43] However, Manasse did not put the slices of Doppler shift together with the range rings. The originator of that idea was Paul Green.

Green remembers how the idea came to him: "I was sitting in my living room wondering what the relationship was between the two of them. I also had noticed that Ben Yaplee had actually measured those things."[44] Ben Yaplee and others at the Naval Research Laboratory were using range rings to refine the Earth-Moon distance. They discovered that details in the structure of the return echoes could be correlated with lunar topography.[45]

However, they did not develop planetary radar range-Doppler imaging. "It was simply an unevaluated measurement," Green explained. "There was no attempt to know what deeper message might be behind that. I was just thinking, 'Hey! Wait a minute! That's kind of an interesting thing to do.' Maybe it was obvious, but might there not be something deep behind it?"[46]

Soon after formulating range-Doppler mapping, Green discovered that classified military radar research at the University of Michigan had led to the conception of a similar process but with significant differences. The military process involved imaging the Earth from aircraft and relied on developing a radar "history" of the target to create an image, while planetary range-Doppler mapping created a "snapshot" of a planetary surface from a ground-based radar. Because of the similarities in the two methods, Green was careful to call his *planetary* range-Doppler mapping. The University of Michigan radar effort, as we shall see in the next chapter, eventually had a profound impact on planetary radar astronomy.

Paul Green first presented his ideas on range-Doppler mapping at the pioneering Endicott House Conference on Radar Astronomy, then at the URSI workshop on radar astronomy that followed immediately afterward in San Diego. An abstracted form of that paper and the others presented at the workshop soon appeared in the *Journal of Geophysical Research*.[47]

---

43.   Roger Manasse, *The Use of Radar Interferometer Measurements to Study Planets*, Group Report 312–23 (Lexington: Lincoln Laboratory, March 1957).

44.   Green 20 September 1993.

45.   See Ch. 1, note 69, and Ch. 3, note 14.

46.   Green 20 September 1993.

47.   Green 20 September 1993; Leadabrand, "Radar Astronomy Symposium Report," pp. 1111-1115. Earlier, a more complete exposition of the theory appeared as Green, *A Summary of Detection Theory Notions in Radar Astronomy Terms*, Group Report 34–84 (Lexington: Lincoln Laboratory, 18 January 1960). See, also, Ch. 3, note 22.

Green did not apply his theory to actual radar mapping of the planets. Instead, it was his Lincoln Laboratory colleague Gordon Pettengill who used it beginning in 1960. Initially, Pettengill explored the surface of the Moon with the Millstone radar. The result was an image that barely resembled the lunar surface. Pettengill concluded, "It is obvious that much patient work lies ahead before detailed correlation with optical photographs may be attempted."[48]

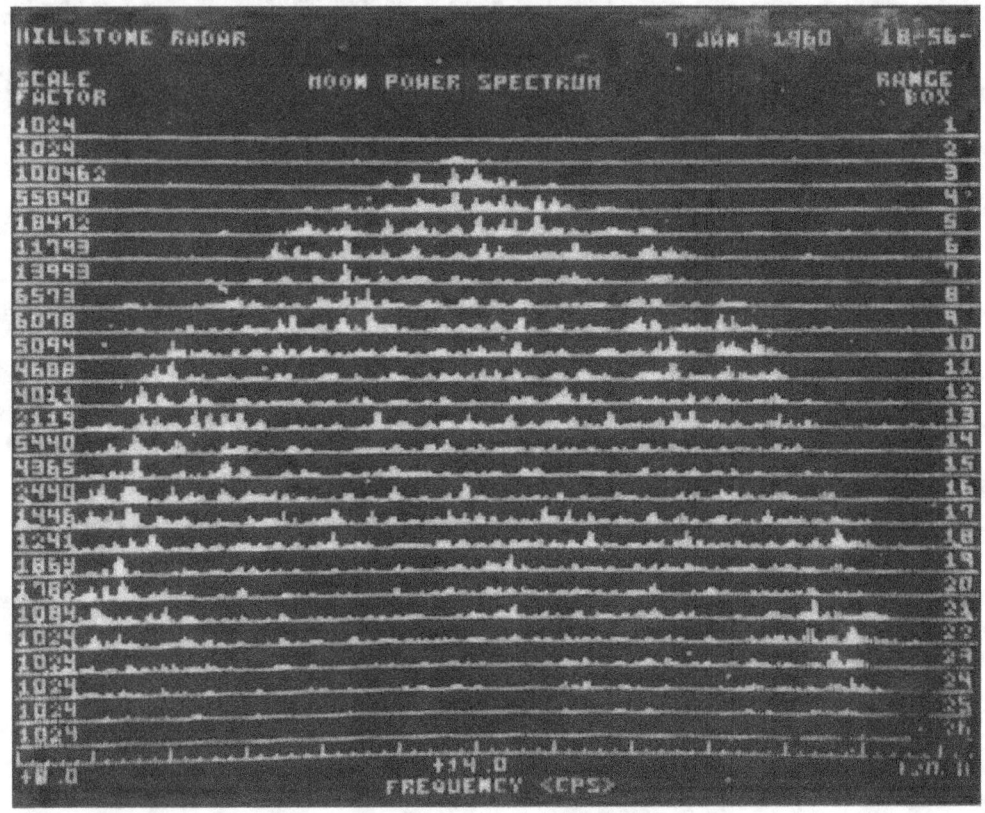

Figure 16

The first range-Doppler image of the Moon, 7 January 1960, made by Gordon Pettengill, using techniques developed by his Lincoln Laboratory colleague Paul Green. The top of the image (shown in range box 2) represents the point on the lunar surface closest to the radar. Pettengill, as the first associate director of the Arecibo Ionospheric Observatory, as it was then called, later guided range-Doppler imaging of the Moon and planets at Arecibo as well as at the Haystack Observatory. (Courtesy of MIT Lincoln Laboratory, Lexington, Massachusetts, photo no. 261209-1D.)

48. Pettengill, "Measurements of Lunar Reflectivity Using the Millstone Radar," *Proceedings of the IRE* 48 (May 1960): 933–934.

Lunar Radar Mapping

Pettengill made a second attempt at lunar radar mapping in June 1961, again using the Millstone radar. Those and the previous Pettengill radar images had what radar astronomers call north-south ambiguity. The nature of range-Doppler mapping is to create an uncertainty (called north-south ambiguity), such that the observer does not know from which hemisphere the echoes are returning. The range-Doppler technique creates two points, one in the northern hemisphere and the other in the southern hemisphere, with exactly the same range and Doppler values. The radar data cannot distinguish the hemisphere of origin of the return echo and thus presents a confusing picture of the target's features.

Pettengill had no technique yet for distinguishing northern-hemisphere echoes from southern-hemisphere echoes. He knew that the youngest, and therefore the roughest, large feature on the lunar surface visible from Earth was the crater Tycho. During a full Moon, this crater appears to have rays emanating across the lunar surface. When Pettengill looked at the echo spectra, he found anomalously high spikes that were consistent from run to run. He assumed that they were in the southern hemisphere (the location of Tycho) and found they matched the crater's location.[49]

For the first time, a lunar surface feature and a radar return had matched. However, the north-south ambiguity problem stood in the way of refining range-Doppler mapping into a useful tool for exploring the solar system. One solution appeared when the Arecibo Ionospheric Observatory began operation in November 1963. There, at the instigation of Pettengill, who was now Associate Director of Arecibo, Thomas W. Thompson, then a Cornell graduate student, and Rolf Dyce began range-Doppler mapping of the Moon.

In contrast to Millstone, the Arecibo radar antenna had a narrow beamwidth relative to the angular size of the Moon. The Moon has a diameter of a half degree or 30 minutes of arc. The width of the Arecibo antenna beam was 10 minutes. Instead of aiming the antenna at the center of the lunar disk facing Earth, Thompson and Dyce aimed it at a point 10 minutes of arc south from the center. The Arecibo telescope received echoes, therefore, only from the lower or "southern" part of the Moon. The technique assuaged the problem of north-south ambiguity, but was applicable to only the Moon. Venus was only a speck, slightly more than one minute of arc, compared to the Moon's 30 minutes of arc.[50]

Using this approach, Thompson and Dyce explored eight regions of the lunar surface and collected data on echo strength. They converted the data into "contour" lines of relative reflectivity. Thompson placed these lines, computed and plotted on a transparent overlay, over lunar maps made from photographs. The resultant radar contour map had a resolution of 20 by 30 km.[51]

Thompson continued to carry out range-Doppler mapping of the Moon by taking advantage of the increasingly narrow beamwidth of the Arecibo antenna. By reducing the beamwidth from 10 to 7 minutes of arc, he succeeded in creating a range-Doppler map of the crater Tycho with surface resolutions between 7 and 10 km. The output from a given radar observation now represented a considerable quantity of data; between 10,000 (10^4) and 100,000 (10^5) values of intensity (or pixels) constituted a single map.

---

49.   Pettengill and John C. Henry, "Enhancement of Radar Reflectivity Associated with the Lunar Crater Tycho," *Journal of Geophysical Research* 67 (1962): 4881–4885. Pettengill's co-author was an MIT electrical engineering graduate student who used the experience in writing his master's thesis. Henry, "An Automated Procedure for the Mapping of Extended Radio Sources," M.S. thesis, MIT, 1965.

50.   Source for the arc measurement of Venus: Goldstein, "Radar Studies of Venus," in Audoin Dollfus, ed., *Moon and Planets* (Amsterdam: North-Holland Publishing Company, 1967), p. 127.

51.   Thompson 29 November 1994; Thompson and Dyce, "Mapping of Lunar Radar Reflectivity at 70 Centimeters," *Journal of Geophysical Research* 71 (1966): 4843–4853.

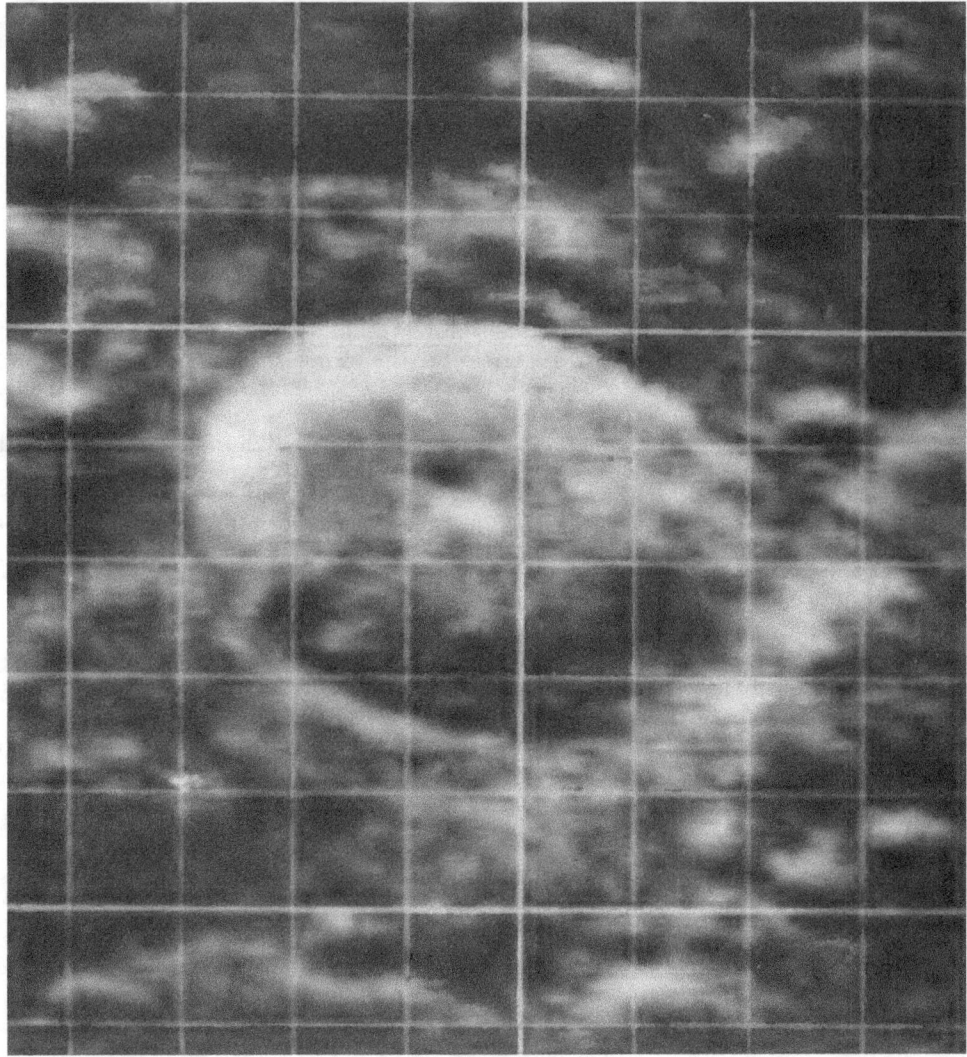

Figure 17

Radar map of the lunar crater Tycho with a resolution of 1 kilometer made with the 3.8-cm (7,750-MHz) Haystack Observatory radar. The grid lines are spaced about 17 km apart. (Courtesy of MIT Lincoln Laboratory, Lexington, Massachusetts, photo no. 242336-1.)

At the same time, Gordon Pettengill guided lunar radar observations at Haystack, which had become available in late 1964. Haystack, moreover, had a narrower antenna beamwidth, only 3 minutes of arc, and the higher operating frequency of Haystack (3.8 cm, X-band) compared to Arecibo (70 cm, UHF) helped Haystack to achieve a much finer resolution on Tycho: between 1 and 2 km. The Haystack radar images now approached the quality of lunar photographs made from Earth. In the words of Pettengill and

Thompson, "The most immediately striking feature of the 3.8-cm [Haystack] observations is their resemblance to the optical photograph...."[52]

The coincidental refinement of lunar range-Doppler imaging and the commitment to place an American on the Moon before the end of the 1960s enhanced the value of the lunar radar work done at both Arecibo and Haystack. NASA Apollo mission staff used the radar images to help select landing sites, and Apollo funded Thompson's dissertation and subsequent radar studies of the Moon. Once the resolution of radar images surpassed the resolution of lunar photographs made from Earth, the value of lunar radar studies to NASA grew even more. Thus, the new technique brought radar astronomy closer to the scientific needs of NASA, increasingly the patron of radar astronomy.

At Arecibo, Tommy Thompson and Rolf Dyce undertook radar mapping of the Moon at both 40 MHz (7.5 meters) and 430 MHz (70 cm) under a supplementary grant from NASA. A joint report with Lincoln Laboratory compared the Arecibo results with those carried out at Haystack by Stan Zisk with additional NASA funding under a contract between MIT and the Manned Spacecraft Center in Houston. NASA funded lunar studies at both telescopes until 1972, when the Apollo program came to an end.[53]

Venus Radar Mapping

In 1964, as Thompson and Zisk were starting their lunar mapping activities, Roland Carpenter and Dick Goldstein analyzed spectra from Venus and discovered the first features on that planet's surface. The Goldstone Venus radar lacked sufficient sensitivity to apply range-Doppler mapping to Venus. However, once the Mars Station became available, Goldstein continued his exploration of Cytherean surface features using range-Doppler techniques, but without resolving the north-south ambiguity. Thus began one of the most long-lived and extensive activities of planetary radar astronomers. This scientific niche for radar resulted from that planet's opaque atmosphere which barred exploration with optical methods.

When Dick Goldstein observed Venus during the 1964 inferior conjunction, he looked only at the structure of the spectra returned from the planet. This was the same technique that Roland Carpenter had used earlier to discover the retrograde motion of Venus; it was not range-Doppler mapping. A few topographic features were visible as details in the return spectra. Goldstein found two features represented as peaks. They moved slowly across the spectrogram, a graph plotting echo power density versus frequency, from the high-frequency side to the low-frequency side, in synchronization with the planet's rotation.

Goldstein then placed his two features on a coordinate system with the first feature, named Alpha (α), located on his zero degree meridian in the southern hemisphere. The second feature, named Beta (β), Goldstein placed in the northern hemisphere. His coordinate system was somewhat arbitrary out of necessity, as astronomers generally had not agreed upon any Cytherean coordinate system. Additional analysis of the 1964 data revealed three more features around the equator. Goldstein named them Gamma (τ), Delta (δ), and Epsilon (ε).

52.   Thompson 29 November 1994; Pettengill and Thompson, "A Radar Study of the Lunar Crater Tycho at 3.8-cm and 70-cm Wavelengths," *Icarus* 8 (1968): 457–471, esp. 464.
53.   The research was conducted under NASA grant NGR-33-010-024. NEROC, *Semiannual Report of the Haystack Observatory*, 15 July 1972, p. ii. See, also, Ch. 4, note 15.

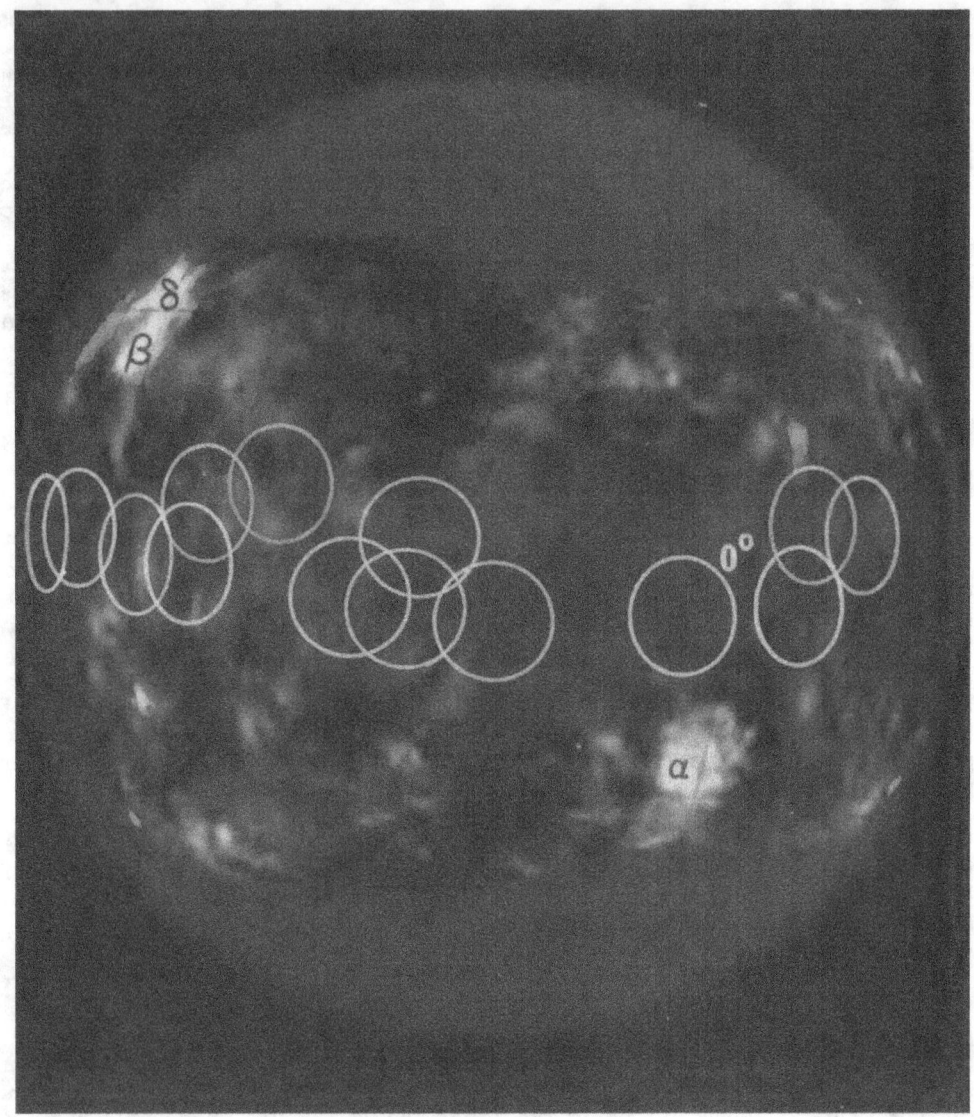

Figure 18

One of the earliest range-Doppler images of Venus made by Richard Goldstein of JPL with the Goldstone radar. The notation "0°" indicates the meridian in Goldstein's coordinate system. Visible are the first surface features identified by Goldstein: Alpha (α), on the meridian in the southern hemisphere, Beta (β), in the far west of the northern hemisphere, and Delta (δ), just to the north of Beta. Gamma (τ) and Epsilon (ε), two additional features identified by Goldstein, are not labelled. The radar names Alpha and Beta were retained when astronomers began naming the surface features of Venus. (Courtesy of Jet Propulsion Laboratory, photo no. 331-4849AA.)

Although he judged that these features were probably mountain ranges, Goldstein had insufficient evidence. What were they? "Venus is still a mystery planet," Goldstein concluded. "However, it may no longer be viewed as featureless, but rather as an exciting object for further study."[54]

Using the data taken with the newly operational Mars Station during the 1967 Venus inferior conjunction, Goldstein studied the Beta region in more detail, attempting to determine its size and character, rather than searching for new features. The Mars Station, moreover, provided sufficient sensitivity to attempt range-Doppler mapping. Goldstein observed Beta, Delta, and an unnamed region at (his) 40° South latitude and made a crude radar image of the ß region. Still, Goldstein lacked sufficient data to determine whether Beta was a mountain range or another type of feature.[55]

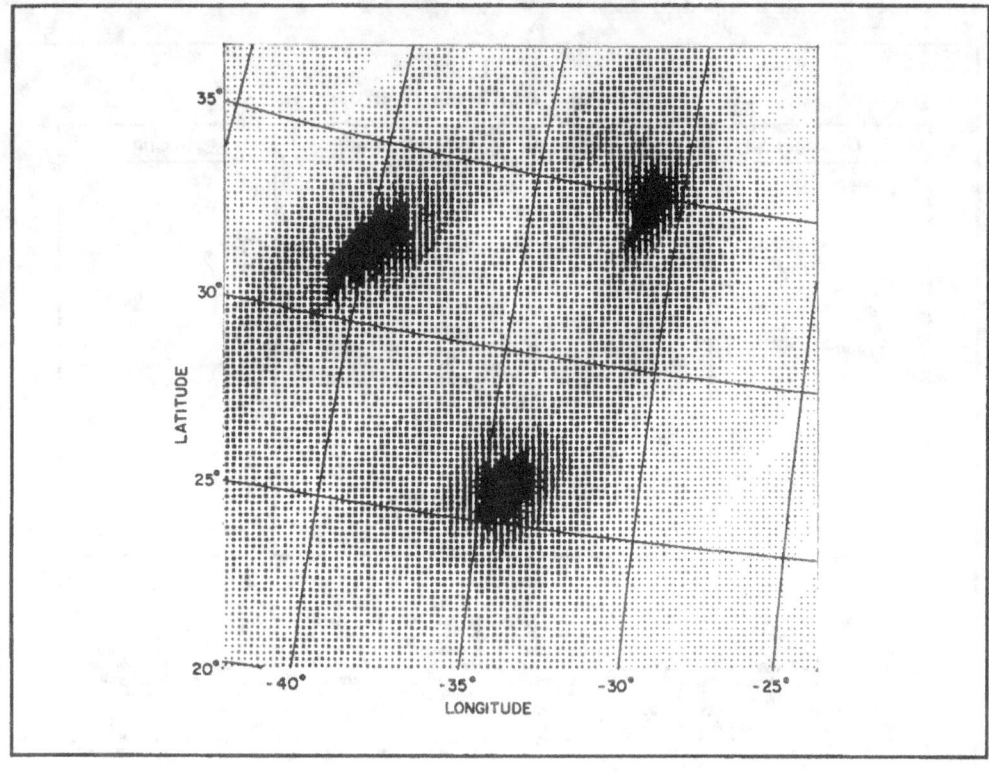

Figure 19

A detailed radar view of the Beta region of Venus, 1967, made by Dick Goldstein of JPL using the Goldstone radar. It exemplifies the limits of resolution available in some of the earliest radar images of that planet. (Courtesy of Jet Propulsion Laboratory, photo no. P-8882.)

54.   Goldstein, "Preliminary Venus Radar Results," *Journal of Research of the National Bureau of Standards, Section D: Radio Science* 69D (1965): 1623–1625; Goldstein, "Radar Studies of Venus," in Dollfus, *Moon and Planets,* pp. 126–131. This article also appeared as Goldstein, *Radar Studies of Venus,* Technical Report 32–1081 (Pasadena: JPL, 1967).

55.   Goldstein and Shalhav Zohar, "Venus Map: A Detailed Look at the Feature ß," *Nature* 219 (1968): 357–358; Goldstein, "A Radar View of the Surface of Venus," *Proceedings of the American Philosophical Society* 113 (June 1969): 224–228. Goldstein's co-author, Shalhav Zohar, was a fellow JPL employee who developed much of the software used in the experiment.

Meanwhile, Roland Carpenter, who was now both a JPL employee and an instructor in the Department of Astronomy of the University of California, Los Angeles, had analyzed 1964 Venus inferior conjunction radar data. Carpenter found two distinct peaks in the return spectra that persisted day after day and moved slowly with time. On closer examination, the first peak appeared to have three components, which he hesitated to interpret because he felt their nature could not be determined with the available data.

Using Goldstein's coordinate system, Carpenter began to identify the most pronounced features with letters of the alphabet from A to G. He labeled less probable locations as numerical extensions of nearby features, e.g., B1, C1, C2, D2, and D3. Correlations between Carpenter's and Goldstein's features began to emerge. Carpenter's feature F had the same location as Goldstein's α, and Carpenter's group B, C, and D corresponded to Goldstein's β (Table 2).[56]

Table 2
Radar Features of Venus

Goldstein	Carpenter	Haystack	Arecibo
Alpha	F	Haystack I	Faraday
Beta	B		
Beta	C	Haystack IV	
Beta	D		
Beta		Haystack II	Gauss
Beta		Haystack III	Hertz
Gamma			
Delta	D2	Haystack A (later Haystack VI)	
Epsilon			
	A		
	B1	Haystack B	
	C1		
	C2	Haystack V	
	D3		
	E		
	G		
		Haystack C	
		Haystack D	
			Maxwell

Sources

R.M. Goldstein, "Radar Studies of Venus," in Audoin Dollfus, ed., *Moon and Planets* (Amsterdam: North-Holland Publishing Company, 1967), pp. 126–131; R.M. Goldstein and H.C. Rumsey, Jr., "A Radar Snapshot of Venus," *Science* 169 (1970): 974–977; R.L. Carpenter, "Study of Venus by CW Radar: 1964 Results," *The Astronomical Journal* 71 (1966): 142–152, especially pp. 148–151; A.E.E. Rogers, T. Hagfors, R.A. Brockelman, R.P. Ingalls, J.I. Levine, G.H. Pettengill, and F.S. Weinstein, *A Radar Interferometer Study of Venus at 3.8 cm*, Technical Report 444 (Lexington: Lincoln Laboratory, 14 February 1968); A.E.E. Rogers, R.P. Ingalls, and G.H. Pettengill, "Radar Map of Venus at 3.8 cm Wavelength," *Icarus* 21 (1974): 237–241; D.B. Campbell, R.F. Jurgens, R.B. Dyce, F.S. Harris, and G.H. Pettengill, "Radar Interferometric Observations of Venus at 70-Centimeter Wavelength," *Science* 170 (1970): 1090–1092; R.F. Jurgens, "Some Preliminary Results of the 70-cm Radar Studies of Venus," *Radio Science* 5 (1970): 435–442; and R.F. Jurgens, "A Study of the Average and Anamalous Radar Scattering from the Surface of Venus at 70 Cm Wavelength," PhD diss., Cornell University, June 1968, also published internally as CRSR Research Report no. 297 (Ithaca: CRSR, May 1968).

Carpenter dropped out of radar astronomy and pursued a teaching career, while Goldstein continued to explore Venus. The 1969 inferior conjunction of Venus provided an opportunity to use range-Doppler mapping. Goldstein combined the 1969 data with earlier data, then applied a mathematical method devised by fellow JPL employee Howard C. Rumsey, Jr., which involved the construction of a large matrix of range and Doppler values.

The mapping process divided the surface of Venus into small cells 1/2° square in latitude and longitude. A column vector (X) consisted of the unknown reflectivities of these cells, while a second column vector (S) contained all the processed data from 17 days of

56. Carpenter, "Study of Venus by CW Radar: 1964 Results," *The Astronomical Journal* 71 (1966): 142–152, especially pp. 148–151.

observations. Already, Goldstein and Rumsey were dealing with a large amount of data; vector X had about 40,000 components, vector S about 120,000 components. They expressed the relationship between vectors S and X as the equation:

$$AX = S,$$

in which A was a matrix whose components could be computed from known parameters and the motion of Venus and Earth. Matrix A consisted of 120,000 by 40,000 components.

As the authors wrote, "Obviously, we cannot compute every component of a matrix with over 10^9 entries." The matrix was "so big," Goldstein recalled, "that we couldn't even read it into the computer except one line at a time."[57] Despite the difficulty of handling the gargantuan matrix, Goldstein produced a number of somewhat unambiguous images of Venus. Once the 1969 data had been converted into a range-Doppler map, in which each resolution cell represented an area on the planet's surface, Goldstein made cumulative maps by adding earlier data. The north and south areas of the cumulative maps were similar, but not identical; however, the images suffered serious flaws, including the "runway" strip running more or less along the planet's equator. Nonetheless, Goldstein succeeded in resolving α for the first time on a map. It was a roundish feature, about 1,000 km across.[58]

Goldstein continued to map Venus with Rumsey's mathematical approach, adding data taken during the 1970 inferior conjunction to that acquired in 1969. The 1970 data were better, being less noisy, because the Deep Space Network had increased the transmitter power of the Goldstone Mars Station from 100 to 400 kilowatts. The total system noise temperature stood at a low 25 K. Regions α and β remained the dominant features of the JPL radar map.[59]

Meanwhile, at Arecibo and Haystack, radar astronomers were creating Venus images with their own techniques. At Arecibo, Cornell University doctoral student Ray Jurgens, with support from an NSF Faculty Fellowship, undertook the analysis of radar data taken during the 1964 inferior conjunction. Dyce and Pettengill had made the radar observations to supply the Planetary Ephemeris Program data base, not to make a range-Doppler map of Venus.[60]

In correlating the radar data with the Cytherean surface, Jurgens abandoned his own zero degree meridian in favor of a modified version of Carpenter's coordinate system that incorporated the latest pole position and rotation rate supplied by Irwin Shapiro from the PEP. Consequently, Goldstein's α and Carpenter's F were not at the zero meridian but closer to 5° longitude. Jurgens identified the features he found by latitude and longitude (e.g., 20°,-102°, in which "-" indicated South latitude or West longitude), then compared his features with those discovered by Goldstein and Carpenter.

Jurgens gave particular attention to Goldstein's β region (Carpenter's group B, C, D, to which Jurgens added E), and he managed to locate most of Carpenter's features. In addition, Jurgens spotted a new feature near Goldstein's Beta. Borrowing from Tommy Thompson's lunar radar mapping work, Jurgens interpreted the feature as a ring structure, specifically a crater, and argued that such a crater might be caused by meteoric impact. Jurgens admitted that "although the evidence for a ring structure is not as strong as one might desire, it at least raises the question of whether such structures would be expected on Venus."[61] Indeed, it was one of the first attempts to relate radar observations and geological interpretation.

---

57. Goldstein 14 September 1993.
58. Goldstein and Rumsey, "A Radar Snapshot of Venus," *Science* 169 (1970): 974–977.
59. Goldstein and Rumsey, "A Radar Image of Venus," *Icarus* 17 (1972): 699–703.
60. Jurgens 23 May 1994.
61. Jurgens, "A Study of the Average and Anomalous Radar Scattering," pp. 71 and 87–110.

Before completing his dissertation, Jurgens observed Venus during the 1967 inferior conjunction, when the Arecibo antenna had an improved receiver system, better data acquisition procedures, and a lower receiver noise temperature. Jurgens combined the 1967 data with additional observations made during the subsequent 1969 conjunction. In order to mitigate the north-south ambiguity problem, he compared observations made a few weeks apart, thereby taking advantage of the changing Doppler geometries between Earth and Venus.

Jurgens continued to explore the β region, in particular, as well as new areas of the planet's surface. On the urging of Tommy Gold, he named his features after scientists famous for their work in electromagnetism: Karl Friedrich Gauss (1777–1855), Heinrich Rudolph Hertz (1857–1894), Michael Faraday (1791–1867), and James Clerk Maxwell (1831–1879). Gauss and Hertz both corresponded strongly to Goldstein's β region. Faraday was Goldstein's α. However, Maxwell, discovered during the 1967 conjunction, had no match among previous citings of Cytherean surface features.[62] It was an original and enduring contribution to Venus mapping.

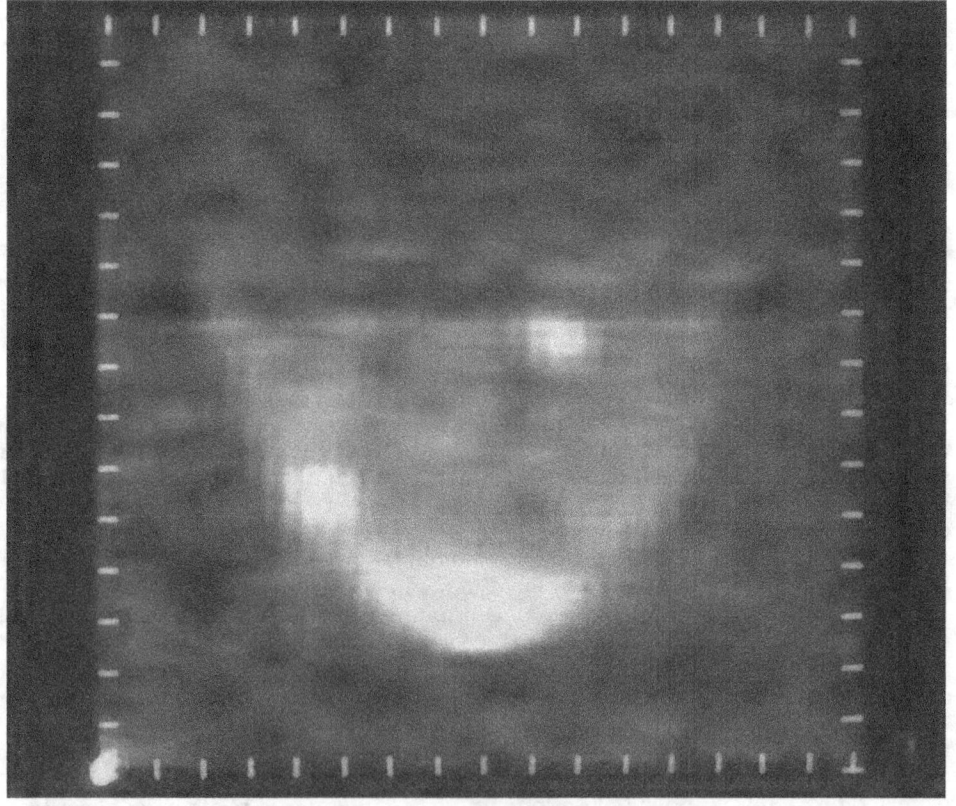

Figure 20

Ray Jurgens discovered a new Venus surface feature, named Maxwell, from these range-Doppler images made at the Arecibo Observatory on 4 September 1967 during inferior conjunction. The bright spot at the leading edge of the image is the subradar point, while the spot closest to the subradar point is the Beta region. Maxwell is the spot farther from the planet's leading edge. (Courtesy of Ray Jurgens.)

---

62. Jurgens, "Some Preliminary Results of the 70-cm Radar Studies of Venus," *Radio Science* 5 (1970): 435–442; AIO, *Research in Ionospheric Physics*, Research Report RS 74 (Ithaca: CRSR, 31 July 1968), pp. 84–85.

Investigators at Haystack Observatory also observed Venus during the 1967 conjunction, but they used a unique technique they pioneered called radar interferometry. It resolved the problem of north-south ambiguity in a superior fashion. An optical interferometer is an instrument for analyzing the light spectrum by studying patterns of interference, that is, how light waves interact with each other. Martin Ryle and other radio astronomers had been designing interferometers since the late 1950s. These radio interferometers used two or more radio telescopes arranged along a straight line (called the base line) and allowed astronomers to "synthesize" observations at higher resolutions than possible with a single antenna.[63]

The inventor of the radar interferometer was Alan E. E. Rogers, then an electrical engineering graduate student at MIT. MIT Prof. Alan H. Barrett was recruiting students to participate in his radio astronomy work on the newly discovered OH spectral line. Alan Rogers joined him and did his masters and doctoral theses on the OH line. As part of his doctoral thesis research, Rogers helped to develop a radio interferometer that linked the Millstone and Haystack radars.

After graduating and spending a year home in Africa, Rogers returned to Lincoln Laboratory, where he was hired to work in the radar group with Gordon Pettengill. Although trained as a radio astronomer, Rogers rapidly became absorbed in planetary radar work and proposed a radar interferometer to eliminate the problem of north-south ambiguity that was typical of range-Doppler mapping.[64] This was not the first time that a radar astronomy technique derived from radio astronomy.

The X-band (7,840 MHz; 3.8 cm) radar interferometer linked the Haystack and Project Westford antennas, which are 1.2 km apart, in the so-called Hayford configuration. In the interferometry experiments, Haystack transmitted a continuous-wave signal to Venus, and both the Haystack and Westford antennas received. Technicians working under Dick Ingalls of Haystack reduced and analyzed the echoes to create a range-Doppler map. The size of the resolution cell on the planet's surface was about 150 km square.

63.   See Bracewell, "Early Work on Imaging Theory," pp. 167–190 and Scheuer, "Aperture Synthesis at Cambridge," pp. 249–265 in Sullivan.
64.   Rogers 5 May 1994.

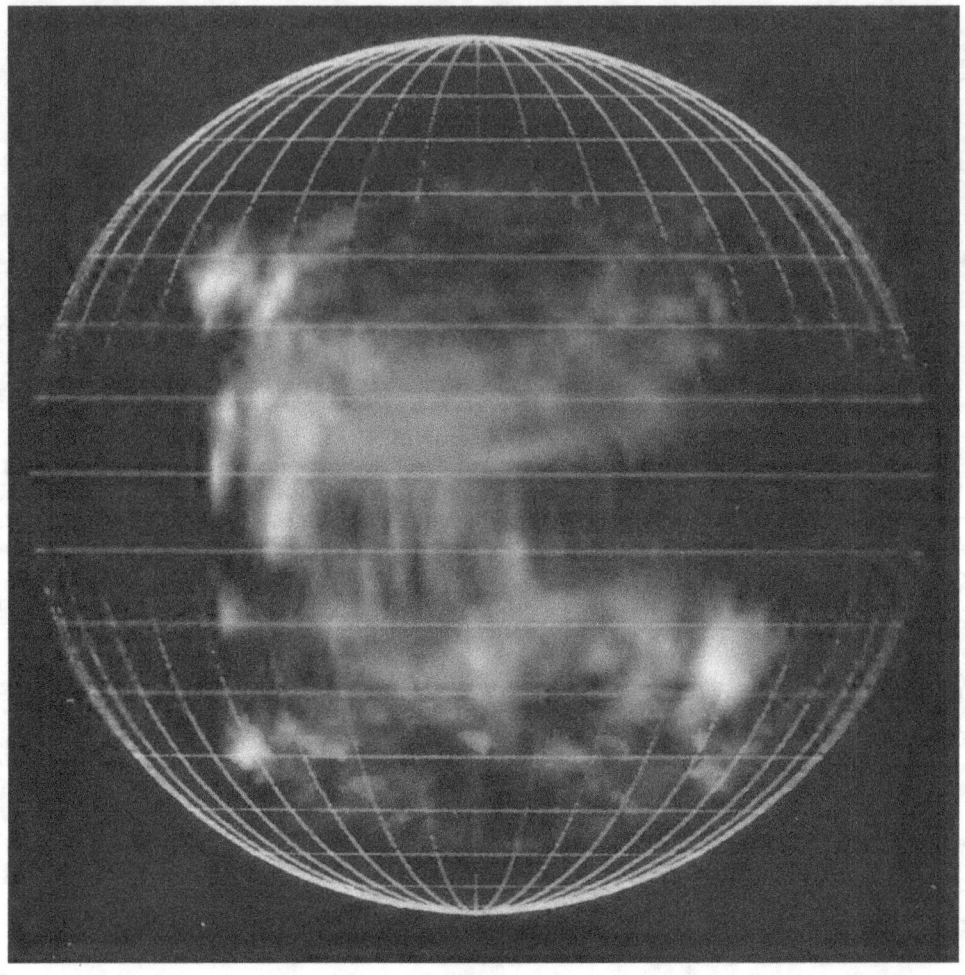

Figure 21

One of the first range-Doppler images of Venus made with a radar interferometer, the Haystack and Westford antennas in tandem, in 1967. Not only are the Alpha and Beta regions discernible, but the complexity of Beta is revealed. (Courtesy of Alan E. E. Rogers.)

Next, Rogers and Ingalls combined the signals from the two antennas to obtain the fringe amplitude and phase for each range-Doppler cell. In an elaborate computer procedure, they rotated the fringe pattern so that the lines of constant phase were normal to the axis of apparent rotation of the planet. The lines of constant phase now were perpendicular to the slices of equal Doppler value. Although each pair of resolution cells that exhibited north-south ambiguity had the same range and Doppler shift values, one could distinguish the north and south cells because they had opposite phases.[65]

One of the first applications of this radar interferometer was to the lunar work being carried out at Haystack for the NASA Manned Spacecraft Center by Stan Zisk. The lunar topographic maps that Zisk created with the Hayford interferometer were carried out under the name "Operation Haymoon" until December 1972, when the Apollo mission ended.[66] Tommy Thompson carried out a similar interferometric study of the Moon using the 40-MHz (7.5-meter) radar at Arecibo.

Alan Rogers and Dick Ingalls also studied Venus during the 1967 inferior conjunction with the Hayford interferometer and identified eight surface regions. Just as each previous radar astronomer had invented his own nomenclature, they labeled features with Roman numerals and letters. The features of which they were certain became Haystack I through Haystack IV. The probable regions were Haystack A through Haystack D. Five of these eight regions corresponded to features already observed by either Goldstein or Carpenter. Haystack I appeared to be Goldstein's α and Carpenter's F, while Haystack II matched Goldstein's ß (Table 2). Jurgens' Arecibo results had not yet been published.[67]

Alan Rogers and Dick Ingalls then published a map of Venus showing the correlation of Haystack and JPL features in a 1969 issue of *Science*.[68] The ß region now appeared to be large and complex. The Hayford radar interferometer confirmed and extended the observations of Goldstein and Carpenter. With interferometer data taken during the 1969 and 1972 conjunctions using an instrument with a lower system noise temperature, Alan Rogers and Dick Ingalls refined their map of Venus; the data continued to indicate agreement among the Haystack and JPL features.[69]

The waxing tide of links between Lincoln Laboratory and Arecibo set in motion by the appointment of Gordon Pettengill as associate director of Arecibo facilitated the transplanting of radar interferometry to Arecibo. In fact, investigators at Arecibo built two additional antennas to study the Moon and Venus with the new technique. The lunar interferometer used the 40-MHz antenna, while the planetary radar interferometer used the 430-MHz antenna. NASA continued to underwrite Tommy Thompson's lunar radar work through a supplementary grant.[70]

----

65.   For a description of the radar interferometer, see Rogers, Hagfors, Brockelman, Ingalls, Levine, Pettengill, and Weinstein, *A Radar Interferometer Study of Venus at 3.8 cm*, Technical Report 444 (Lexington: Lincoln Laboratory, 14 February 1968).

66.   Rogers 5 May 1994; Documents in 44/2/AC 135; "Haystack Operations Summary, 8/11/69–5/18/70," 37/2/AC 135; "Funding Proposal, "Programs in Radio Astronomy at the Haystack Observatory," NSF, 10/1/72–9/30/73," 28/2/AC 135, MITA; NEROC, *Semiannual Report of the Haystack Observatory*, 15 July 1972, p. ii; NEROC, *Final Progress Report Radar Studies of the Planets*, 29 August 1974, p. 1. A number of techniques for extracting lunar topography from interferometric data were devised. Delay-Doppler stereoscopy was developed by Irwin Shapiro and independently by Thomas Thompson and Stan Zisk. Another technique, called delay-Doppler interferometry, was suggested by Shapiro and developed by Zisk and Rogers; Thompson pointed out the strength of the Hayford interferometer for this application. Shapiro, Zisk, Rogers, Slade, and Thompson, "Lunar Topography: Global Determination by Radar," *Science* 178 (1972): 939-948, esp. notes 19 and 21, p. 948.

67.   Thompson, "Map of Lunar Radar Reflectivity at 7.5-m Wavelength," *Icarus* 13 (1970): 363-370.

68.   Brockelman, Evans, Ingalls, Levine, and Pettengill, *Reflection Properties of Venus at 3.8 cm*, Report 456 (Lexington: Lincoln Laboratory, 1968), especially pp. 34–35, 44, and 49–50; Rogers and Ingalls, "Venus: Mapping the Surface Reflectivity by Radar Interferometry," *Science* 165 (1969): 797–799.

69.   Rogers and Ingalls, "Radar Mapping of Venus with Interferometric Resolution of the Range-Doppler Ambiguity," *Radio Science* 5 (1970): 425–433; Rogers, Ingalls, and Pettengill, "Radar Map of Venus at 3.8 cm Wavelength," *Icarus* 21 (1974): 237–241.

70.   AIO, *Research in Ionospheric Physics*, Research Report RS 75 (Ithaca: CRSR, 31 March 1969), pp. 2 and 11–12; *Annual Summary Report, Center for Radiophysics and Space Research, July 1, 1968—June 30, 1969*, 30 June 1969, p. 4.

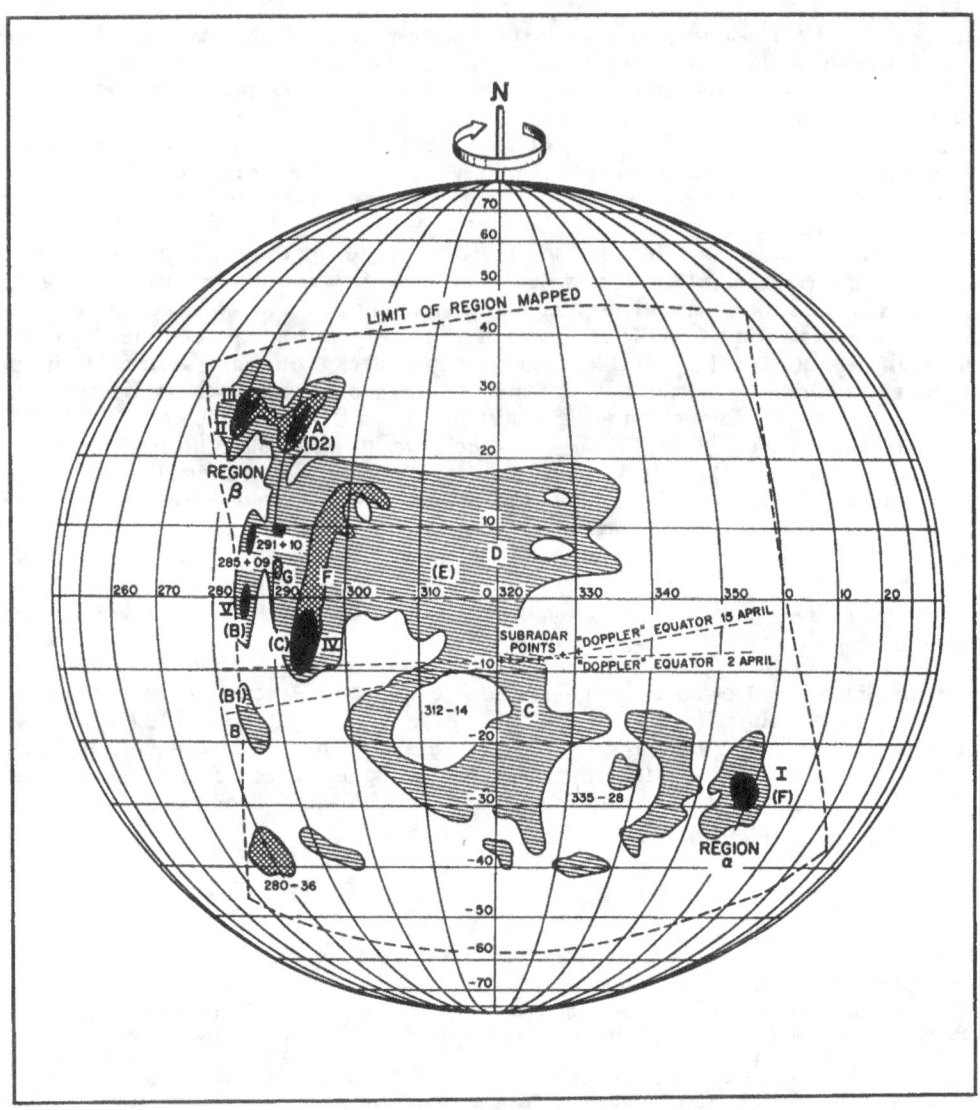

Figure 22

Diagram of Venus surface features made with the Haystack-Westford interferometer. Features observed with the Haystack-Westford interferometer are indicated variously by capital letters, Roman numerals, and coordinate numbers. Goldstein's Alpha and Beta regions are indicated (Region α and Region β), while the labels given by Carpenter are shown in parentheses. (Courtesy of Alan E. E. Rogers.)

Figure 23

The antenna built by Arecibo Observatory employee and radio amateur Sam Harris and located at Higuillales about 10 km from the main dish. Harris and his antenna are a reminder of the important role self-taught engineers and radio amateurs have played in the design and construction of scientific instruments, particularly in the field of astronomy. (Courtesy of Ray Jurgens.)

Undertaking radar interferometric observations of Venus at Arecibo was Cornell graduate student Don Campbell. Campbell came to Cornell from Australia, his native country, where he had studied radio astronomy at the University of Sydney, though not through the agreement between the two universities.[71] His observations of Venus in 1969 with the radar interferometer formed the basis of his doctoral thesis. Located about 10 km from the Arecibo Observatory at Higuillales near Los Caños, the auxiliary interferometer antenna was a square parabolic section, 30 meters by 30 meters (100 ft by 100 ft) with a movable offset feed that allowed tracking up to 10° from the zenith.[72]

---

71.   Campbell 7 December 1993.
72.   Donald B. Campbell, "Radar Interferometric Observations of Venus," Ph.D. diss., Cornell, July 1971; AIO, *Research in Ionospheric Physics*, Research Report RS 75 (Ithaca: CRSR, 31 March 1969), pp. 12–13; Campbell, Jurgens, Dyce, F. Sam Harris, and Pettengill, "Radar Interferometric Observations of Venus at 70-Centimeter Wavelength," *Science* 170 (1970): 1090–1092.

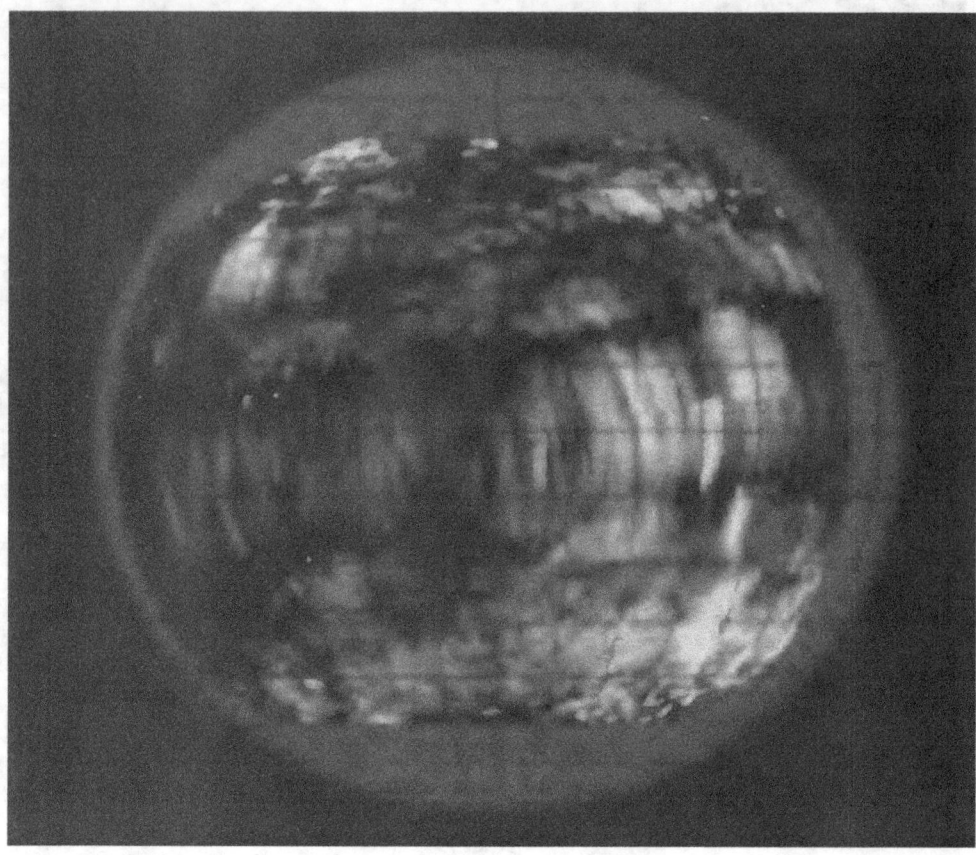

Figure 24
Radar interferometric image of Venus made by Don Campbell for his 1971 doctoral dissertation, which was a study of Venus using the Arecibo Observatory and Higuillales antennas as a radar interferometer. The resolution is about 150 km. The Alpha region can be seen in the lower right corner, and Beta Regio is visible in the upper left corner. (Courtesy of D. B. Campbell, Cornell University.)

As Don Campbell remembered, the original antenna was owned by Sam Harris, an Arecibo employee, who used it for his backyard amateur radio Moon bounces. Harris was a self-taught engineer well known in the "ham" community for his Moon-bounce work and had a column in the popular ham journal *QST* for many years. "He was a real character," Campbell reflected. "I always enjoyed working with him and in getting this interferometer to work over the year or so that it took."[73]

The 430-MHz radar interferometer went into operation in March 1969. Jurgens, Campbell, and Dyce made interferometric observations of Venus between 20 March and 27 April 1969. Unfortunately, because the interferometer antenna was so small, the radar sensitivity fell sharply, and they achieved a surface resolution of only 300 km. The north-south ambiguity had been resolved, but at the loss of resolution. From the data, nonetheless, Campbell deduced that the Faraday region was the same as Goldstein's α and

---

73.   Campbell 7 December 1993.

Carpenter's F. He concluded, "Despite the considerable advance that the radar interferometer represents over other methods in mapping the surface scattering of Venus at radio wavelengths, we still know very little about the actual nature of the surface."[74] In other words, the images really said nothing about the planet's geology.

Campbell returned to Cornell, wrote his thesis, and graduated in July 1971. He then returned to Arecibo as a Research Associate employed by the NAIC. An improved line feed promised better observations during the next Venus inferior conjunction in 1972. Although delays in manufacturing the new analog-to-digital converters, as well as power outages, caused lost observing time, Campbell mapped Venus with the radar interferometer and achieved a resolution of about 100 km. "That was the last fling prior to the upgrade," Campbell recalled.[75]

Campbell also derived Venus topographical (relief or surface height) information from the 1972 data. The most notable result was the discovery of what appeared to be a mountainous zone located at a longitude of 100° and having a peak height of about 3 km. Although not at the same location as Jurgens's suspected crater, which still remained noted only in his dissertation, these mountains became the second clearly identified topographical feature on the surface of Venus, following a pioneering study by Smith and other Lincoln Laboratory and MIT investigators at Haystack published two years earlier.[76]

Dick Goldstein also observed Venus in 1972 with a radar interferometer that combined the Mars Station with a nearby 26-meter antenna. These were Goldstein's first Venus observations with an interferometer. Because he did not suffer the obstacles thrown at Don Campbell, Goldstein was able to update his large-scale, low-resolution map of Venus, which now had a resolution of 10–15 km. He also assembled his first altitude map. A gray scale of only five levels, with each level representing a set of altitude values, indicated the degree of relief. The map showed a large crater about 160 km in diameter about 36° West longitude and 2° South latitude. Goldstein estimated the height of the crater rim to be about 500 meters above the crater floor. This was the first distinguishable crater Goldstein found in his radar data; several years earlier, though, Ray Jurgens had identified a crater in his Cornell dissertation.[77]

By the 1972 inferior conjunction of Venus, the combination of range-Doppler mapping and radar interferometry was beginning to reveal a general overview of the planet's major surface features. Although Venus still looked like a strange fish bowl in radar images, lunar range-Doppler images looked more like photographs. These initial tentative steps, whatever their drawbacks, began to set in motion a shift in the planetary radar paradigm from astronomy to geology. Like the far more successful (because they looked like and had greater resolution than ground-based photographs) lunar radar images, Venus radar images showed that planetary radar astronomy could tell scientists useful information about distant surface formations. These images were not the only techniques radar astronomers had for describing planetary surface conditions. Coincidental with the gradual evolution of planetary radar toward these geological problems, NASA was turning from Apollo to planetary missions.

74. Campbell 7 December 1993; AIO, *Research in Ionospheric Physics*, Research Report RS 75 (Ithaca: CRSR, 31 March 1969), pp. 12–13; *Ibid.*, Research Report RS 76 (Ithaca: CRSR, 30 September 1969), p. 23; Campbell, Jurgens, Dyce, Harris, and Pettengill, "Radar Interferometric Observations," pp. 1090–1092.
75. Campbell 7 December 1993; NAIC QR Q2/1972, pp. 3–4, and Q3/1973, pp. 3–4.
76. Campbell, Dyce, Ingalls, Pettengill, and Shapiro, "Venus: Topography Revealed by Radar Data," *Science* 175 (1972): 514–516. Smith, Ingalls, Shapiro, and Ash, "Surface-Height Variations on Venus and Mercury," *Radio Science* 5 (1970): 411–423, presented an earlier topographical study of Venus made from Haystack data taken over a period of years. That study was confined to the planet's equator and found a 2-km feature. It was remarkable for the variety of radar techniques used, as well as for its discovery of the first topographical feature on Venus.
77. Rumsey, Morris, R. Green, and Goldstein, "A Radar Brightness and Altitude Image of a Portion of Venus," *Icarus* 23 (1974): 1–7; Jurgens, "A Study of the Average and Anomalous Radar Scattering," pp. 87–110.

Chapter Six

Pioneering on Venus and Mars

Range-Doppler mapping and radar techniques for determining the roughness, height variations, and other characteristics of planetary surfaces came into their own in the early 1970s and shaped the kinds of problems planetary radar could solve. Radar techniques and the kinds of problems they solved were cross-fertilizing forces in the evolution of planetary radar astronomy. In the early 1970s, NASA was shifting gears. The landing of an American on the Moon, the zenith of the Apollo program, was history when in December 1972 Apollo 17 became the last to touch down on the Moon. Now the unmanned exploration of the planets began in earnest.

The usefulness of radar to planetary exploration had been argued by radar astronomers as early as the 1959 Endicott House conference. However, not everyone shared their enthusiasm. Smith and Carr, for example, in their 1964 book on radio astronomy, wrote: "It is inevitable that the importance of the exploration of the planetary system by radar will diminish as instruments and men are carried directly to the scene by space vehicles. However, that time is still to come. In the meantime, the information that radar provides will be vital in man's great effort to conquer space."[1] Soviet radar astronomers B. I. Kuznetsov and I. V. Lishin expressed similar sentiments in 1967: "Certainly, radar bombardment of the planets gives less information than a direct investigation of them with spaceships and interplanetary automatic stations." However, they did foresee that information about planetary surfaces would "help designers in the development of spaceships intended for making a 'soft' landing on the planets."[2]

As NASA came to fund planetary radar research, experiments and NASA missions became linked. Goldstone antenna time depended on mission approval, while Haystack radar funding was tied to specific, mission-oriented tasks. It is not surprising, then, that planetary radar in the 1970s evolved in point and counterpoint to the NASA space program, at first modestly to correct data returned from Soviet and American missions to Venus, next to help select a Mars landing site, and then to image Venus from a spacecraft. This evolution followed from the precedent established by NASA's funding of lunar radar imaging for the Apollo program. The Pioneer Venus radar imaging and altimetry missions took radar astronomy off the ground and into space. Again, just as ground-based radar astronomy had piggybacked itself onto Big Science radio astronomy facilities, so the Pioneer Venus radar attached itself to a larger mission to explore the planet's atmosphere.

The new techniques and problem-solving activities drew radar astronomers into closer contact with planetary scientists from a variety of disciplines who were not necessarily familiar with radar or the interpretation of radar results. It was one thing for radar astronomers to determine a spin rate for a planet or the value of the astronomical unit; astronomers easily grasped those discoveries. However, when radar astronomers described planetary surfaces in such abstract terms as *root-mean-square slope* to geologists, whose discipline rests heavily on hands-on field knowledge, a communication problem arose and serious misinterpretations and misunderstandings of radar results ensued.

---

1. Smith and Carr, pp. 130–131.
2. Kuznetsov and Lishin, p. 201.

The Radar Radius of Venus

On 18 October 1967, the Soviet Venera 4 space probe entered the atmosphere of Venus and began to transmit data back to Earth. From that data, Soviet scientists calculated a value for the radius of Venus, 6,079 ± 3 km, on the assumption that the break in the probe's transmissions indicated that it had reached the planet's surface. On the following day, Mariner 5 passed within 4,100 km of Venus and conducted a series of experiments. From the data beamed back to Earth, Mariner scientists at JPL calculated a value for the radius of Venus that was compatible with that determined by their Soviet colleagues, 6,080 ± 10 km.

The data from Venera 4 and Mariner 5 were consistent with each other and with the latest optical data, which yielded a value of 6,089 ± 6 km. However, the space and optical values differed markedly from the size of the radius, 6,056 ± 1.2 km, determined by Irwin Shapiro, Bill Smith, and Michael Ash with the Lincoln Laboratory radars as part of the Planetary Ephemeris Program.[3]

If the spacecraft and optical measurements were correct, then the radar data or its analysis were in error. The radius of Venus was a critical radar measurement; its value, for example, could serve to study the planet's topography. Radar astronomers associated with MIT and the Haystack Observatory, Gordon Pettengill, Irwin Shapiro, Dick Ingalls, Michael Ash, and Marty Slade, and those at the Arecibo Observatory, Rolf Dyce, Don Campbell, Ray Jurgens, and Tommy Thompson, took up the challenge in collectively authored papers that appeared in *Science* and the *Journal of the Atmospheric Sciences*. The publications embraced both a general audience and atmospheric specialists.

In addition to data collected previously at Millstone, Haystack, and Arecibo, the MIT-Arecibo radar astronomers added data from fresh radar observations made in 1966 and 1967 as well as optical observations from the U.S. Naval Observatory from the period 1950 through 1965. The magnitude of the data base was impressive and convincing. The Arecibo and MIT investigators analyzed their data separately and obtained radii of 6,052 ± 2 km and 6,048 ± 1 km, respectively. They concluded that Mariner 5 had misjudged its distance from the planet's center by about 10 km, and that "the simple possibility that Venera 4 underestimated its altitude by about 35 km cannot yet be ruled out."[4]

Dewey Muhleman, now professor of planetary science at the California Institute of Technology, with Bill Melbourne and D. A. O'Handley of JPL, made observations of Venus between May 1964 and October 1967 with the Goldstone Mars Station. Because their data were reported only in internal JPL reports, Lincoln Laboratory did not use that data. Consequently, they asserted, their observations constituted "an entirely independent data source." Muhleman and his JPL colleagues determined a value for the radius of Venus of 6,053.7 ± 2.2 km, in strong agreement with the MIT and Arecibo results.[5]

Arvydas Kliore and Dan L. Cain, two JPL scientists on the Mariner mission, saw the agreement between the Caltech-JPL and the Arecibo-MIT values and realized that "the consistency between reductions from data taken by different radars and reduced by different investigators cannot be ignored." They discovered that the different timing systems

3.   C. W. Snyder, "Mariner 5 Flight past Venus," *Science* 158 (1967): 1665–1669; Arvydas Kliore, Gerald S. Levy, Dan L. Cain, Gunnar Fjeldbo, S. Ichtiaque Rasool, "Atmosphere and Ionosphere of Venus from the Mariner 5 S-band Radio Occultation Experiment," *Science* 158 (1967): 1683–1688; Gerard H. de Vaucouleurs and Donald H. Menzel, "Results of the Occultation of Regulus by Venus, July 7, 1959," *Nature* 188 (1960): 28–33; Ash, Shapiro, and Smith, *Astronomical Journal* 72 (1967): 338–350.

4.   Ash, Campbell, Dyce, Ingalls, Jurgens, Pettengill, Shapiro, Martin A. Slade, and Thompson, "The Case for the Radar Radius of Venus," *Science* 160 (1968): 985–987; Ash, Campbell, Dyce, Ingalls, Jurgens, Pettengill, Shapiro, Slade, Smith, and Thompson, "The Case for the Radar Radius of Venus," *Journal of the Atmospheric Sciences* 25 (1968): 560–563; Shapiro 1 October 1993.

5.   William G. Melbourne, Muhleman, and D. A. O'Handley, "Radar Determination of the Radius of Venus," *Science* 160 (1968): 987–989.

used by the Deep Space Network to acquire Mariner 5 data, namely Station Time and Ephemeris Time, had introduced an error into their calculations. The amount of that error, 8.85 km, brought the Mariner 5 value for the radius of Venus in line with the radar results.

To explain what was now the anomalous Soviet value for the radius of Venus, Kliore and Cain concluded that either the Venera 4 capsule landed on a peak or plateau that was about 25 km high and not detected by planetary radar or the capsule stopped transmitting before reaching the solid surface of Venus. The problem with Venera 4, Don Campbell ventured, "was tied up in an ambiguity difficulty in their own radar system, which was a pulsed altimeter radar. I think, frankly, that the scientists who reported the results did not know how it worked. It was a military radar altimeter. They were just provided the answer, essentially. Although I don't know, and probably didn't know at the time either, what exactly the circumstances were, that was the impression that one got."[6]

"A Little Radar Knowledge is a Dangerous Thing."

Well before radar astronomers began collaborating with geologists, misinterpretations of radar data occurred. In fact, radar astronomers themselves were not immune to misconstruing radar results, as the case of the radar brightness of Mars illustrates. In initial observations of that planet, radar astronomer Dick Goldstein assumed a relationship between radar brightness and optical darkness. Arecibo observations appeared to confirm that relationship, which snowballed among planetary astronomers into a hypothesis that correlated radar brightness and topography (continental blocks and dry ocean basins). A reconsideration of evidence showed no such correlation.

When Dick Goldstein made his pioneering radar observations of Mars in 1963, he discovered what he thought was a relationship between radar "brightness," that is, the average amount of power returned in the echo from a given surface area of the planet, and the optical darkness of that same surface area. Goldstein constructed what he called a radar map of Mars, which showed variations in radar brightness. He noted, for example, that the Syrtis Major region appeared bright to the radar, but dark to visual observations. Because radar brightness is a function of surface roughness, he argued, the brightest radar areas were regions of flatness, while dark radar areas were topographically rough.[7]

In 1965, Goldstein observed Mars at the next opposition and again looked at the radar brightness of the planet's surface, this time at latitude 21° North. The average power returned (radar brightness) reached a maximum in the region of Trivium Charontis (an optically dark area), then dropped off abruptly when the neighboring area of Elysium (optically bright) was the radar target. Based on the known relationship between surface roughness and radar brightness, Goldstein concluded the existence of a very smooth, strongly reflecting area extending 20° to 30° in longitude and having an unknown latitudinal extent in the region of Trivium Charontis.[8]

During the same opposition, Gordon Pettengill, Rolf Dyce, and Don Campbell observed Mars with the UHF radar at Arecibo. When they compared their results with an optical map of Mars, the Arecibo investigators found a general tendency for weak echoes to correlate with the (optically) lighter areas of Mars, such as Arabia, Elysium, Tharsis, and

6.   Kliore and Cain, "Mariner 5 and the Radius of Venus," *Journal of Atmospheric Sciences* 25 (1968): 549-554; Campbell 7 December 1993. Murray, pp. 90–91, provides further anecdotal accounting of Soviet embarrassment over the incident.
7.   Goldstein and Gillmore, "Radar Observations of Mars," *Science* 141 (1963): 1172.
8.   Goldstein, "Mars: Radar Observations," *Science* 150 (1965): 1715–1717. His results were reported also in Goldstein, "Preliminary Mars Radar Results," *Radio Science* 69D (1965): 1625–1627.

Amazonis, and a tendency for strong echoes to correspond with visually darker features, such as the regions near Trivium Charontis and Syrtis Major. They did, however, note that the correlation between radar brightness and optical lightness was not perfect. For instance, the peak radar echo near Trivium Charontis occurred at 201° longitude, which is on one edge of the visually dark region. Likewise, the visually darkest region of Syrtis Major corresponded to a local minimum in echo strength.[9]

The Arecibo results were rather convincing. Not only had they been obtained from roughly the same area (22° North latitude) that Goldstein had studied, but the Arecibo and Goldstone observations had been made at two different frequencies (UHF vs. S-band). The persistence of the correlation between optical darkness and radar brightness at both frequencies was persuasive.

Astronomers Carl Sagan and James B. Pollack, then at the Smithsonian Astrophysical Observatory, and Richard Goldstein carried out a lengthy and detailed analysis of the JPL 1963 and 1965 radar data. They maintained and extensively documented the correlation between high radar reflectivity and optical darkness, despite some exceptions. Not only did radar bright and optically dark areas correlate; they claimed that topography and radar brightness also were related. Dark areas were elevations similar to continental blocks; bright areas were comparable to dry ocean basins.[10] The notion that Martian dark areas were elevated land masses rapidly gathered support from other planetary astronomers in the United States and Britain.[11]

Nonetheless, Pettengill, who had participated in the earlier effort at Arecibo, now opposed the correlation of visual darkness and radar brightness and undertook observations at Haystack, during the 1967 opposition, specifically in order to oppose the prevailing hypothesis that now correlated topography and radar brightness. Pettengill conducted a series of straightforward, precise range measurements to establish the topographical variations along latitude 22° North. Then he compared those range measurements with the average planetary radius taken from the planetary ephemeris data. He also plotted echo power over longitude along that same latitude.

Pettengill found no significant correlation between radar brightness and topography. A direct comparison between the radar results and a map of visible Martian surface features revealed no clear one-to-one association between bright or dark areas and topographical extremes. What others had observed as variations in radar brightness, Pettengill argued, resulted from the deviant properties of relatively small regions of the surface near the subradar point. Moreover, he pointed out, arguments for the hypothetical correlation between elevation extremes and brightness had been based largely on conclusions drawn from a range of disparate isolated locations. Further Haystack observations of Mars carried out under Pettengill's direction reinforced the conclusion that no correlation existed between regions of high radar reflectivity and optically dark areas.[12]

Perhaps one of the most notorious examples of misinterpreted radar results is that of Thomas Gold of Cornell University. Gold had been developing theories about the lunar surface since the 1950s. Long before he ever saw any radar data, Gold favored a meteoric

9.   Dyce, Pettengill, and Sánchez, "Radar Observations of Mars and Jupiter at 70 cm," *The Astronomical Journal* 72 (1967): 771–777; Campbell 7 December 1993.

10.   Carl Sagan, James B. Pollack, and Goldstein, "Radar Doppler Spectroscopy of Mars: 1. Elevation Differences between Bright and Dark Areas," *The Astronomical Journal* 72 (1967): 20–34. This article appeared earlier as Sagan, Pollack, and Goldstein, *Radar Doppler Spectroscopy of Mars: 1. Elevation Differences between Bright and Dark Areas*, Special Report 221 (Cambridge: SAO, 6 September 1966).

11.   See, for example, D. G. Rea, "The Darkening Wave on Mars," *Nature* 210 (1964): 1014–1015; R. A. Wells, "Evidence that the Dark Areas on Mars are Elevated Mountain Ranges," *Nature* 207 (1965): 735–736. Rea was at the University of California at Berkeley, and Wells at University College, London.

12.   Pettengill, Counselman, Rainville, and Shapiro, "Radar Measurements of Martian Topography," *The Astronomical Journal* 74 (1969): 461–482; Pettengill, Rogers, and Shapiro, "Martian Craters and a Scarp as Seen by Radar," *Science* 174 (1971): 1324.

explanation for lunar craters and developed an explanation for the presence of vast flat level surfaces that did not require the deposition of volcanic lava. His hypothesis was that these flat expanses consisted of dust from meteoric impacts. Gold interpreted radar observations of the Moon as supporting the existence of a surface layer of fine rock powder several meters deep, which a seismic experiment carried out by Apollo 12 allegedly supported. The implications for landing an American on the Moon were obvious; an astronaut might sink several centimeters into the powder or even "wallow" in it.[13]

Many scientists greeted Gold's prediction of a deep layer of powder with disbelief. As Don E. Wilhelms wrote, "Four Surveyor and six Apollo landings established the strength, thickness, block content, impact origin, and paucity of meteoric material in the Moon's regolith. There is fine pulverized soil, but it is weak only for a few centimeters of its thickness. Yet Thomas Gold is still fighting the battle. Still believing radar more than geological sampling..."[14] Wilhelms went so far as to state, "A little radar knowledge is a dangerous thing."[15] Gold later defended himself by insisting that although the "Gold dust" (as it has come to be called) would be many meters thick, the idea of sinking in it was a "total misconception."[16]

The Apollo program started the process of bringing together radar astronomers and geologists. The lunar radar images created by Tommy Thompson and Stan Zisk from data gathered at Arecibo and Haystack contributed not inconsequentially to America's exploration of the Moon. On occasion, nonetheless, radar astronomers misinterpreted lunar landing sites. In one instance, a landslide was mistaken for a field of boulders at the Apollo 17 landing site, while in another radar astronomers incorrectly characterized the roughness of the Apollo 14 Cone Crater site. These problems, however, arose not from mistaken readings of radar images, but from misinterpretations of the root-mean-square slope and dielectric constants of the surface.[17]

Landing on Mars

During the preparation for the Viking mission to Mars, radar astronomers encountered the challenge of making radar data understandable to NASA mission personnel unfamiliar with the interpretation of radar results. Until Congress funded the Voyager mission to Jupiter and Saturn, Viking was NASA's biggest and most expensive program for planetary exploration. Viking was to land on that planet, and NASA needed a landing site that was both safe for the lander and interesting to scientists. Radar astronomers collected and interpreted data to help with the selection of candidate sites.

The selection of the Viking lander site also brought together ground-based planetary radar astronomy and the Stanford bistatic radar approach under the aegis of NASA. Ground-based planetary radar astronomy had distinguished itself from "space exploration" (the Stanford approach), but the boundary between ground-based planetary radar astronomy and "space exploration" softened, as radar astronomers played an expanding role in NASA missions of planetary exploration and as Stanford investigators extended their field of applications.

13.   Gold, "The Lunar Surface," *Monthly Notices of the Royal Astronomical Society* 115 (1955): 585–604; Malcolm J. Campbell, Juris Ulrichs, and Gold, "Density of the Lunar Surface," *Science* 159 (1968): 973; Gold and Steven Soter, "Apollo 12 Seismic Signal: Indication of a Deep Layer of Powder," *Science* 169 (1970): 1071-1075; Gold, "The Moon's Surface," in Wilmot N. Hess, Menzel and John A. O'Keefe, eds., *The Nature of the Lunar Surface* (Baltimore: Johns Hopkins University Press, 1966), pp. 107–121; Gold, "Conjectures about the Evolution of the Moon," *The Moon* 7 (May–June 1973): 293–306.

14.   Don E. Wilhelms, *To A Rocky Moon: A Geologist's History of Lunar Exploration* (Tucson: The University of Arizona Press, 1993), p. 347.

15.   Wilhelms, p. 299.

16.   Gold 14 December 1993.

17.   Schaber 27 June 1994; Thompson 29 November 1994.

Images of Mars from earlier missions provided a clue in selecting candidate Viking landing sites. As early as 1965, Mariner 4 had flown past Mars and snapped 22 pictures of about one percent of the planet's surface. Mariner 6 and Mariner 7 took about 200 images of around 10 percent of the surface in 1969. The goal of Mariner 9, to make a complete photographic map of Mars was thwarted; when the spacecraft arrived at its destination, a planet-wide dust storm concealed most of the surface. Once the storm appeared to subside, Mariner 9 began to transmit images to Earth in early 1972, and the study of Martian topography began in earnest.[18]

Unlike the Mariner flybys, Viking was to study Mars by landing on its surface. A pair of orbiters was to focus on atmospheric studies, while a pair of landers studied the surface, if all went well. If the Viking landers were to touch down on a large rock or precariously on an edge, the entire mission might be lost. The clearance under the lander body was only 23 cm (nine inches), so a relatively smooth landing surface was a prime mission requisite.

NASA selected landing and backup sites for two landers. The sites had to be around 25° North latitude; at any other latitude, the orbiter solar panels would not receive sufficient solar energy to keep the spacecraft's batteries charged. That power was critical to the transmission of telemetry to Earth.

A major criteria for selecting candidate landing sites was the potential availability of water. Water meant the possibility of finding life, which was a major mission objective. Chryse, located at 19.5° North and 34° West, was scientifically interesting, because it is located at the lower end of a valley where the largest group of Martian channels diverges. The site may have been a drainage basin for a large portion of equatorial Mars and, therefore, would have collected deposits of a variety of surface materials.[19]

Despite the scientific interest in Chryse as the prime Viking landing site, the high-resolution Mariner 9 images lacked sufficient resolution to determine the site's safety. As Don Campbell recalled: "NASA was very concerned about how rough the surface was at the landing site. None of the Mariner 9 imagery had any hope of giving information at scales of 10 cm to a meter, which was the amount of surface roughness that they cared about."[20] Mariner 9 images had a resolution of about 100 meters, roughly the size of a football field, and simply did not show objects small enough to jeopardize the touchdown of the lander, which had a clearance of only 23 cm. The radar data, in contrast, were capable of indicating surface roughness down to objects only a few centimeters across. Once again, radar was going to try to solve a problem left unresolved by optical methods.

The Stanford Center for Radar Astronomy

In order to help select candidate Viking landing sites, NASA turned to radar astronomy and its ability to appraise gross and fine surface characteristics. The chief advocate for the use of radar data was Carl Sagan. Sagan was concerned about the possibility that the first lander might disappear in quicksand at one of the equatorial sites. In general, he believed that too much stress had been placed on visual images with a resolution of only 100 meters and not enough on radar, which could indicate surface irregularities at the

18.   Corliss, *The Viking Mission to Mars*, NASA SP-334 (Washington: NASA, 1974), pp. 6–8; Thomas A. Mutch, Raymond E. Arvidson, James W. Head, III, Kenneth L. Jones, R. Stephen Saunders, *The Geology of Mars* (Princeton: Princeton University Press, 1976).

19.   Martin Marietta Aerospace, *The Viking Mission to Mars* (Denver: Martin Marietta, 1975), pp. III-21 to III-23; Edward Clinton Ezell and Linda Neuman Ezell, *On Mars: Exploration of the Red Planet, 1958–1978*, NASA SP-4212 (Washington: NASA, 1984), p. 298.

20.   Campbell 8 December 1993.

10-cm scale. Sagan urged further study of the meaning of the radar data, so that the properties of the Martian soil could be better evaluated.

In response to Sagan's urging, on 1 March 1973, Tom Young and Gerald Soffen, Viking science integration manager and project scientist, respectively, met with Von Eshleman and Len Tyler of the Stanford Center for Radar Astronomy. Both already were investigators on Viking with a radio scattering experiment. Young and Soffen asked Tyler to acquire, analyze, and interpret radar data and to set up a radar study team for the selection of Viking landing sites. Tyler agreed.[21]

The Viking Project Office probably approached the Stanford Center for Radar Astronomy because Eshleman and Tyler already were Viking investigators, but also because of the Center's experience in interpreting Doppler spectra from the lunar surface. The Stanford Center for Radar Astronomy (SCRA) was a joint venture of Stanford University and the Stanford Research Institute (SRI) created in 1962 to foster scientific and engineering efforts and to provide graduate student training in radar astronomy and space science. It was the umbrella organization for Eshleman and his program of bistatic radar astronomy. A NASA grant underwrote the Center itself, while additional military and civilian awards supported a range of theoretical and experimental radio and radar research on space, ionospheric, and communication theory topics.[22]

Len Tyler, as did his Stanford colleague Dick Simpson, brought considerable knowledge of radar techniques to the effort. A graduate of Georgia Institute of Technology, Tyler had been at the SCRA since 1967, when he received his doctorate in electrical engineering from Stanford under Von Eshleman. Tyler invited Dick Simpson to work on the Viking data. Simpson, a graduate of the MIT electrical engineering program, had joined the SCRA in 1967 as a research assistant while working on his MS and Ph.D. in electrical engineering.[23]

Later, during the 1978 Mars opposition, Simpson and Tyler conducted 29 bistatic radar observations using the Viking 1 and 2 orbiter spacecraft in conjunction with the DSN antennas at Goldstone and Tidbinbilla (near Canberra, Australia) to study Mars surface roughness and scattering properties, and Simpson made ground-based monostatic radar observations of Mars, not associated with the Viking project, at Arecibo.[24] Their radar work, however, began much earlier, during the Apollo era.

For his doctoral thesis, Tyler had developed a method for creating two-dimensional surface images of the Moon using an Earth-based transmitter and a spacecraft receiver and based on theoretical work laid out earlier by another SCRA investigator, Gunnar Fjeldbo (now known as Lindal).[25] Tyler first applied his bistatic imaging method on Explorers 33 (which missed the Moon) and 35, the first U.S. spacecraft to orbit the Moon,

21.   Tyler 10 May 1994; Ezell and Ezell, pp. 309 and 320–321; "VOIR, Proposal to the NASA Management Section, 2/79," Box 13, JPLMM.

22.   SCRA, *Research at the Stanford Center for Radar Astronomy*, semi-annual status report no. 2 for the period 1 July–31 December 1963 (Stanford: RLSEL, February 1964), pp. 3–4; *Ibid.*, no. 4 for the period 1 July – 31 December 1964 (Stanford: RLSEL, January 1965), pp. 2–3; *Ibid.*, no. 5 for the period 1 January–30 June 1965 (Stanford: RLSEL, July 1965), pp. 5–6; *Ibid.*, no. 6 for the period 1 July–31 December 1965 (Stanford: RLSEL, January 1966), p. 4; *Ibid.*, no. 7 for the period 1 January–31 June 1966 (Stanford: RLSEL, August 1966), p. 5; *Ibid.*, no. 9 for the period 1 January–30 June 1967 (Stanford: RLSEL, 9 July 1967), pp. 6–8; John E. Ohlson, *A Radar Investigation of the Solar Corona*, SU-SEL-67- 071, Scientific Report 21 (Stanford: RLSEL, August 1967).

23.   Simpson 10 May 1994.

24.   Richard A. Simpson and G. Leonard Tyler, "Viking Bistatic Radar Experiment: Summary of First-Order Results Emphasizing North Polar Data," *Icarus* 46 (1981): 361–389; Simpson and Tyler, "Radar Measurement of Heterogeneous Small-Scale surface Texture on Mars: Chryse," *Journal of Geophysical Research* 85 (1980): 6610–6614; Simpson 10 May 1994.

25.   Fjeldbo, "Bistatic-Radar Methods for Studying Planetary Ionospheres and Surfaces," Ph.D. diss., Stanford University, 1964, especially pp. 64–82. Later published as Fjeldbo, *Bistatic-Radar Methods for Studying Planetary Ionospheres and Surfaces*, SR 2 (Stanford: RLSEL, 1964).

and obtained crude meter-scale measurements of surface roughness and radar brightness.[26] With Simpson, Tyler performed bistatic radar experiments on the Moon using the Apollo 14, 15 (at 13 and 116 cm), and 16 (at 13 cm only) command service modules while those vehicles were in lunar orbit; at the same time, they were receiving the S-band (13 cm) signals at Goldstone and the VHF (116 cm) signals at the Stanford 46-meter (150-ft) dish. However, Tyler and Simpson did not do imaging; they were more concerned with scattering mechanisms.[27]

Mars Radar

Tyler and Simpson began working on the Viking landing site selection problem by surveying and re-analyzing the available data. Radar data from several oppositions already were available, and those data obtained during the 1965 opposition were from the latitude of the preferred Viking landing sites, around 20° North. Radar studies of Mars made during the 1969 opposition provided useful topographical and surface roughness measurements, though not at latitudes interesting to the Viking mission. Haystack observed a swath of the planet's surface near the equator (latitudes 3° and 12° North), while Goldstone took observations at three latitudes (3°, 11°, and 12° North).[28]

26.   Tyler 10 May 1994; Tyler, *The Bistatic Continuous-Wave Radar Method for the Study of Planetary Surfaces*, SU-SEL-65-096, Scientific Report 13 (Stanford: RLSEL, October 1965), which later appeared as Tyler, "The Bistatic, Continuous-Wave Radar Method for the Study of Planetary Surfaces," *Journal of Geophysical Research* 71 (1966): 1559-1567; Tyler, *Bistatic-Radar Imaging and Measurement Techniques for the Study of Planetary Surfaces*, SU-SEL-67-042, Scientific Report 19 (Stanford: RLSEL, May 1967); Tyler and Simpson, *Bistatic-Radar Studies of the Moon with Explorer 35: Final Report Part 2*, SR 3610-2, SU-SEL-70-068 (Stanford: RLSEL, October 1970); Tyler and Simpson, "Bistatic Radar Measurements of Topographic Variations in Lunar Surface Slopes with Explorer 35," *Radio Science* 5 (1970): 263-271; SCRA, *Proposal to the National Aeronautics and Space Administration for Bistatic Radar Astronomy Studies of the Surface and Ionosphere of the Moon based upon Transmission from the Earth and Reception in a Surveyor Orbiter*, Proposal RL 21-62 (Stanford: RLSEL, 7 September 1962), Eshleman materials.
27.   Tyler 10 May 1994; Simpson 10 May 1994; Simpson and Tyler, "Radar Scattering Laws for the Lunar Surface," *IEEE Transactions on Antennas and Propagation* AP-30 (1982): 438-449; Simpson, "Lunar Radar Echoes: An Interpretation Emphasizing Characteristics of the Leading Edge," Ph.D. diss., Stanford University, 1973.
28.   Tyler 10 May 1994; Rogers, Ash, Counselman, Shapiro, and Pettengill, "Radar Measurements of Surface Topography and Roughness of Mars," *Radio Science* 5 (1970): 465-473; Goldstein, Melbourne, Morris, George S. Downs, and O'Handley, "Preliminary Radar Results of Mars," *Radio Science* 5 (1970): 475-478.

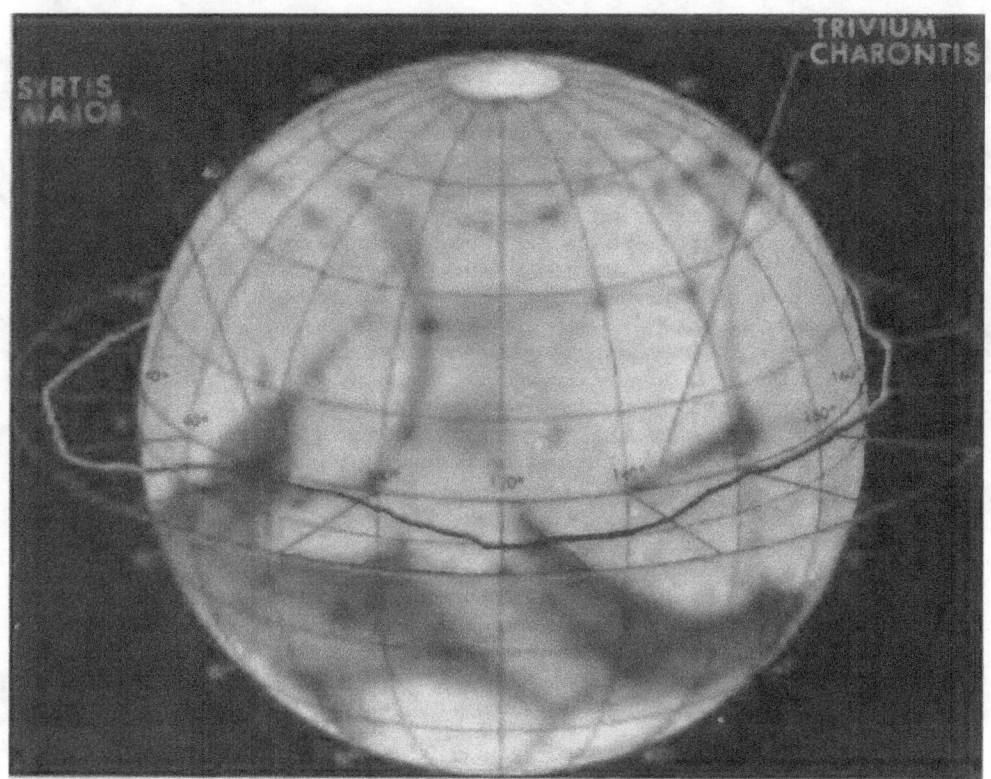

Figure 25

Outline of Mars topography at 8° north of the equator released by JPL in July 1969. The outer white circle indicates a six-mile-high scale. The inner irregular line traces topographical variations found by radar. Syrtis Major and Trivium Charontis were found to be long slopes. The correlation of radar topographic data with known features in Mars photographic images aided geologists' ability to interpret the physical and historical geology of the planet. (Courtesy of Jet Propulsion Laboratory, photo no. 331-4539.)

The 1969 and earlier data, moreover, were too noisy to be of any use in sorting out a Viking landing site. The best data had been collected during the 1971 opposition, when Mars came closer to Earth than it would again for 17 years. Goldstone achieved its highest resolutions to date; results showed a rugged terrain, with elevation differences greater than 13 km from peak to valley. Altitude profiles showed heavy cratering, including several large craters 50 to 100 km in diameter and 1 to 2 km deep. Haystack also made high-resolution Mars radar observations during the 1971 opposition, measured surface heights with relative errors down to about 75 meters, and correlated craters detected by radar with those in images taken by Mariner.[29]

29.   Goldstein 14 September 1993; Downs, Goldstein, R. Green, Morris, "Mars Radar Observations, A Preliminary Report," *Science* 174 (1971): 1324–1327; Downs, Goldstein, R. Green, Morris, and Reichley, "Martian Topography and Surface Properties as Seen by Radar: The 1971 Opposition," *Icarus* 18 (1973): 8–21; Pettengill, Shapiro, and Rogers, "Topography and Radar Scattering Properties of Mars," *Icarus* 18 (1973): 22–28; Pettengill, Rogers, and Shapiro, "Martian Craters and a Scarp as Seen by Radar," *Science* 174 (1971): 1321–1324.

Nonetheless, even that high-resolution data was not useful to the selection of a Viking landing site. Because of the geometries of the Earth and Mars during that opposition, planetary radar astronomers observed the southern hemisphere of the planet. The Goldstone radar observed Mars at latitude 16° South. Haystack observations during the 1971 opposition also examined southern latitudes.[30] The best candidates for the Viking mission were in the northern hemisphere.

Thus, in 1973, when Tyler undertook the interpretation of Mars radar data for the selection of the Viking landing sites, radars had not observed the preferred Viking landing area near 20° North since 1967, nor any of the backup sites near the equator prior to 1975. The Viking Project Office funded a round of Mars radar observations in 1973 at the Haystack, Arecibo, and Goldstone radar telescopes at UHF, S-band, and X-band frequencies. Don Campbell and Rolf Dyce provided the Arecibo data, while Dick Goldstein and George Downs took the Goldstone data, and Gordon Pettengill furnished the Haystack data.[31]

The 1973 Haystack Mars data was placed in the same format as that obtained at Arecibo in order to facilitate their comparison. Although Haystack provided an abundance of radar data, its signal-to-noise ratio was generally too low for a detailed study of surface characteristics. The Haystack klystron was acting up,[32] and Haystack ceased to participate in the Viking mission; shortly thereafter Haystack stopped all planetary radar experiments.

The 1973 Viking Mars data provided no direct information on potential landing sites. The orbital geometries of Earth and Mars meant that the subradar points of the three telescopes swept areas in the southern hemisphere, between latitudes 14° and 22° South far from either the main or backup landing sites.[33] The 1973 data, nonetheless, provided an opportunity to better understand the radar properties of the Martian surface and for Tyler, in particular, to begin the difficult task of explaining the surface roughness of Mars in terms of root-mean-square (rms) slope to an audience unacquainted with the interpretation of radar data.

Radar Tutorials

Mariner images made the surface of Mars obvious to everyone. Radar data on surface roughness was not at all obvious and required expert interpretation. The "rift between believers in radar and believers in photography," in the words of Edward Clinton Ezell and Linda Neuman Ezell, first appeared at a meeting of the Viking landing site working group on 25 April 1972,[34] well before Tyler and radar became a part of the site selection process.

---

30.   Pettengill, Shapiro, and Rogers, "Topography and Radar Scattering Properties of Mars," *Icarus* 18 (1973): 22–28; Pettengill, Rogers, and Shapiro, "Martian Craters and a Scarp," pp. 1321–1324.

31.   Simpson, Tyler, and Belinda J. Lipa, *Analysis of Radar Data from Mars*, SR 3276-1, SU-SEL-74-047 (Stanford: SCRA, October 1974).

32.   Ingalls 5 May 1994; Simpson 10 May 1994.

33.   Memorandum, Sebring to Distribution, 9 December 1970, 44/2/AC 135; "Applications of High Power Radar to Studies of the Planets, NASA, 7/1/69- 6/30/70," 67/2/AC 135; "Radar Studies of the Planets, NASA, 7/1/72-6/30/73," 68/2/AC 135, MITA; NEROC, *Final Progress Report Radar Studies of the Planets*, 29 August 1974, pp. 1–2; NEROC, *Semiannual Report of the Haystack Observatory*, 15 July 1972, p. ii; Simpson, Tyler, and Lipa, "Mars Surface Properties Observed by Earth-Based Radar at 70-, 12.5-, and 3.8-cm Wavelengths," *Icarus* 32 (1977): 148. For the radar results themselves, see Pettengill, John F. Chandler, Campbell, Dyce, and D. M. Wallace, "Martian Surface Properties from Recent Radar Observations," *Bulletin of the American Astronomical Society* 6 (1974): 372; Downs, Goldstein, R. Green, Morris, and Reichley, "Martian Topography and Surface Properties as Seen by Radar: The 1973 Opposition," *Icarus* 18 (1973): 8–21; Downs, Reichley, and R. Green, "Radar Measurements of Martian Topography and Surface Properties: The 1971 and 1973 Oppositions," *Icarus* 26 (1975): 273–312.

34.   Ezell and Ezell, p. 298.

The key radar information on surface characteristics was not expressed visually, but mathematically. The abstract results were neither visual nor directly accessible by any of the senses. Moreover, the transformation of raw radar data into information on surface characteristics involved the interpretation of the data in terms of scattering laws and their expression as degrees of rms slope. The number of degrees of rms slope indirectly but reliably described the planet's surface roughness.

When a radar wave strikes an irregular planetary surface covered by boulders or other material with multiple sides, a complex scattering process takes place. Some power returns to the radar, some power is deflected away from the radar return path, while some power scatters among the boulders. The rougher the surface, the less power returns to the radar and the flatter is the return power spectra.

Because each radar target has a different surface makeup, its scattering behavior varies. Radar astronomers have sought general laws that describe scattering behavior. These scattering laws are mathematical descriptions of how much power is reflected back towards the radar at different angles of incidence. They are important tools for interpreting planetary radar data. At Haystack and Arecibo, radar investigators used what had become known as the "Hagfors Law," named after the Cornell University ionosphericist and radar astronomer.

The Hagfors Law mathematically expresses the general roughness of a planetary or lunar surface in terms of average slope. The root-mean-square is a specific type of mathematical average for the expression of these average slopes. When using the Hagfors Law, the value for the slope varies up to 3°, the upper theoretical limit for the validity of the assumptions underlying the Hagfors Law, although in practice much higher slope values are normal. The 1973 slope estimates for Mars ranged from 0.5° to at least 3°, suggesting that some areas, those closest to 0.5°, were suitable for a Viking landing. However, none of the 1973 radar experiments had observed areas of potential Viking landing sites.[35]

Tyler and the members of his radar study group presented their results to the landing site working group meeting at Langley on 4 November 1974. Tyler announced that his study group had learned a great deal: overall, the Martian surface was very heterogeneous; Mars tended to have greater variation in surface reflectivity than Earth or the Moon; and the planet appeared smoother than the Moon to the radar. However, he concluded, data acquired in the southern hemisphere could not be applied to northern latitudes without variation. Also, correlation between radar features and Mariner 9 imagery was poor.

Both Tyler and Gordon Pettengill "laced their presentations strongly with tutorial material which greatly enhanced the ability of the group to understand and correctly interpret their findings," reported Edward and Linda Ezell.[36] After all, geologists would rather think about rocks than about Hagfors' Law, rms slopes, or dielectric constants, and those in charge of making the landing site selection had no knowledge of radar.[37] The abstract nature of the radar data, as well as its complex and difficult interpretation, had an impact on the actual use of radar in the selection of the Viking landing site.

---

35.   Simpson, Tyler, and Lipa, "Mars Surface Properties Observed by Earth-Based Radar at 70-, 12.5-, and 3.8-cm Wavelengths," *Icarus* 32 (1977): 156.

36.   Ezell and Ezell, p. 322.

37.   Simpson 10 May 1994; Schaber 27 June 1994; Shoemaker 30 June 1994; Soderblom 26 June 1994; Gold 14 December 1993.

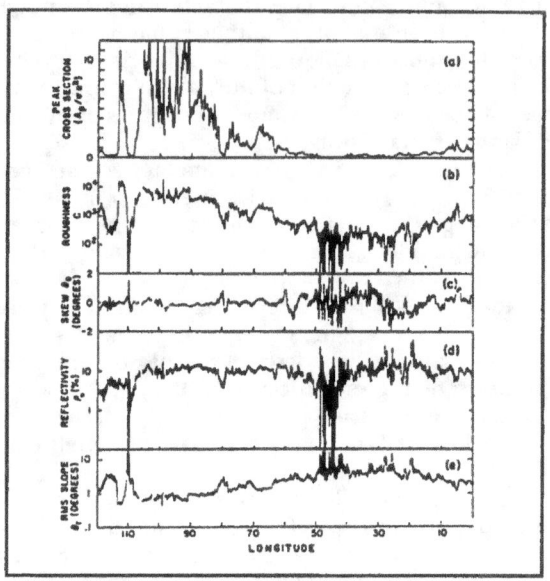

Figure 26

The radar data used to help select candidate landing sites for Viking often were expressed in degrees of rms slope. This illustration depicts the abstract nature of that radar data. Above, (e) is the rms slope derived from the roughness data obtained near latitude -16° and shown in (b). (a) through (c) were obtained by fitting the Hagfors scattering law to measured angular power spectra, while (d) was surface reflectivity derived from data in (a) and (b). (Courtesy of Jet Propulsion Laboratory.)

The Landing Site

The selection of the final landing site of Viking 1 was a long, tedious, and dramatic process.[38] The sites under consideration at the last minute, literally during the two weeks that Viking orbited Mars, all had been either observed by radar or photographed. Part of the problem was the lack of overlap between the radar and optical data. Areas with good radar data had poor photographic documentation, while sites with good photographic views had poor radar data. Everybody wanted a safe landing; nobody wanted to take a chance with a site confirmed by only radar or only photographs. The indecision foiled an earlier interest in landing Viking on the Fourth of July in honor of the country's bicentennial.

In order to acquire additional information on the candidate landing sites, the Viking Project Office commissioned another round of radar observations by Goldstone and Arecibo. Observing conditions were not ideal, because Earth and Mars were not in opposition. However, the Arecibo S-band (2380 MHz; 13 cm) 400-kilowatt radar had just come on line, and Tyler recommended making additional Arecibo observations with it. Earlier Mars radar studies had been conducted only when Mars-Earth distances were less than one astronomical unit. Signal strengths during the August 1975–February 1976 equatorial observations were good, but the Earth-Mars distance reached 2 astronomical units in May-July 1976.[39]

---

38.   For a full discussion, see Ezell and Ezell, pp. 317-346, as well as Downs 4 October 1994.
39.   Simpson, Tyler, and Lipa, *Analysis of Radar Data from Mars*; Simpson, Tyler, and Campbell, "Arecibo Radar Observations of Mars Surface Characteristics in the Northern Hemisphere," *Icarus* 36 (1978): 156–157.

Arecibo observed Mars between August 1975 and July 1976 over the latitudes between 12° South and 24° North. The results between 12° South and 4° North were relevant to potential alternate (i.e., backup) Viking sites. Between October 1975 and April 1976, Goldstone observed the two regions Syrtis Major and Sinus Meridiani, particularly a number of proposed Viking landing sites, including the prime site (called A1) near longitude 34° and latitude 19.5° North. As a result of the radar data, the A1 site was rejected on 26 June 1976, while other sites came under consideration.[40]

Simpson, Tyler, and Campbell made additional Arecibo observations for the Viking Project Office near 20° North latitude, the latitude of the landing site, particularly the Viking Chryse and Tritonis Lacus (the A2 site, first alternate to A1) landing areas. The search for a suitable site then moved toward the northwest where a region designated A1NW was tentatively selected because of its apparent smoothness as seen from orbit. The A1NW site was finally abandoned because of its questionable radar properties. It was toward the west that the Viking site selection and certification teams moved after turning down A1NW.[41]

Did Radar Help?

Had radar observations and expressions of rms slope actually helped in the selection of the final Viking 1 landing site? Certainly, Tyler's reports to the landing site working group did not go totally unheeded, and radar turned down some potential but suspect landing sites. As NASA official John E. Naugle wrote in November 1976, "The choice of the actual landing site was eventually based on a combination of the S-band [Arecibo] radar data and high resolution photography obtained from the Viking 1 Orbiter."[42] However, not everyone was as diplomatic as Naugle; some doubted the utility of the radar data.

Tom Young, Viking science integration manager, believed that radar data eventually played a role, although when the project selected initial landing site candidates, he admitted that, "radar played no role, because we weren't smart enough to know how to use it." On the other hand, James Martin, Viking Project Manager, remained skeptical about the utility of the radar data. Radar provided no useful information, he felt, although it was "an input and a source of information that [we] could not ignore." Frankly, he admitted, "The fact that it [radar] was so different scared me off."[43]

It was that difference, the general unfamiliarity with radar data, that raised a barrier to the use of radar results. "People didn't quite know what to make of us," Tyler explained. "People were willing to listen, but it was clear that they didn't like the answer!" Farther to the cynical side was the judgement of Dick Simpson: "I've always said that the radar contribution to picking landing sites on Mars probably came out with a net result of zero....If we'd never been involved, they probably would have had the same end result, but we got to play in the game and sometimes that's part of it." George Downs, who analyzed the Mars radar data at JPL, was convinced that project personnel simply looked for a site as Viking 1 orbited Mars, ignoring the radar data entirely. The attitude of many, he felt, was that the radar astronomers were getting their answers as if from a ouija board.[44]

---

40.   Simpson, Tyler, and Campbell, "Arecibo Radar Observations of Martian Surface Characteristics Near the Equator," *Icarus* 33 (1978): 102–115; Downs, R. Green, and Reichley, "Radar Studies of the Martian Surface at Centimeter Wavelengths: The 1975 Opposition," *Icarus* 33 (1978): 441–453.

41.   Simpson, Tyler, and Campbell, "Mars Surface Characteristics in the Northern Hemisphere," pp. 153–173.

42.   John E. Naugle to H. Guyford Stever, 8 November 1976, NHOB.

43.   Ezell and Ezell, p. 357.

44.   Tyler 10 May 1994; Simpson 10 May 1994; Downs 4 October 1994.

The radar data presented was indeed quite different; it was degrees of rms slope, rather than images universally understood. Perhaps if range-Doppler mapping of Mars had been possible, the difference would not have been so great. Still, the episode illustrated the kinds of challenges that radar astronomers would have to confront as they played an increasing role in planetary exploration and sought to share their results with scientists who lacked an understanding of radar. It was simply not enough to meet with planetary geologists and other scientists; radar astronomers had to communicate their results in a way understandable by other scientists.

The availability of the Mars radar data at JPL was the catalyst for the kind of interdisciplinary communication and collaboration that interpreting the radar results demanded. George Downs struck up an alliance with Ladislav Roth at JPL and Gerald Schubert at UCLA. Roth and Schubert saw value in the radar data; that is, the topographical information, not the surface roughness measurements. Roth, in fact, had approached Downs to collaborate in interpreting the radar topographical data, and several studies grew out of that collaboration.[45]

A Venus Radar Mapper?

Concurrently with Viking preparations, NASA planned a mission to Venus. Pioneer Venus marked a significant departure for radar astronomy. Don Campbell and Tor Hagfors had distinguished planetary radar astronomy from space exploration, in particular, the bistatic radar work done at Stanford University. Pioneer Venus challenged that distinction; it was no longer ground-based planetary radar astronomy, and it marked a significant entree into a new area of Big Science.

Instead of Big Science providing a large, Earth-based dish, like the Arecibo radar, spacecraft missions furnished the opportunity, but not the hardware, to do planetary radar astronomy from a point just above the target, not millions of kilometers away. Like piggybacking radar astronomy onto an Earth-based facility, placing a radar experiment and its necessary hardware on a spacecraft demanded participating in the politics of Big Science. Radar astronomy aboard Pioneer Venus remained Little Science, though, conducted by a single investigator, Gordon Pettengill, who carried out the entrepreneurial burden of placing the radar instrument on the spacecraft and who brought fellow ground-based radar astronomers into the project as analyzers of the radar data.

Pioneer Venus also facilitated the shift of planetary radar toward serving the planetary geology community. Within the working groups established by NASA space missions, planetary radar astronomers and planetary geologists worked together. Behind this shift, too, was the ability of radar astronomers to solve problems of interest to geologists. If planetary radar astronomy had focused solely on refining planetary orbital parameters, the prime users of planetary radar results would have remained astronomers. Radar techniques that described planetary surfaces, in contrast, solved problems of interest to geologists, especially those geologists at the United States Geological Survey (USGS) interested in planetary geology, or what the USGS called astrogeology. The shift to geology was an educational experience for both geologists and radar investigators, and it eventually manifested itself in the journals and professional societies attended by planetary radar astronomers and culminated in the Magellan mission to Venus.

The idea of using radar to image Venus from a probe predated the Pioneer Venus project. The official history of Pioneer Venus dates the beginning of the project to

45.   Downs 4 October 1994; Ladislav E. Roth, Downs, Saunders, and Gerald Schubert, "Radar Altimetry of South Tharsis, Mars," *Icarus* 42 (1980): 287–316; Roth, Saunders, Downs, and Schubert, "Radar Altimetry of Large Martian Craters," *Icarus* 79 (1989): 289–310.

October 1967, shortly after the Venera 4 and Mariner 5 spacecraft visited Venus. Three scientists, Richard M. Goody (Harvard University), Donald M. Hunten (University of Arizona; Kitt Peak National Observatory), and Nelson W. Spencer (Goddard Space Flight Center) formed a group to consider the feasibility of exploring the Cytherean atmosphere from a spacecraft. The group's formation led to a study published in January 1969 by the Goddard Space Flight Center.[46]

The idea of mapping Venus with a radar started much earlier. As early as 1959, NASA contracted with the University of Michigan to design a Venus radar. In 1961, NASA let out three more grants and contracts to develop radars for a future Venus mission to map the planet's surface to investigators at the University of New Mexico, MIT, and Ohio State.[47] In 1961, for example, NASA funded a study under J. F. Reintjes, Director of MIT's Electronics Systems Laboratory, "to perform an investigation of radar techniques and devices suitable for the exploration of the planet Venus." NASA awarded the funds because the space agency saw radar as an attractive technique for exploring the surface of Venus and as "a logical experiment for a Venus flyby or orbiter."

Developing a radar system appropriate for space travel presented numerous problems. The equipment had to meet certain weight, space, and reliability criteria. The MIT goal was to design and build a space radar that required fewer than 100 watts and weighed no more than 50 pounds. After completion of an engineering model by Reintjes and the MIT Electronics Systems Laboratory, in October 1967, tests aboard an aircraft, the Convair CV-990 owned by NASA Ames Research Center, began.[48]

Throughout the 1960s, then, and well before the formation of the Goddard study group in 1967, the idea of imaging Venus with a spacecraft-borne radar was already "in the air." But before a spacecraft could carry a radar to Venus, NASA had to formulate and fund a voyage of exploration to the planet. In June 1968, a Space Science Board study on planetary exploration urged NASA to send a space probe to Venus, though without recommending inclusion of a radar experiment.[49] By June 1970, the NASA program of planetary exploration still contained no significant Venus missions. The planned flyby of Venus and Mercury was essentially a Mercury mission with only a small contribution to Venus science. In contrast, NASA had a robust plan for exploring Mars and an ambitious program for investigating the outer planets.[50]

In June 1970, to address the lack of a serious Venus mission, the NASA Lunar and Planetary Missions Board and the Space Science Board brought together 21 scientists to study the scientific potential of a mission to Venus (Table 3). Richard Goody and Donald M. Hunten, who had helped start the Goddard study, co-chaired the meeting. Their report, known as the Purple Book because of the color of its cover, recommended that exploration of Venus should be prominent in the NASA program for the 1970s and 1980s. The group presented its recommendations to NASA management, and the Space Science Board endorsed them.[51]

Significantly, the Purple Book study brought together a planetary radar astronomer, Gordon Pettengill, then Director of the Arecibo Ionospheric Observatory, and a planetary geologist, Harold Masursky of the USGS. Pettengill's participation in the Purple Book

46.   Richard O. Fimmel, Lawrence Colin, and Eric Burgess, *Pioneer Venus*, NASA SP-461 (Washington: NASA, 1983), pp. 14–15; Colin, "The Pioneer Venus Program," *Journal of Geophysical Research* 85 (1980): 7575.

47.   Tatarewicz, pp. 150–151.

48.   Memorandum, Oran W. Nicks, 10 March 1966, and Memorandum, Brunk, 29 November 1966, NHOB; J. F. Reintjes and J. R. Sandison, *Venus Radar Systems Investigations Final Report* (Cambridge: MIT, Electronic Systems Laboratory, Department of Electrical Engineering, March 1970), Pettengill materials.

49.   Space Science Board, *Planetary Exploration, 1968–1975* (Washington: National Academy of Sciences, 1968).

50.   Space Science Board, *Venus: Strategy for Exploration* (Washington: National Academy of Sciences, June 1970), p. 3.

51.   C. H. Townes, Preface, Space Science Board, *Venus: Strategy for Exploration*, n.p.

Table 3
Purple Book Scientists

Scientist	Institution
Richard M. Goody, Chair	Harvard University
Donald M. Hunten	Kitt Peak National Observatory
Don L. Anderson	California Institute of Technology
W. Ian Axford	University of California, San Diego
Alan H. Barrett	Massachusetts Institute of Technology
Leverett Davis, Jr.	California Institute of Technology
Thomas M. Donahue	University of Pittsburgh
John C. Gille	Florida State University
Seymour Hess	Florida State University
Garry E. Hunt	Atlas Computer Laboratory
Robert G. Knollenberg	University of Chicago
John S. Lewis	Massachusetts Institute of Technology
Michael B. McElroy	Kitt Peak National Observatory
Gordon H. Pettengill	Arecibo Ionospheric Observatory
Robert A. Phinney	Princeton University
S. Keith Runcorn	University of Newcastle
Verner E. Suomi	University of Wisconsin
Patrick Thaddeus	Columbia University
G. Leonard Tyler	Stanford University
James A. Weiman	University of Wisconsin
George W. Wetherill	University of California, Los Angeles

study marked his initial involvement in Pioneer Venus.[52] By then, Pettengill, the future Professor of Planetary Physics in the MIT Earth and Planetary Sciences Department, had acquired stature in his field, having been one of the radar astronomy pioneers at Lincoln Laboratory, but also as Associate Director, then as Director, of the prestigious Arecibo Observatory.

Masursky had joined the USGS after graduating from Yale in 1947. After a number of years as a general geologist, Masursky joined the USGS's Branch of Astrogeologic Studies. In 1967, he became chief of the astrogeology branch, then starting in 1971 and until his death, chief scientist of that branch. Masursky was a science investigator on almost every NASA flight project to the Moon and the planets, including the Ranger, Lunar Orbiter, Surveyor, Apollo, Mariner 9, and Viking missions.[53]

The Purple Book meeting thus was a first step in planetary radar's shift toward geology, providing an initial setting for planetary radar and geology to interact and to develop a common approach for the study of Venus's surface, within the broader context of NASA-sponsored research of the planet's atmosphere. As the Purple Book itself noted, the space missions of the 1960s had given rise to new fields of study: "Very rapidly studies of planetary meteorology, planetary aeronomy, planetology, and planetary biology emerged which involved, in the main, research workers from the parallel terrestrial disciplines. Earth and planetary studies suddenly merged and simultaneously diverged from astronomy. In some major universities, departmental and research center organization was changed to meet this development."[54]

Images sent back from space had encouraged geologists, like Hal Masursky, to become interested in planetary surfaces and in the processes that shaped them. However, ground-based radar images of Venus had yet to find their audience among planetary geologists.[55]

---

52.   Pettengill 28 September 1993.
53.   *V-Gram* no. 12 (July 1987): 15; "Harold Masursky," in R. R. Bowker, comp., *American Men and Women of Science*, 18th edition (New Providence, NJ: R. R. Bowker, 1992), vol. 5, p. 275.
54.   Space Science Board, *Venus: Strategy for Exploration*, p. 4.
55.   Campbell 9 December 1993.

As far as ground-based planetary radar was concerned, the Purple Book applauded its success. "Virtually all our present knowledge of the radius, rotation, and surface of Venus has been obtained using ground-based radars," the Purple Book proclaimed. With resolutions ranging from 100 to 500 km, radar had revealed features, and even the lack of topographic relief, in the equatorial region of Venus. Including a radar system on a Venus probe would yield "maps similar in appearance and usefulness to photographic maps of the same region."

Ironically, the Purple Book cautioned against imaging Venus with a spacecraft radar. It pointed out that such radar images would be "directly competitive with ground-based observations and would provide similar data." That point resurfaced later during planning for Magellan. Although a spaceborne radar could cover more of the planet's surface, the Purple Book concluded that "it is not yet clear whether the high cost of the additional information could be justified." Not only would a spacecraft radar require "great weight and complexity" in order to compete with the resolution already achieved by ground-based radars, but the "rapidly improving capabilities of radar observatories on the earth to image Venus" made radar mapping of the planet from orbit "less important at the present time." The report reflected the anticipated benefits of the planned upgrade of the Arecibo telescope.

While technological and cost constraints militated against an orbiting radar, a viable alternative, according to the Purple Book, was the Stanford bistatic radar method, specifically that mode in which a radar on Earth transmitted and a receiver on the spacecraft collected echoes. The Purple Book concluded that "bistatic-radar experiments, in conjunction with ground-based observations, can provide a significant insight into the details of the surface structure and electromagnetic properties of Venus."

The recommendation to conduct a bistatic experiment was not surprising; Len Tyler of the SCRA was one of the 21 Purple Book scientists. Although Tyler planned to do some bistatic observations with Pioneer Venus, those plans fell by the wayside. Later, as Pettengill was writing a proposal for Pioneer Venus and was looking for scientists to join him, he invited Tyler. Tyler turned down the invitation because of his heavy commitment to the Voyager project.

In addition to the bistatic experiment, the Purple Book recommended using a radar altimeter to measure surface relief. Radar altimeter readings would complement the equatorial topographic information available from ground-based radar observations, and a simple, low-power orbiting radar could measure vertical relief over those portions of the planet not covered by ground-based radars.[56] Using the altimeter to gather relief measurements was the cheapest and technologically least complicated alternative. In the end, a modified version of this approach was to fly on Pioneer Venus.

After Venera 7 succeeded in transmitting data from the surface of Venus for 23 minutes on 15 December 1970, a special panel reviewed the Purple Book conclusions. Their recommendation, to make no changes in the Purple Book, opened the door for NASA to issue an Announcement of Opportunity in July 1971 for scientists to participate in defining the Venus program.[57]

NASA established the Pioneer Venus Science Steering Group in January 1972, in order to enlist widespread participation of the scientific community in the early selection of the science requirements for the Pioneer Venus project. The Science Steering Group met with Pioneer Venus project personnel between February and June 1972. The Group developed in great detail the scientific rationale and objectives for several voyages to Venus and outlined candidate payloads.[58]

---

56.   Tyler 10 May 1994; Simpson 10 May 1994; *Venus: Strategy for Exploration*, pp. 58–62.
57.   Fimmel, Colin, and Eric, pp. 17–18.
58.   Fimmel, Colin, and Burgess, p. 18.

The search for mission objectives stirred radar astronomer Gordon Pettengill to propose a radar experiment for the mission. Pettengill recalled: "I remember doing a calculation literally on the back of an envelope. I realized that if we could get even a tiny little
antenna into a reasonable orbit around Venus, we could do an awful lot in terms of
measuring the altitude and the reflecting properties of the surface....By going around
Venus in a polar, rather than an equatorial orbit, we could get a totally new view of Venus.
We could detail the whole surface, instead of just the equatorial band that we observed at
Arecibo."

Pettengill then began "beating the drums" to include a radar experiment in the
Venus program. "The Science Working Group studied the concept. I didn't think I was
going to survive that," he recalled. "Pioneer Venus was strictly an atmospheric mission. A
radar experiment to study the surface stood out like a sore thumb." Nonetheless, NASA
awarded Pettengill funds to conduct a feasibility study of a radar to image Venus.[59]

Above all else, the prime mission of Pioneer Venus was to study the planet's atmosphere. An article published in 1994 in *Scientific American*[60] evaluated the scientific achievements of Pioneer Venus and emphasized its contributions to atmospheric science, but
failed to mention the radar experiment. Peter Ford, who collaborated with Pettengill on
the Pioneer Venus radar experiment, pointed out that the *Scientific American* article's
emphasis on the atmospheric science balanced the record; during the first three years of
the Pioneer Venus mission, most publicity had focused on the radar imaging.[61]

Next, the Science Steering Group published its comprehensive report, called the
Orange Book. Among the 24 areas of research advocated, only one was related to the plan-
et's surface. As the project evolved, Pioneer Venus matured into a single-opportunity mission with a multiprobe and an orbiter. In September 1972, NASA disbanded the Science
Steering Group and issued an Announcement of Opportunity for scientists to participate
in the multiprobe mission. Not until August 1973 did NASA issue an Announcement of
Opportunity for the orbiter. Over the ensuing months, the NASA Instrument Review
Committee evaluated proposals for orbiter scientific payloads, including Pettengill's radar
experiment, then presented its recommendations to NASA Headquarters in May 1974.
When NASA selected the final orbiter payloads on 4 June 1974, the radar experiment was
among them.

The radar was only one of 12 scientific instruments on the orbiter. In contrast to the
spaceborne radar initially developed at MIT, which was to consume no more than 100
watts and weigh less than 50 pounds, the Pioneer Venus radar required only 18 watts and
weighed 9.7 kilograms (21.3 pounds). "You could literally put the thing under one arm
and carry it," as Pettengill characterized it. Compared to other instruments on Pioneer
Venus, though, 9.7 kilograms was an appreciable load; it accounted for 22 percent of the
total weight (45 kilograms) of all 12 orbiter scientific instruments.[62]

Although the radar experiment, in Pettengill's words, "stood out like a sore thumb,"
NASA Headquarters wanted to see the surface features of Venus through its white, sulfu-
ric-acid clouds. The information was a vital part of planning for a future mission to Venus
to map the planet's surface, known eventually as Magellan. The only reason the radar
experiment stayed on Pioneer Venus, according to Pettengill, was that Advanced
Programs at NASA Headquarters wanted it, even though its inclusion made life "a little
uncomfortable for the other experiments."[63]

59.   Pettengill 28 September 1993.
60.   Janet G. Luhmann, Pollack, and Colin, "The Pioneer Mission to Venus," *Scientific American* 270 (April
1994): 90–97.
61.   Ford 3 October 1994.
62.   Pettengill 28 September 1993; Fimmel, Colin, and Burgess, pp. 18–21, 38 and 58.
63.   Pettengill 28 September 1993.

The key individual in NASA's Office of Advanced Programs who supported Pettengill and the radar experiment was Daniel H. Herman. Before joining NASA in 1970 as head of Advanced Programs in the Office of Lunar and Planetary Programs, Herman had worked at Northrup on the development of surveillance synthetic aperture radar (SAR) mappers for the Navy, specifically investigating the feasibility of transmitting reconnaissance data in real time. At NASA, his job was to develop new missions and to "sell" them through the NASA hierarchy and ultimately to the President and Congress. Danny Herman's job, then, was to sell the Pioneer Venus mission. In Pettengill's words, Herman was "an eminence grise" and "a supersalesman." As early as 1972, Danny Herman also began to put together and push the Magellan project.[64]

Unlike Magellan, Pioneer Venus strictly speaking did not have a synthetic aperture radar; instead, the radar altimeter had a mapping mode. The most valuable data returned from the Pioneer Venus radar experiment would be the extensive topographical information acquired by the altimeter. The mapping mode did generate crude, low resolution images of portions of the planet's surface.

Far more impressive were the images generated by synthetic aperture radars (SARs) mounted on aircraft and regularly utilized by geologists to study the geology and topography of Earth. The use of SARs in Earth geology was but one part of a long and complex history that stretched from the interpretation of aerial photographs to the emergence of remote sensing, an all-encompassing term which has came to involve the interpretation of infrared, ultraviolet, microwave, gamma ray, and x-ray images, as well as optical photographs.

Radar Geology

Radar geology, as the study of geologic surface features from radar maps has come to be called, had its roots in the military surveillance radar research of the 1950s. It began to find a home in NASA during the 1960s and found a common bond with planetary radar astronomy in the 1970s, thanks largely to Pioneer Venus and Viking. The trickle of astrogeologists converted to planetary radar images by Pioneer Venus and Viking swelled through purposeful steps taken in the planning of Magellan to bring together planetary geology and planetary radar investigators.

By World War I, aerial photography had become a key tool in gathering military intelligence. The scientific applications of photointerpretation grew after the war, particularly during the 1930s. Government agencies, such as the Agricultural Adjustment Administration, the Forestry Service, and the Tennessee Valley Authority, began to use aerial photographs, and the USGS entered the field of photogrammetry, the making of maps from photographs, with a series of geologic and topographic maps constructed from aerial photographs.[65]

After World War II, the military sponsored research on two types of Side-Looking Airborne Radar (SLAR) used in remote sensing and especially for surveillance. One type, known as real-aperture or incoherent radar, relied on transmission of a narrow beam to provide fine image resolutions in the direction parallel to the flight of the aircraft. The other type, known as synthetic aperture radar (SAR), relied on coherent data processing to synthesize a very large effective aperture in the direction of motion and, thereby, to provide a very narrow corresponding antenna beam. Continuously operating SARs achieve a surface resolution that is independent of wavelength and approximately equal to their along-orbit physical antenna dimension. Normally, in real-aperture radars,

---

64.   Pettengill 28 September 1993; Daniel H. Herman, telephone conversation, 20 May 1994.
65.   William A. Fischer, "History of Remote Sensing," in Robert G. Reeves, *Manual of Remote Sensing* (Falls Church, Virginia: American Society of Photogrammetry, 1975), [2 volumes] vol. 1, pp. 27-39.

resolution is better the shorter the wavelength. In order to achieve high resolution, SARs replace the need for a large aperture with a large amount of data processing.[66]

The military branches developed SARs in the 1940s and 1950s under highly classified conditions in corporate and university laboratories, such as those at the Goodyear Aircraft Corporation, the Philco Corporation, the University of Illinois Control Systems Laboratory, and the University of Michigan Willow Run Research Center. By the late 1950s, a number of experimental SAR systems emerged, such as the one built by Texas Instruments for the Army. In 1961, under Air Force contract, Goodyear built the first operational SAR system; it had a resolution of about 15 meters. Throughout the 1960s, Goodyear and other firms began to commercialize SAR applications.[67]

A series of symposia underwritten by the Office of Naval Research (ONR) and held at the University of Michigan, where a great deal of SLAR work took place under contract with the ONR, greatly stimulated and advanced radar geology.[68] The University of Michigan symposia series grew out of a study initially recommended by a subcommittee of the National Academy of Sciences (which soon formed the Committee on Remote Sensing of Environment) and the Geography Branch of the ONR. A group from the ONR and the National Academy of Sciences met in January 1961 to discuss the need for more advanced and efficient data acquisition techniques in the Earth sciences. Although University of Michigan faculty dominated the first symposium, held in February 1962, subsequent symposia participants reflected the spreading commercial importance of SAR systems in studying the Earth. By the third symposium, held in October 1964, the emphasis had shifted to remote sensing from weather and other satellites.[69]

During the third University of Michigan symposium, held in October 1964, R. F. Schmidt of the Avco Corporation, Cincinnati, presented a theoretical study on the feasibility of imaging Venus's surface with a radar. Schmidt failed, however, to address such practical questions as weight and power requirements.[70] Nonetheless, it was clear that those interested in remote sensing, and in radar imaging in particular, were open to the idea of imaging Venus from a spaceborne radar.

Meanwhile, commercial applications of SARs to geology and topography expanded. The successful radar mapping of Panama in 1967–1968 by Westinghouse in Project RAMP, considered to be one of the major achievements in radar geology, further stimulated commercial radar mapping. In late 1971, Westinghouse surveyed the entire country of Nicaragua, and that same year the Aero Service/Goodyear RADAM Project (RADar of the AMazon), initially intended to cover only 1.5 million square kilometers, eventually covered the entire country of Brazil, over 8.5 million square kilometers. RADAM was considered the most impressive radar mapping program ever conducted.[71]

66.   For a discussion of space SARs by one of its leading practitioners, see Charles Elachi, *Spaceborne Radar Remote Sensing: Applications and Techniques* (New York: IEEE Press, 1987).

67.   Fischer, pp. 42–43; Allen M. Feder, "Radar Geology, the Formative Years," *Geotimes* vol. 33, no. 11 (1988): 11–14. See also John J. Kovaly, *Synthetic Aperture Radar* (Dedham, MA: Artech House, Inc., 1976), Chapter One. I am grateful to Louis Brown for this last reference.

68.   Feder, p. 12.

69.   *Proceedings of the First Symposium on Remote Sensing of Environment* (Ann Arbor: University of Michigan Institute of Science and Technology, March 1962). Of the 72 participants, 37 of them, or 51%, were University of Michigan faculty. *Proceedings of the Second* (Ann Arbor: University of Michigan Institute of Science and Technology, February 1963); *Proceedings of the Third Symposium* (Ann Arbor: University of Michigan Institute of Science and Technology, October 1964); *Proceedings of the Fourth Symposium* (Ann Arbor: University of Michigan Institute of Science and Technology, June 1966).

70.   Peter C. Badgley, "The Applications of Remote Sensors in Planetary Exploration," in *Proceedings of the Third Symposium*, pp. 9–28; R. F. Schmidt, "Radar Mapping of Venus from an Orbiting Spacecraft," *ibid.*, pp. 51–61.

71.   Feder, p. 13; H. MacDonald, "Historical Sketch: Radar Geology," pp. 23–24 and 27–28 in *Radar Geology: An Assessment* Publication 80–61 (Pasadena: JPL, 1 September 1980). This was a report of the Radar Geology Workshop, held at Snowmass, Colorado, 16–20 July 1979.

Parallel with the development of SAR mapping of Earth was the rise of astrogeology within the USGS in the late 1950s in response to a shortage of funds and a surplus of geologists within the Survey. Following the discovery of an abundant supply of uranium ore in New Mexico, the USGS uranium project closed down in 1958. Eugene Shoemaker, a geologist who moved to the USGS Pacific Coast Regional Center at Menlo Park, California, following the closure of the uranium project, suggested lunar geologic mapping as one way to help alleviate the money and personnel problems.

Shoemaker, who did his dissertation on Meteor Crater, sold lunar geologic mapping to NASA, which in contrast to the USGS had funding but too few geologists. The result was the creation of the Astrogeologic Studies Group, USGS, Menlo Park, on 25 August 1960. Later, Shoemaker led a group of astrogeologists to a new location in Flagstaff, Arizona.[72] In 1963, geologist Peter C. Badgley came to NASA from the Colorado School of Mines. Badgley was interested in techniques for observing Earth from space, particularly to support the Apollo program. He let out contracts to firms, such as Westinghouse, and universities, especially the University of Michigan, to carry out radar geologic studies from aircraft. Moreover, Badgley continued to shift NASA money to the USGS to fund lunar and planetary geology.[73] Thus, the evolution of the NASA space program and the USGS astrogeology branch marched forward in tandem.

During the Apollo program, certain USGS astrogeologists began collaborating with radar astronomers Stan Zisk and Tommy Thompson. Among them were Henry John Moore II, Shoemaker's former field assistant and part of the Menlo Park Astrogeologic Studies Group, and Gerald G. Schaber, UCLA and USGS Flagstaff.[74] These early lunar efforts involved radar mapping and topographical data collected from ground-based radars, not abstract data on rms slope and dielectric constant. Schaber collaborated with Tyler on interpreting lunar bistatic radar results, which were expressed in abstract mathematical terms. Schaber admitted, "I never really did much with the interpretation of bistatic radar, because it is kind of a theoretical interpretation I don't really understand too much."[75]

The launch of SEASAT in the summer of 1978 began the era of satellite radar imagery. SEASAT demonstrated the feasibility of radar observations of Earth on a global basis, and initial examination of the SEASAT radar data indicated that one could fruitfully apply the data to a variety of problems in geology, agriculture, hydrology, and oceanography, as well as to planetary exploration.[76]

In order to assess the application of radar imaging to terrestrial geologic problems and to make recommendations to NASA, JPL sponsored the Radar Geology Workshop in Snowmass, Colorado, 16–20 July 1979, with funding from NASA. Among those on the organizing committee were Harold Masursky, USGS, and R. Stephen Saunders, JPL, who later played a role on Magellan. The workshop focused on radar observations of Earth, not the planets.[77]

Thus, by the launch of SEASAT in 1978, the year also of Pioneer Venus's launch, a good number of geologists were familiar with and could interpret radar images of Earth made from aircraft. But those geologists were more interested in terrestrial than extraterrestrial geology. On the other hand, through the pioneering efforts of Gene Shoemaker, the USGS Astrogeologic Studies Group already had embraced lunar radar geology. The

---

72.   Wilhelms, pp. 37–40, 43, 46, 48 and 77.
73.   Pamela E. Mack, *Viewing the Earth: The Social Construction of the Landsat Satellite System* (Cambridge: The MIT Press, 1990), pp. 46–49; MacDonald, pp. 26 and 28–29.
74.   Wilhelms, p. 47; Thompson 29 November 1994. See, for instance, Shapiro, Stanley H. Zisk, Rogers, Slade, and Thompson, "Lunar Topography: Global Determination by Radar," *Science* 17 (1972): 939–948.
75.   Schaber 27 June 1994.
76.   John P. Ford, "Seasat Orbital Radar Imagery for Geologic Mapping: Tennessee-Kentucky-Virginia," *American Association of Petroleum Geologists Bulletin* 64 (1980): 2064–2094; *Radar Geology: An Assessment*, p. 1.
77.   *Radar Geology: An Assessment*, passim.

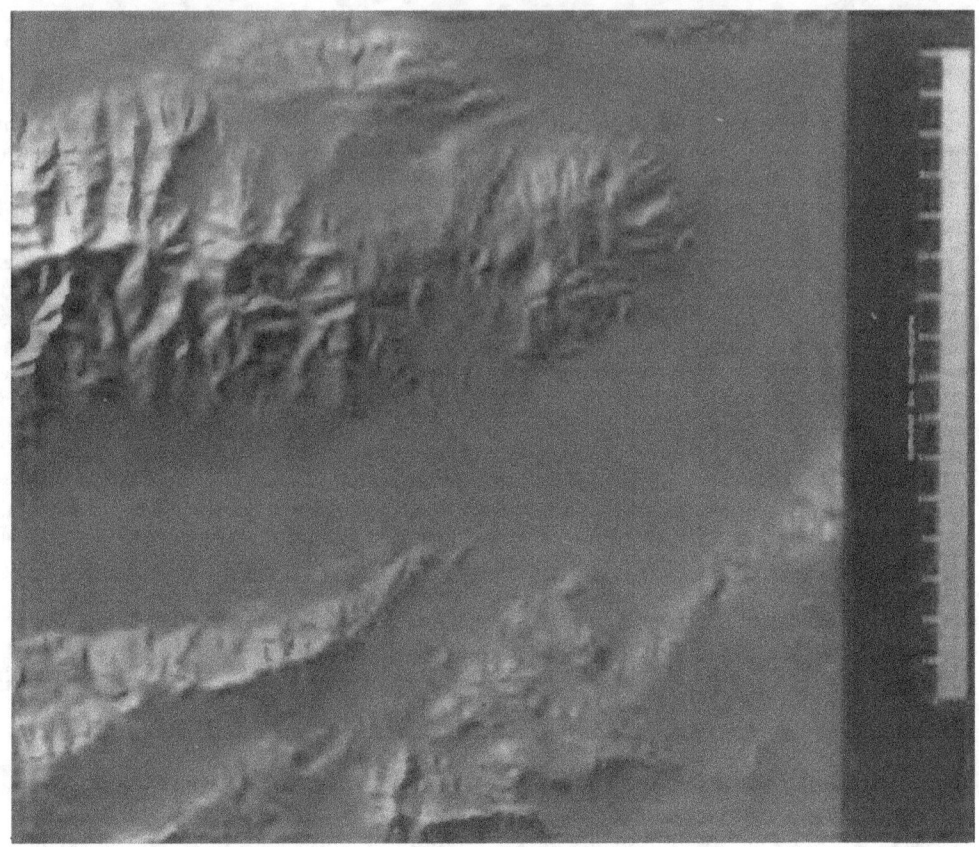

Figure 27

SEASAT image of Death Valley, Earth. The launch of SEASAT in 1978 began the era of satellite radar imagery. The resolution of images made by military surveillance satellites was much finer, however. Utilization of SEASAT technology was a basic strategy adopted by JPL in the planning of VOIR (later Magellan). (Courtesy of Jet Propulsion Laboratory, photo no. P-30224.)

potential for planetary geologists and planetary radar astronomers to work together already had been realized in the Apollo program through the work of Stan Zisk and Tommy Thompson. The NASA Pioneer Venus working committees brought together additional radar astronomers and geologists.

Once NASA decided the Pioneer Venus payloads and science experiments in June 1974, the space agency created the Orbiter Mission Operations Planning Committee. Among its members were USGS astrogeologist Hal Masursky and radar astronomer Gordon Pettengill. They also worked together closely in the Surface-Interior Working Group, one of the six mission Working Groups responsible for developing key scientific questions. Hal Masursky chaired that Working Group (Table 4).[78]

---

78.   Fimmel, Colin, and Burgess, pp. 22 and 218.

Table 4
Pioneer Venus Surface/Interior Working Group

Scientist	Institution
Harold Masursky	US Geological Survey
C.T. Russell	University of California, Los Angeles
Gordon H. Pettengill	Massachusetts Institute of Technology
William M. Kaula	University of California, Los Angeles
George E. McGill	Massachusetts Institute of Technology
Roger J. Phillips	Lunar and Planetary Institute
Irwin I. Shapiro	Massachusetts Institute of Technology

Pioneer Venus

Without the radar experiment, Pioneer Venus would not have brought together planetary geologists and radar astronomers. Attending meetings of the Working Groups, as well as all mission meetings, was vital to the survival of the radar on a project whose prime objectives were atmospheric. As Pettengill explained: "It was a very demanding project that had to be watched closely. I had to make sure that we did not lose radar capability. We were fighting with 11 other Principal Investigators on Pioneer Venus. It was very important that I never miss a meeting. If I missed one meeting, those guys might come to some decision that would compromise the experiment."

The Pioneer Venus atmospheric experiments competed with the radar experiment for spacecraft parameters. The atmospheric scientists wanted a different set of orbits and a different allocation of down link data bits. "It was a jungle out there!" Pettengill recalled. "You had to have a certain number of bits, or you could not do your work. If you turned your back, literally if you missed one meeting, they could make a decision to allocate 20 percent of that particular format to some experiment instead of only 10 percent. Then you have lost that 10 percent. In 1975, especially, all of this was coming together. I couldn't miss a meeting. It really was taking up my time."[79]

The data handling system on the orbiter integrated all analog and digital telemetry data into formats for transmission back to Earth. Telemetry storage, playback, and real-time rates varied. The orbiter had a total of 14 telemetry formats; some were used during periapsis, others during apoapsis. The radar was a heavy user of two formats designed for use at periapsis, and in fact it used more of those two formats than any other experiment.

NASA procured scientific instruments for Pioneer Venus in a variety of ways. Normally, the principal investigator was responsible for an instrument's design and construction. Either his own laboratory or a subcontractor built the instrument. NASA used a different procurement method for the Pioneer Venus radar. The project office at Ames Research Center built it for a radar team headed by Pettengill. Carl Keller, an Ames Research Center engineer, had overall decision-making responsibility, and the instrument prime contractor was the Hughes Aircraft Company Space and Communications Group, El Segundo, California, as a result of an open bid procurement. Pettengill characterized Carl Keller as an engineer from "the old school, a seat-of-the-pants, no nonsense teutonic. He would look at all the details. He was the right guy for the job. I enjoyed working with him. Not everybody did."[80]

---

79.   Pettengill 28 September 1993.
80.   Pettengill 28 September 1993; Fimmel, Colin, and Burgess, pp. 22, 41 and 43.

Both MIT's Center for Space Research, with which Pettengill was associated, and JPL competed for the Pioneer Venus radar contract. The rivalry between MIT and JPL was tense, "a real shootout" in Pettengill's words. At JPL, Walter Brown had been working on a Venus orbiter radar since the 1960s. His approach, however, differed considerably from that of MIT.

Walter Brown's radar proposal involved placing a 100-MHz (3-meter) transmitter on the Pioneer Venus orbiter, while Pettengill and the MIT Center for Space Research proposed a 1,757-MHz (17-cm) system. The MIT antenna was directive, so that when the spacecraft rotated, it took data for only a fraction of each 12-second rotation of the spacecraft.[81]

As Pettengill reflected: "Meanwhile, JPL thought they had the inside track. They were a NASA center, after all, and this was a NASA project. If I have a fault to lay on JPL, it is that they think that there is no place else in the world that does things as well as they do. They think they deserve the first cut of everything, because they are so much better than everybody else. They don't take kindly to new ideas that are not in-house; not invented here is very much a JPL hallmark. Irwin Shapiro has fought this on the Planetary Ephemeris Program. We fought it on radar work, and Stanford has fought it. It has been difficult over time. JPL is so institutionalized into thinking that no one else can do anything but them. It has been an uphill battle over the years. It has put grey hairs on Von Eshleman's head; it certainly put a few on mine."

JPL lost the radar battle. Their design would have bathed the whole spacecraft, even the solar panels, in radiation from the radar. The antenna extended all around the spacecraft, so that as the spacecraft rotated, the radar always was transmitting. To Walter Brown, that was the advantage, but it made the electronics engineers nervous.

In the end, neither MIT nor JPL built the Pioneer Venus radar, but it is typical of the kinds of fights for hardware contracts that mark NASA space missions. The winner of the contract was Hughes. Hughes devised a method by which the radar altimeter could image the planet's surface at low resolution with a small, 38-cm-diameter antenna. The electronics of the MIT design were clumsy, Pettengill admitted, whereas the Hughes proposal was "very clever and efficient."

"If we had done the experiment," he mused, "it probably would not have stayed in. I have to hand Hughes some credit for that. They really had a flash of insight into a clever way of instrumenting it....They had a good team, and so did we. The main reason Hughes won was that they were willing to take a loss." For Hughes, taking a loss on the Pioneer Venus radar contract was a gambit to gain leverage on the Magellan radar contract, which they ultimately won. "At the time," Pettengill recalled, awarding Hughes the radar mapper contract "hurt a bit. I was hoping to get the hardware here at the Center for Space Research."

In August 1974, Congress approved Pioneer Venus as a new start for fiscal 1975, and in November 1974, NASA made the final contract award to Hughes Aircraft Company. By 1975, only three years away from launch, Pettengill recalled, "it all came together. With the Hughes contract, we started cutting metal."[83] On 20 May 1978, the orbiter left Cape Kennedy, followed atop a second Atlas-Centaur rocket by the multiprobe on 8 August 1978. Both reached Venus in early December 1978.[84]

The radar was a complicated instrument capable of operating in one of two modes, altimeter or mapper. It was a 1,757-MHz (17-cm) radar with a peak output of 20 watts and utilized relatively long pulses to improve the signal-to-noise ratio. Such a radar could not

81.   Pettengill 28 September 1993; Memorandum, Brunk, 29 November 1966, NHOB.
82.   Pettengill 28 September 1993.
83.   Pettengill 28 September 1993.
84.   Fimmel, Colin, and Burgess, pp. 27 and 35.

have performed planetary radar astronomy experiments from Earth, but reducing the distance to the target made all the difference.

Shortly after its encounter with Venus, the orbiter began making altimeter measurements of surface relief. The altimeter measured the distance from the orbiter to the planet's surface. In order to ascertain the height and depth of surface features, that distance was subtracted from the spacecraft's orbital radius, that is, the distance between the spacecraft and the planet's center of mass. The Deep Space Network, while maintaining two-way communications with the spacecraft, generated radiometric data from which JPL accurately calculated its orbit, and the MIT group then used both the orbital and radar data to determine the radius of the planet at discrete positions on the surface.

In the radar mapper mode, the instrument compensated for the complex motion of the spacecraft. Because the orbiter spun on its own axis about five revolutions per minute, radar observations took place only periodically, about one second out of each 12-second spin of the orbiter. The radar mapper also automatically compensated for the Doppler shift caused by the motion of the orbiter relative to the planet.

The instrument took altimeter data, whenever the orbiter was below 4,700 km, and imaging data, when the orbiter was below 550 km, subject to competition with other experiments for the limited telemetry capacity. In order to minimize telemetry requirements, the orbiter processed the radar echoes on board the spacecraft. The radar mapper achieved its best resolution, a footprint 23 km long and 7 km wide, at periapsis. The radar data also provided information on surface roughness and electrical conductivity.[85]

The radar mapper's first sweeps showed a region of Venus previously unexplored by ground-based radar. With the exception of a deep trench near the equator, the surface of Venus appeared relatively flat, similar to the Earth's surface and quite different from the rough, cratered surfaces of Mars, Mercury, and the Moon. Pioneer Venus continued to complete one orbit per day, when on the 14th orbit, the radar mapper began to malfunction; data was lost. Scientists and engineers failed to find a remedy. Mission control turned off the radar for about two weeks around Christmas and the New Year. When mission control turned on the radar mapper, they discovered that it worked, though not quite normally.

The problem, eventually traced to a timing malfunction that resulted from a differential "aging" rate in two interconnected semiconductor devices, appeared when the instrument operated longer than ten hours. Pettengill, the experiment team leader, and project personnel decided to operate the radar mapper intermittently. After about 10 days of intermittent operation, the instrument started to function normally on 20 January 1979 (orbit 47).[86]

Somehow, though, the mission had to recover the lost data. Data recovery was not possible during the first extended mission (September 1979), because the Deep Space Network was handling communications with Pioneer 11 at Saturn, so it took place during the second extended mission, April–May 1980. The 10 other scientific instruments (the infrared experiment failed after a few months and never ran again) continued to transmit data back to Earth; the radar mapper, however, was turned off as planned after Orbit 834 on 19 March 1981.[87]

---

85.   Fimmel, Colin, and Burgess, pp. 58-59 and 113–115; Pettengill, D. F. Horwood, and Carl H. Keller, "Pioneer Venus Orbiter Radar Mapper: Design and Operation," *IEEE Transactions on Geoscience and Remote Sensing* GE-18 (1980): 28–32; Pettengill, Peter G. Ford, and Stewart Nozette, "Venus: Global Surface Radar reflectivity," *Science* 217 (1982): 640–642.

86.   Pettengill, Ford, Walter E. Brown, William M. Kaula, Carl H. Keller, Harold Masursky, and George E. McGill, "Pioneer Venus Radar Mapper Experiment," *Science* 203 (1979): 806–808; Colin, "The Pioneer Venus Program," *Journal of Geophysical Research* 85 (1980): 7588–7589; Fimmel, Colin, and Burgess, p. 107; Pettengill, Ford, Brown, Kaula, Masursky, Eric Eliason, and McGill, "Venus: Preliminary Topographic and Surface Imaging Results from the Pioneer Orbiter," *Science* 205 (1979): 91–93.

87.   Colin, pp. 7589 and 7590; Fimmel, Colin, and Burgess, p. 191.

The processing and interpretation of Pioneer Venus altimeter and mapper data sets by the MIT group again brought together planetary geologists and radar astronomers. Peter G. Ford, Pettengill's colleague in the MIT Department of Earth and Planetary Sciences, was a central player in the MIT effort. A native of Britain, Peter Ford initially came to MIT to work in VLBI (Very Long Baseline Interferometry) radio astronomy with Irwin Shapiro. From 1977 to 1985, he worked on various aspects of the Pioneer Venus orbiting radar experiments, including their geologic interpretation, although his training was in nuclear physics. The USGS processed some of the data to create a three-dimensional effect which graphically revealed depressions and mountains. Key among the planetary geologists were Hal Masursky and Gerald Schaber of the USGS, Flagstaff, and George E. McGill, University of Massachusetts at Amherst.

The collaboration of radar astronomy and planetary geology resulted in many important discoveries about the surface of Venus, although preliminary analysis showed that much more could be learned about the planet's geological history. The altimeter data was used to create a number of maps, including a topographic contour map, a shaded relief map, and a map showing relative degrees of surface roughness. The altimeter and radar mapper data sets were assembled and placed in position by computer; however, variations from orbit to orbit were edited by hand then smoothed out by computer.

In preparation for the mission, a preliminary map was compiled from ground-based images and used by mission operations for planning. For this map, Goldstone radar images were computer mosaicked, and images obtained at Arecibo were mosaicked from photographic copy. The scale of this map was 1:50,000,000. Once the Pioneer Venus data were in hand, the map was updated to combine the spacecraft and ground-based data.

The radar altimeter yielded a topographic map covering 93 percent of the Venus globe, with a linear surface resolution of better than 150 km. Vertical measurement accuracy exceeded 200 meters. Relief was expressed as a center-of-mass-to-surface radius. Extremes went from a low of 6,049 km to a high of 6,062 km. Despite these impressive extremes of surface height and depth, the Pioneer Venus data confirmed and greatly expanded previous Earth-based observations on the global smoothness of Venus relative to the Moon, Mars, and Earth. Only about five percent of the observed surface was elevated more than two km above the mean radius, $6{,}051.5 \pm 0.1$ km.

Radar astronomer Gordon Pettengill processed and interpreted Pioneer Venus altimeter and mapper data sets. Don Campbell at Arecibo, and Dick Goldstein and Howard C. Rumsey, Jr., at JPL, supplied ground-based radar images and digital tapes, many before publication. The Arecibo and JPL radar images were compiled into a mosaic for the Pioneer Venus Planning Chart that was used in mission operations. Their high-resolution, Earth-based radar-imaging data also was essential for the interpretation of the spacecraft images and altimetry data. Thus, ground-based radar astronomers were brought into the Pioneer Venus project, and association with the project facilitated radar astronomers' access to the Goldstone radar.

The radar brightness and elevation extremes dominated the imaging and topographic maps of the highlands province, which included Ishtar Terra, Aphrodite Terra, and Beta Regio. The two highland regions, Ishtar and Aphrodite terrae, resembled terrestrial "continents" because they were high and had areas comparable to continents on Earth. Ishtar and Aphrodite appeared to be the size of continents, roughly equivalent to Australia and Africa, respectively. Beta, a much smaller feature initially detected with ground-based radar, appeared to differ from Ishtar and Aphrodite in roughness characteristics and possibly in age and chemical composition. Ishtar Terra was the most elevated region found on Venus. It included three topographic elements: Lakshmi Planum, a western plateau area; Maxwell Montes, the central mountainous area previously studied with Earth-based radars; and a complex eastern region. The highest point found on Venus was the summit of Maxwell Montes. Standing 11.1 km above the planet's average radius (in Earth terms, above sea level), Maxwell Montes was higher than Mount Everest, which reaches 8.8 km above sea level. The lowest point found on Venus was a rift valley or trench named Diana Chasma.[88]

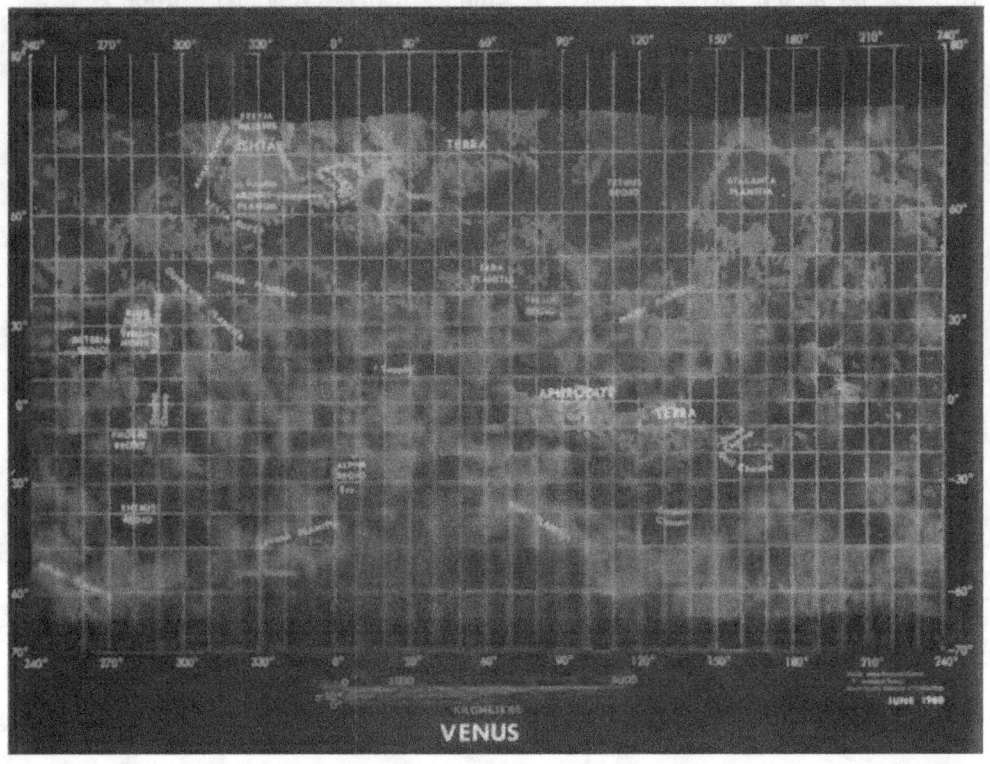

Figure 28

Pioneer Venus map of Venus, 1980, showing Alpha Regio and Maxwell Montes, along the planet's meridian, and Beta Regio at longitude 280°. Diana Chasma is at longitude 160°. Compare this map with the Venus mosaic made from Arecibo Observatory radar observations (Fig. 30). (Courtesy of Jet Propulsion Laboratory, photo no. P45744.)

The planetary radar and geology collaboration yielded a host of new topographical names. In order to systematically standardize the names of Venus surface features, as well as those discovered earlier on Mars and the Moon, on an international level, the International Astronomical Union (IAU) created the Working Group for Planetary System Nomenclature (WGPSN) during its 15th General Assembly at Sydney, 21–30 August 1973. The IAU established the WGPSN because of the recent rapid advance in knowledge of the topography and surfaces of planetary bodies, as well as the necessity of coordinating the approved systems of nomenclature among the different planets and their satellites.

---

88.   Fimmel, Colin, and Burgess, p. 154; Masursky, Eliason, Ford, McGill, Pettengill, Gerald G. Schaber, and Schubert, "Pioneer Venus Radar Results," *Journal of Geophysical Research* 85 (1980): 8232–8260; Pettengill, Eliason, Ford, George B. Loriot, Masursky, and McGill, "Pioneer Venus Radar Results: Altimetry and Surface Properties," *Journal of Geophysical Research* 85 (1980): 8261–8270; *V-Gram* no. 10 (January 1987): 20.

Unlike most other IAU working groups, the WGPSN did not report through any commission or group of commissions, but was responsible to only the IAU Executive Committee. The WGPSN was charged with formulating and coordinating all topographic nomenclature on the planetary bodies of the solar system and had certain powers of action in the interval between General Assemblies. Radar astronomer Gordon Pettengill was a member of the WGPSN. The Task Group for Venus Nomenclature, responsible for compiling the detailed material presented to the WGPSN, included Gordon Pettengill, chair, JPL radar astronomer Dick Goldstein, USGS geologist Hal Masursky, and the Soviet scientist M. Ya. Marov.

Although the first meeting of the WGPSN, held in Ottawa, 27–28 June 1974, did not concern itself with the naming of surface features on Venus, at the second meeting, held in Moscow, 14–18 July 1975, the WGPSN named three valleys on Mercury Arecibo, Goldstone, and Haystack after the radar observatories and established two themes for naming Venus features. The first theme was the "feminine mystique long associated with Venus." Hence, for example, the continent-sized features Ishtar and Aphrodite were named for the Babylonian and Greek goddesses of love, respectively.

The second theme arose from the "extensive and opaque cloud cover which surrounds the planetary sphere" which "requires the use of radio and other techniques in order to study and map the surface." Therefore, the WGPSN proposed "to assign the names of deceased radio, radar and space scientists to topographic features." One exception, Alpha, was admitted. Alpha was one of the first Cytherean features to be observed "and which has served to help define the origin of the official IAU system of longitude for the planet." During subsequent meetings of the WGPSN, held in Grenoble, 30–31 August 1976; Washington, 1–2 June 1977; Innsbruck, 2 June 1978; and Montreal, 13–15 August 1979, the WGPSN approved not only Alpha, but Beta and Maxwell as well.[89] Thus, the feature names first given by ground-based radar astronomers were fixed on the map of Venus.

Pioneer Venus awakened more planetary geologists to the value of radar data, especially radar images. Pioneer Venus also was a new taste of Big Science that would lead to the Magellan mission. In turn, Magellan culminated the linking of planetary geology with radar astronomy and further blurred the distinction made earlier in the history of planetary radar astronomy between ground-based radar and space exploration.

---

89. *Transactions of the International Astronomical Union* 17A (1979): 113–114; "Working Group for Planetary System Nomenclature," *Ibid.* 16B (1977): 321–369; "Working Group for Planetary System Nomenclature," *Ibid.* 17B (1980): 285–304.

Chapter Seven

Magellan

Magellan culminated the shift of radar astronomy toward planetary geology kindled by Apollo and fostered by Viking and Pioneer Venus with the creation of workshops and microsymposia. The workshops attempted to bridge the gap between radar and geologic knowledge among practitioners, while the microsymposia provided annual opportunities for U.S. and Soviet geology and radar scientists interested in Venus to exchange research results. This shifting of the planetary radar paradigm toward geology also manifested itself in articles co-authored with planetary geologists, publication in new journals, especially the *Journal of Geophysical Research*, and attendance at American Geophysical Union meetings.

Furthermore, the close relationship between NASA missions and ground-based planetary radar astronomy that had developed at Haystack, Arecibo, and Goldstone since 1970 continued with Magellan. The Arecibo and Goldstone radars observed Venus throughout the two decades spanned by Pioneer Venus and Magellan, and their data contributed to the success of those missions. In addition, the range-Doppler images created from that data also drew geologists to planetary radar astronomy.

Magellan, like Pioneer Venus, was not ground-based planetary radar astronomy; it was space exploration. By carrying out imaging from a spacecraft, radar astronomer Gordon Pettengill had erased that distinction. That distinction no longer seemed to describe the field, as Len Tyler and Dick Simpson joined the Magellan radar team. Tyler and Simpson had not abandoned bistatic radar and radio occultation experiments; they had simply added Magellan radar science to their wide range of research interests.

Unlike the Pioneer Venus mission, or the Goldstone and Arecibo facilities, Magellan was not a case of radar astronomy "Little Science" piggybacking onto a Big Science facility. Magellan *was* Big Science. Moreover, its single scientific instrument was a radar. The Smithsonian push to have Congress fund the NEROC 440-ft (134-meter) dish never reached the floor. With Magellan, then, Congress considered for the first time underwriting construction of a facility dedicated primarily to planetary radar astronomy, albeit one whose lifetime was rather limited. Magellan illustrates the range of factors that influence the scientific conduct and outcome of a Big Science project. The change of administration in 1980, Cold War politics, and the Challenger accident, as well as ongoing and changing budgetary and technological constraints largely shaped the scale and scope of the Magellan mission and its science.

Allez VOIR!

As a mission concept, Magellan began in 1972, when Danny Herman, the head of NASA Advanced Programs, convened an informal meeting of scientists, including Gordon Pettengill, NASA engineers, and representatives of several key aerospace companies at the

Denver division of Martin Marietta Aerospace, to discuss putting a synthetic aperture radar on a spacecraft to Venus.[1]

Subsequently, two NASA laboratories, Ames Research Center and JPL, organized study groups and began planning the mission and appropriate spacecraft parameters. At Ames, Byron Swenson and John S. McKay put together a study group that worked closely with Martin Marietta Aerospace in planning a Venus SAR mapping mission. They initially proposed a variation on Pioneer Venus with an elliptic orbit. During the period 1972 through 1974, Ames Research Center, Martin Marietta Aerospace, and the Environmental Research Institute of Michigan (ERIM), which had been involved in the development of airborne SAR systems for the military as early as the 1950s, carried out a preliminary evaluation of data handling problems and techniques. The 1974 joint report of Martin Marietta Aerospace and the ERIM defined the project's science requirements and argued in favor of a circular orbit.[2]

At the same time, a similar study was underway at JPL under Louis D. Friedman. In order to distinguish their Venus project from that of the Ames group, Friedman and Al Laderman named the JPL project the Venus Orbiting Imaging Radar (VOIR). Laderman had played a key role in the development of the SEASAT SAR. He and Friedman intended the acronym to connote the sense of the French verb "voir," meaning to see. VOIR was going to "see" the (optically) unseen surface of Venus. The JPL group included science, mission, and radar people. R. Stephen Saunders was the principal study scientist. Later, he became Magellan Project Scientist, as well as an investigator in the radar group. Saunders had served on the Viking Mission to Mars, before carrying out NASA-funded research in planetary geology and participating on the Shuttle Imaging Radar (SIR-A) project.

The JPL group decided, mainly on the urging of Friedman, to use a circular orbit. A circular orbit would simplify the radar imaging process. The radar system always dealt with the same parameters, because it was always at the same height above the surface. Friedman felt that simplifying the radar versus the increased propulsion required to achieve a circular orbit was a good trade-off. Although the added propulsion needed to achieve a circular orbit would increase the overall cost of the mission, at least it was understood. The synthetic aperture radar was a new technology; an elliptical orbit presented a host of radar and data processing problems. Jim Rose's study group, charged with planning the spacecraft, proposed a vehicle based on the Mariner system. The radar study group specified a radar system compatible with the 3-meter (10-ft) antennas built for the Pioneer missions to the outer solar system. Already, the goal was to economize by using existing technology.[3]

Many of the initial assumptions concerning look angle, number of looks, various resolution assumptions, number of bits, and other radar system parameters came under criticism by scientists familiar with optical data, but nonetheless responsible for interpreting the radar data. Those criticisms led JPL to redesign away from the Mariner approach and to exploit internal strengths in synthetic aperture radars gained through the SEASAT program. Ultimately, the SEASAT experience gave JPL an edge over its competitors.

---

1.   Herman, telephone conversation, 20 May 1994.

2.   Memorandum, Louis D. Friedman to J. C. Beckman, "VOIR, Archeology, 10/79," Box 14 [hereafter Friedman-Beckman Memorandum]; VOIR Historical Perspective, "VOIR, VOIR Mission, Briefing to NASA Code S, 5/78," Box 8; "VOIR, A Study of an Orbital Radar Mapping Mission to Venus, Vol. 1, 9/73," Box 10; "VOIR, Report, A Study of an Orbital Radar Mapping Mission to Venus, Vol. 2, 9/73," Box 14; "VOIR, Report, A Study of an Orbital Radar Mapping Mission to Venus, Vol. 3, 9/73," Box 14; and "VOIR, (NASA) Correspondence VOIR Mission Study Books, 10/77," Box 10, JPLMM.

3.   Friedman-Beckman Memorandum; VOIR Historical Perspective; "VOIR, Report, A Study of an Orbital Radar Mapping Mission to Venus, Vol. 2, 9/73," and "VOIR, Report, A Study of an Orbital Radar Mapping Mission to Venus, Vol. 3, 9/73," Box 14, JPLMM; V-Gram no. 9 (October 1986): 3; Campbell 8 December 1993.

SEASAT was an Earth-orbiting satellite equipped with a SAR and designed for oceanographic research. In its 1977 mission and systems study, JPL proposed the SEASAT SAR as the potential design base for the VOIR. JPL argued that SEASAT already had converted the concept of a spacecraft imaging radar into a reality. SEASAT used much of the conceptual and system design contained in the original JPL VOIR study, while later VOIR studies borrowed heavily from the SEASAT experience. JPL also contributed SEASAT staff. John H. Gerpheide, SEASAT satellite system manager, became VOIR/Magellan project manager. Anthony J. Spear, sensor manager on SEASAT, became VOIR/Magellan deputy manager.[4]

When the Science Working Group convened at NASA Headquarters in November 1977, NASA already had selected the JPL study. NASA charged the Science Working Group with defining the major scientific objectives and rationale for a Venus orbiter equipped with a radar imager, as well as defining other experiments and defining the radar-imaging requirements of the mission, including coverage, resolution, operating wavelength, telemetry data rate, and data processing. The Science Working Group considered the merit of global coverage at medium resolution and imaging selected areas at high resolution.

The composition of the VOIR Science Working Group drew heavily on Pioneer Venus alumni and from both the planetary radar and geology communities (Table 5). The planetary radar members were Don Campbell (Arecibo), Dick Goldstein (JPL), and Gordon Pettengill (MIT), who chaired the Group. Harold Masursky and Gerald Schaber, both astrogeologists from the USGS, Flagstaff, and both participants in Pioneer Venus, also served on the Science Working Group.

Table 5
VOIR (Magellan) Science Working Group

Scientist	Institution
Gordon H. Pettengill, Chair	MIT
Harry S. Stewart, Executive Secretary	JPL
Donald B. Campbell	NAIC, Cornell
Richard M. Goldstein	JPL
William M. Kaula	UCLA
Michael C. Malin	JPL
Harold Masursky	US Geological Survey
Norman Ness	Goddard Space Flight Center
William L. Quaide, Program Scientist	NASA Headquarters
R. Keith Raney	Canada Center for Remote Sensing
William B. Rossow	Princeton University
R. Stephen Saunders, Project Scientist	JPL
Gerald G. Schaber	US Geological Service
Sean C. Solomon	MIT
David H. Staelin	MIT
A. Ian Stewart	University of Colorado
Robert Strom	University of Arizona
G. Leonard Tyler	Stanford University

The Science Working Group thus became a forum for reinforcing bridges between planetary radar and geology scientists. The geologists were "very helpful in teaching us radar people what it was that turned them on, as it were, while we were helpful to them in terms of optimizing the operation of the radar, so as to provide them with what they wanted," Gordon Pettengill explained. "This interaction shaped the specifications that turned

4.   Friedman-Beckman Memorandum; VOIR Historical Perspective; "VOIR, (NASA) Correspondence VOIR Mission Study Books, 10/77," Box 10, JPLMM; Robert C. Beal, *Venus Orbiter Imaging Radar FY77 Study Report Radar Studies,* Report 660-60 (Pasadena: JPL, 2 May 1977), pp. 5-1 through 5-18; *V-Gram* no. 9 (October 1986): 2. On VOIR's SEASAT legacy, see also Murray, pp. 127–129.

into the VOIR and later the Magellan programs. The process is ongoing. It goes on even today."[5]

NASA was particularly mindful that the Science Working Group "take full account of the anticipated capabilities of Earth-based radar systems as well as the results expected from the Pioneer Venus experiments."[6] The Committee on Planetary and Lunar Exploration (COMPLEX) of the Space Science Board, and in particular its chairman, Caltech professor of geology and geophysics Gerald J. Wasserburg, was behind that request. The request was logical, Herman judged in retrospect. Given the high cost of VOIR, why should NASA and the Congress commit a large sum of money to a space mission, when Arecibo could acquire the same imaging data for far less money? Having the 1977 VOIR Science Working Group assess the science yield from a large ground-based radar telescope, like Arecibo, compared to the science yield from a spacecraft was, in Herman's words, "very necessary to yield off the devil's advocate question."[7]

Herman already had emphasized to the initial JPL study group the need to consider the capabilities of Arecibo for undertaking ground-based radar observations of Venus. The chief weakness in the development of the Venus radar orbiter concept, he explained, was the belief held by some scientists that upgraded ground facilities could provide data that was almost as good at a far lower cost.

By 1977, range-Doppler imaging of Venus at Goldstone and Arecibo had advanced considerably thanks to the refinement of interferometry techniques and the attainment of finer image resolution. At Goldstone, for example, Dick Goldstein used a radar interferometer, the 400-kilowatt Mars Station linked to a 26-meter Goldstone DSN antenna (DSS-13, known also as the Venus site) located about 22 km to the southeast, to observe and image Venus in 1972 for the first time and subsequently during the winter of 1973–1974 and the summer and fall of 1975. Over that period, image resolution fell from 15 to 10 km, although in some instances Goldstein realized resolutions as low as 5 to 9 km in the East-West direction and 7 to 10.8 km North to South. In 1977, Ray Jurgens and Dick Goldstein organized a three-station interferometer; the Mars Station transmitted, then it and two 26-meter Goldstone DSN antennas (DSS-12 and DSS-13, the Echo and Venus sites, respectively) received. The three-station data yielded image resolutions of 10 and even down to 8 km.[8]

Planetary scientists R. Stephen Saunders and Michael C. Malin of the JPL Planetology and Oceanography Section studied the Goldstone Venus images and concluded that they revealed a complex and varied terrain. They found degraded impact craters and evidence for volcanism. In these radar images, Beta now appeared to be a 700-km-diameter region elevated a maximum of about 10 km relative to its surroundings with a 60-by-90-km-wide depression at its summit. Saunders and Malin tentatively identified Beta Regio as a shield volcano.[9]

Meanwhile at Arecibo, the radar upgrade from UHF to S-band increased the resolution of Venus radar images abundantly. In 1969, with the old 430-MHz radar operating in an interferometric mode, Campbell, Ray Jurgens, and Rolf Dyce achieved a resolution of

5.   Pettengill 29 September 1993.
6.   "VOIR, (NASA) Correspondence, VOIR Mission Study Books, 11/78," Box 13, JPLMM.
7.   Herman, telephone conversation, 20 May 1994.
8.   Herman, telephone conversation, 20 May 1994; Rumsey, Morris, R. Green, and Goldstein, "A Radar Brightness and Altitude Image of a Portion of Venus," *Icarus* 23 (1974): 1–7; Goldstein, R. Green, and Rumsey, "Venus Radar Images," *Journal of Geophysical Research* vol. 81, no. 26 (10 September 1976): 4807–4817; Goldstein, R. Green, and Rumsey, "Venus Radar Brightness and Altitude Images," *Icarus* 36 (1978): 334–352; Jurgens, Goldstein, Rumsey, and R. Green, "Images of Venus by Three-Station Radar Interferometry—1977 Results," *Journal of Geophysical Research* vol. 85, no. A13 (30 December 1980): 8282–8294.
9.   Saunders and Michael C. Malin, "Surface of Venus: Evidence of Diverse Landforms from Radar Observations," *Science* 196 (1977): 987-990; ibid., "Geologic Interpretation of New Observations of the Surface of Venus," *Geophysical Research Letters* 4 (1977): 547–550.

only 300 km. An improved line feed brought Venus image resolution down to about 100 km in 1972, the last Venus observations before the S-band upgrade.[10]

Concomitant with the S-band radar upgrade, the NAIC constructed a second antenna, a 30-meter equatorially mounted reflector, at a site about 11 km to the north-northeast of the main 1,000-ft (305-meter) dish. Data taken by Campbell and Dyce in association with Gordon Pettengill during the Venus inferior conjunction of late August and early September 1975 yielded images with surface resolutions approximating those of Goldstone, between 10 and 20 km. Especially interesting was a detailed view of Maxwell.[11]

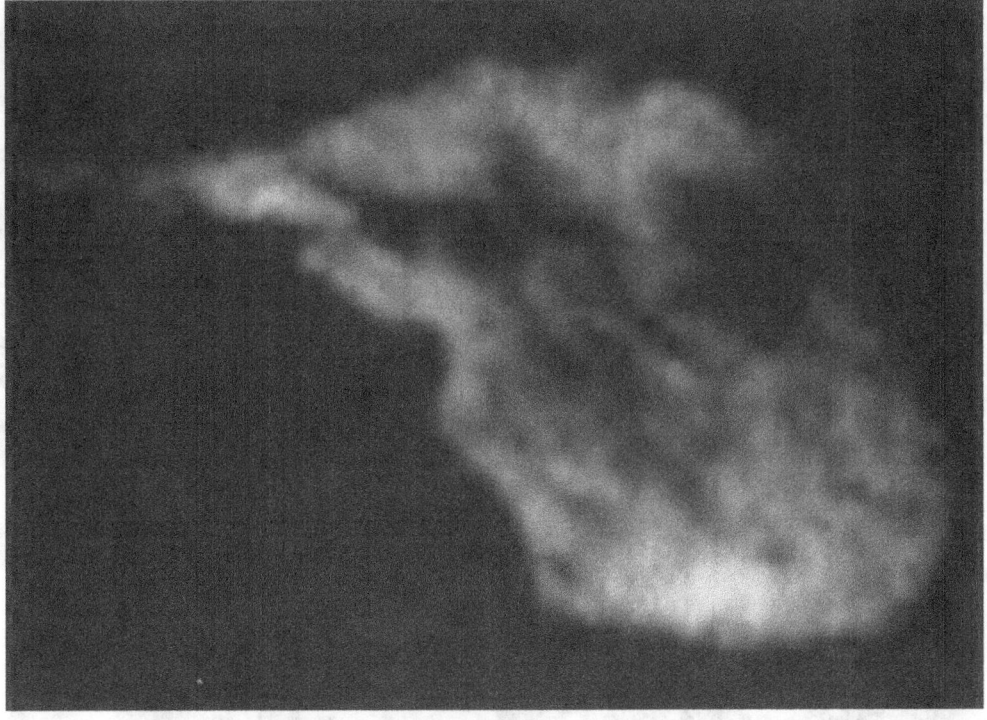

Figure 29

Radar image of Maxwell Montes made at Arecibo. Surface resolution is about 10 kilometers. Maxwell, which measures about 750 kilometers from north to south, includes the planet's highest elevation: 11 kilometers above the planetary mean. (Courtesy of National Astronomy and Ionosphere Center, which is operated by Cornell University under contract with the National Science Foundation.)

Thanks to hardware improvements, Don Campbell and Barbara Ann Burns, his graduate student, increased the resolution of Venus images to five km during the 1977 inferior conjunction. For her doctoral dissertation, Burns used these radar images to study cratering on the planet. She and Campbell identified over 30 circular features in the images and tentatively classified them as craters, but the level of resolution did not permit them to ascertain whether their origin was volcanic or impact.[12] Also, in conjunction with USGS

---

10.   Campbell, Jurgens, Dyce, Harris, and Pettengill, "Radar Interferometric Observations of Venus at 70-Centimeter Wavelength," *Science* 170 (1970): 1090-1092; NAIC QR Q2/1972, pp. 3–4, and Q3/1972, pp. 3–4.

11.   Campbell, Dyce, and Pettengill, "New Radar Image of Venus," *Science* 193 (1976): 1123–1124.

12.   Campbell and Barbara Ann Burns, "Earth-based Radar Imagery of Venus," *Journal of Geophysical Research* vol. 85, no. A13 (30 December 1980): 8271–8281; Burns, "Cratering Analysis of the Surface of Venus as Mapped by 12.6-cm Radar," Ph.D. diss., Cornell University, January 1982.

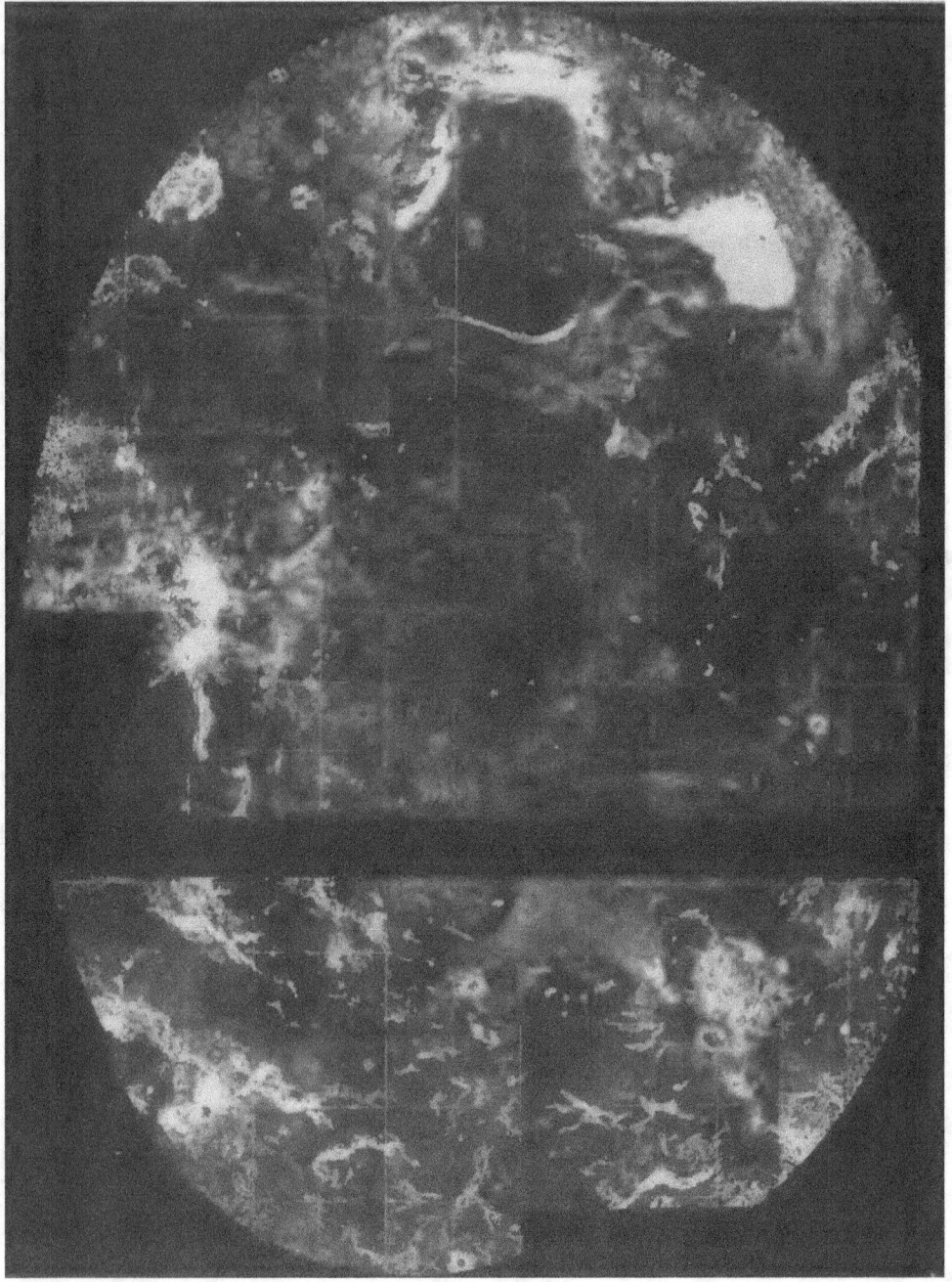

Figure 30

Large mosaic of Venus made from Arecibo radar observations. The image is centered on longitude 320° (see Fig. 28). Maxwell Montes is the large white area in the upper right corner. Left of center is Beta Regio. (Courtesy of National Astronomy and Ionosphere Center, which is operated by Cornell University under contract with the National Science Foundation.)

geologist Hal Masursky, Don Campbell and Gordon Pettengill studied images of Alpha, Beta, and Maxwell made from combined 1975 and 1977 Arecibo observations.[13]

As Campbell and fellow radar astronomers using the upgraded Arecibo telescope achieved resolutions as fine as 5 kilometers on Venus during the 1977 inferior conjunction, the high resolution invited comparison with potential space-based radars. In order to evaluate the capabilities of ground-based radars versus orbiting radars, the JPL study group brought in Thomas Thompson. Thompson had conducted lunar radar work at both Arecibo and Haystack for the NASA Apollo program. As a result of Thompson's advice, as well as the counsel of Danny Herman, Friedman's study group framed a radar orbiter mission that complemented, rather than competed with, ground-based radar observations of Venus.[14]

Thompson judged that the best ground-based facility would be the upgraded Arecibo telescope. He concluded that the Earth-based radar was a very powerful tool for mapping the surface features of Venus. "We should encourage these efforts with great vigor," he wrote. "It seems certain that the Earth-based mapping will show many features that should be mapped in greater detail with the spacecraft. Also, the spacecraft will be needed to map the hemisphere of Venus which is not pointed toward Earth at each inferior conjunction."[15]

The combined revolutions of Venus and Earth around the Sun lead to an interval between inferior conjunctions (known as the synodic period) that nearly matches the spin rate of Venus about its axis, so that Venus presents almost the same hemisphere to Earth observers at inferior conjunction, the only moment when radar astronomers have sufficient signal-to-noise ratio to image the planet.[16] The major argument in favor of a spacecraft imaging mission to Venus was the inability of ground-based radars to image the planet's hidden hemisphere. A major upgrade of the Arecibo (or Goldstone) radar could have enabled it to observe and image Venus at orbital points before and after inferior conjunction. Such an upgrade would have cost less than the Magellan mission, and the improved radar would have been able to carry out radar research on a variety of other solar system targets.

In 1977, NASA asked the VOIR Science Working Group to compare the costs of acquiring the data from a space-based SAR versus from a ground-based radar telescope, like Arecibo. "We knew that NASA did not want to hear that it would be cheaper, even though if you had taken what it actually cost to do Magellan and put it into a ground-based facility, you could have had one beautiful ground-based facility, and you could have endowed a fund to run it for years, forever probably, if you invested the money properly," Gordon Pettengill explained.

Moreover, Pettengill argued, "As an investment in basic research, basic astronomy, a ground-based observatory would be a much wiser investment than sending Magellan out there. But that is not how things work. The money is available for the Space Station, but not available for any ground-based system that perhaps would do some of the same things."[17]

Pettengill assigned the tasks of comparing altimetry and radar imaging capabilities of ground-based versus space-based radars to Don Campbell and Dick Goldstein. They

13.   Pettengill, Campbell, and Masursky, "The Surface of Venus," *Scientific American* 243 (August 1980): 54–65.

14.   Thompson 29 November 1994; Friedman-Beckman Memorandum.

15.   *Venus Orbiting Imaging Radar Study Team Report (Preliminary Draft* (Pasadena: JPL, 31 August 1972), pp. 22–28, and Friedman and J. R. Rose, *Final Report of a Venus Orbital Imaging Radar (VOIR) Study* 760–89 (Pasadena: JPL, 30 November 1973), Pettengill materials.

16.   For an explanation of the relationship between Venus's spin and rotational rates, see Goldreich and Peale, "The Dynamics of Planetary Rotations," *Annual Review of Astronomy and Astrophysics* 6 (1968): 287–320.

17.   Pettengill 28 September 1993.

completed separate reports, with Goldstein considering altimetry and Campbell appraising imaging capabilities. In each case, they compared a feasible radar design (an array located probably in Puerto Rico to have the planet nearly over head) with the current VOIR design requirements and judged whether the radar could achieve the geologic objectives of the Venus mission as well or better than the VOIR design.

Campbell and Goldstein concluded that the radar array could do the VOIR science (almost). The ground-based radar would not observe Venus at the same angles of incidence as VOIR, yet, because it would be able to observe Venus at a distance of 1.5 astronomical units, it could see the side of Venus hidden at inferior conjunction. The 100-meter resolution attainable from Earth was the same as that set for the VOIR mission. Moreover, the radar array could do the job for less. Pettengill decided to not include their conclusions in the Working Group report "for political reasons." He believed that NASA had no interest in the project, and that the conclusions might be embarrassing.[18]

Defining the VOIR

In 1978, VOIR began to come together. The concept and preliminary design studies completed, the time had come to begin implementation studies. Radar development began in 1978 and took place in two stages, called Phase A, lasting from June through August 1978, and Phase B, October 1979 through June 1980. During Phase A, JPL received three proposals to study the VOIR SAR and selected one study contractor, Goodyear Aerospace Corporation. During Phase B, JPL received three proposals and selected two study contractors, Goodyear Aerospace Corporation and Hughes Aircraft Company. Participation in Phase B studies was important for those firms wishing to build the VOIR radar; implementation phase proposals were accepted from only those companies participating in Phase B.[19]

The mission, its spacecraft and radar systems, and its science experiments underwent many revisions, and many of the risks foreseen in 1978 materialized before Magellan left Earth. As planned in 1978, the Space Shuttle would launch the VOIR spacecraft during the period May–June 1983. VOIR would arrive at Venus in November 1983 and spend five months in orbit, reduced from the earlier concept of a 19-month mission. JPL considered the possibility of the launch being delayed until 1984. Such a delay would cause an overlap with Galileo, complicate scheduling the Deep Space Network, and raise costs. A delayed launch also would provide an opportunity for the Soviet Union to obtain Venus SAR images before the United States, thereby making VOIR radar results less interesting, if not inconsequential.

The 1978 version of VOIR also exploited the availability of extant technology. In order to economize and facilitate selling and funding the project, VOIR would use components with proven performance records from other missions. For instance, from Mariner 10 VOIR borrowed its solar array and louvers, from Voyager its spacecraft electronics, from Pioneer Venus its radar altimeter, and from SEASAT its synthetic aperture radar.[20]

JPL hoped to make VOIR an in-house project. NASA had other ideas. In 1979, NASA stipulated that JPL contract out both the radar and the spacecraft to industry. NASA also

18.   Pettengill 3 October 1993.
19.   "VOIR, Venus Orbital Imaging Cost Review, 6/78," Box 5; "VOIR, Venus Orbiting Imaging Radar Review, 4/80," Box 10; and "VOIR, VOIR 88, Viewgraph Presentation to NASA Administrator, 11/87," Box 10, JPLMM.
20.   "VOIR, Status Briefing to Committee on Planetary and Lunar Exploration, NASA Headquarters, 6/78," and "VOIR, VOIR 84, Delayed Launch Option, 6/78," Box 3, JPLMM.

specified that the radar would have a single individual, from NASA, shoulder the responsibility of making it work. The NASA Headquarters decision had an immediate impact on VOIR design. The JPL in-house effort, which came to an end in February 1980, had concentrated on using SEASAT technology. Now an industrial design would serve. In order to economize, JPL had proposed using the Galileo circular 4.8-meter antenna for both communications and the SAR. The weight of the Galileo antenna was significantly less than that of a competing antenna design. Instead, JPL now had to undertake a study of the competing and differing antenna patterns proposed by the contractors Goodyear and Hughes.[21]

In planning the VOIR radar mapper, the Science Working Group took into account the resolution of the images sent back by Mariner 9. Those images had revealed for the first time the diversity of Martian geologic structures, including young volcanoes, liquid cut channels, and large canyons of possible tectonic origin and had led to a fundamentally new understanding of the nature of Mars. The VOIR radar mapper had to have comparable or better resolution than Mariner 9. Steve Saunders, project scientist, and Gerry Schubert, a geophysicist in the Department of Earth and Space Sciences, University of California at Los Angeles, originated the high-resolution design requirement.[22]

By 1978, the Science Working Group had defined four objectives for the 1984 VOIR mission: 1) images at a resolution and image quality equivalent to the Mariner 9 Mars mission (100 percent of the surface at mapping resolution, 600 meters, and a few percent in a high resolution mode, 100 meters); 2) a global topographic map of the planet; 3) a global map of the gravity field; and 4) new investigations of the atmosphere and exosphere. With surface exploration taking pride of place over atmospheric experiments, VOIR would be an inverse of Pioneer Venus.

In October 1978, NASA dissolved the Science Working Group and issued an Announcement of Opportunity requesting proposals for VOIR science experiments in three categories: 1) surface and interior properties of the planet requiring use of the SAR and altimeter, 2) atmospheric and ionospheric and other geophysical experiments requiring flight instruments other than the SAR and altimeter, and 3) other geophysical, atmospheric, and general relativity experiments using existing spacecraft subsystems.[23]

Schooling potential users of Venus radar images became an integral part of the project. Project managers understood the VOIR radar image interpretation community as consisting of 70 investigators plus 130 associates with experience interpreting photographs of the Moon, Mars, and Mercury. In order to inculcate potential users in the interpretation of radar images, JPL planned two radar image interpretation sessions, tentatively scheduled with Goodyear, to take place in 1978 and 1979. NASA and JPL also were to sponsor studies based on the analogy between Venus radar images and radar images from aircraft and Earth satellites.[24]

After the release of the VOIR Announcement of Opportunity, experiment proposals began to arrive at NASA Headquarters. Gordon Pettengill submitted his proposal to use the synthetic aperture radar to image Venus in February 1979. Pettengill defined his radar experiment in such a way as to dovetail radar and geological science. The proposal focused on "those processes that have shaped the surface of Venus and that have led to the evolution of its distinctive atmosphere. A major intermediary in achieving this goal is the preparation of a global map of the surface morphology in sufficient detail to describe and locate the major geological types and processes exhibited by Venus."

21.   Pettengill 28 September 1993; "VOIR, Venus Orbiting Imaging Radar Review, 4/80," and "VOIR, Venus Orbiting Imaging Radar Review, 4/80," Box 10, JPLMM.
22.   A. Gustaferro to W. B. Hanson, 8 May 1979, "Magellan Documentation," NHO; Friedman-Beckman Memorandum; VOIR Historical Perspective.
23.   "VOIR, (NASA) Correspondence, VOIR Mission Study Books, 11/78," Box 13; "VOIR, Status Briefing to Committee on Planetary and Lunar Exploration, NASA Headquarters, 6/78," Box 3; and NASA, Announcement of Opportunity no. OSS-5-78, 12 October 1978, Box 13, JPLMM.
24.   "VOIR, Venus Orbital Imaging Cost Review, 6/78," Box 3, JPLMM.

Pettengill's proposal emphasized the general lack of knowledge about the surface features of Venus. Ground-based observations of Venus, Mariners 2, 5, and 10, the Soviet Venera missions, and Pioneer Venus all provided much information about the planet, but the proposal argued, "This knowledge is heavily weighted toward the atmosphere of Venus and its interaction with the solar wind. Comparatively little is known about the solid surface or the interior of the planet."

Pettengill's proposed team of co-investigators followed closely the membership of the disbanded Science Working Group. Apart from Arecibo radar astronomer Don Campbell, most co-investigators came from MIT's Center for Space Research, Pettengill's home organization, JPL, and the U.S. Geological Survey. Representing the USGS were Pioneer Venus veteran Hal Masursky, Gerald Schaber, then assistant chief of the Branch of Astrogeologic Studies, and Laurence A. Soderblum, chief of the USGS Branch of Astrogeologic Studies. Again, radar and planetary geologists associated in a common endeavor.

Among the geologists who ultimately would be the most influential on the VOIR project was James W. Head, III, an associate professor in the Department of Geological Sciences at Brown University. Head had worked at NASA Headquarters for Bell Communications, a telephone company subsidiary that provided systems analysis and support, including geologic work and landing site selection, to NASA on the Apollo missions. His research interests included comparative planetary geology, and he had been active in the geologic interpretation of radar data from the Moon for some years. More importantly, as we shall see, he was a guest investigator on the Soviet Venera 15 and 16 missions.

Pettengill proposed to organize his co-investigators into Task Groups that would participate in and monitor the design and implementation of all aspects of the SAR instrument, its operation during flight, and the reduction of imaging and ancillary radar data, as well as the subsequent geological and geophysical interpretation of the data.[25]

NASA received several other proposals, but they were not successful for one reason or another. H. MacDonald, a radar geologist at the University of Arkansas, proposed interpreting VOIR data in the form of a radar landform atlas of Venus. The project largely duplicated the mapping contemplated in the Pettengill proposal. Another unsuccessful proposal came from Charles A. Barth, at the Laboratory for Atmospheric and Space Physics, University of Colorado, Boulder, to act as Principal Investigator on the airglow photometer experiment.[26] The airglow photometer was to measure the horizontal and temporal characteristics of the nightside thermospheric circulation. That proposal failed for reasons external to VOIR, as we shall see.

The Stanford Center for Radar Astronomy also submitted a proposal; it succeeded. Proposing radio and radar experiments on NASA space missions was their normal mode of conducting scientific research. Len Tyler, Dick Simpson, and John F. Vesecky proposed to study radar backscatter from the surface of Venus, in order to infer the small-scale physical texture of the surface, and to relate that texture to the large-scale formations visible in the VOIR images. Rather than create a separate investigative group, the Stanford researchers proposed that they participate in the radar group with Pettengill.[27]

25.   "VOIR, Scientific Investigation and Technical Plan, Proposal to NASA, 2/79," Box 13, JPLMM; *V-Gram* no. 11 (April 1987): 16; *V-Gram* no. 13 (October 1987): 14; and *V-Gram* no. 11 (April 1987): 11.

26.   "VOIR, A Proposal to NASA, Submitted by University of Arkansas, 7/79;" "VOIR, Contract Request for Proposal (APE) Airglow Photometer Experiment, 5/81, 12/81;" and "VOIR, Proposal to NASA, for Airglow Photometer Experiment for the VOIR Mission," Box 13, JPLMM.

27.   "VOIR, Proposal to the NASA Management Section, 2/79," Box 13, JPLMM.

The Venus Radar Mapper

Congress already had voted VOIR a new start in the NASA budget, when Ronald Reagan became president in January 1981. As a result of decisions reached in the early months of the new administration, the problems foreseen in 1978—overlap with Galileo, complication of DSN scheduling, escalated costs, and an opportunity for the Soviet Union to obtain Venus SAR images before the United States—all came true. National politics now took its turn in shaping the VOIR mission. Early in the new Republican administration, as a political signal that the new president was serious about cutting the budget, or at least the civilian portion of the budget, the budget czar David Stockman pressured NASA to sacrifice a major project. NASA chose VOIR.[28]

Failure to fund the project until fiscal 1984, when VOIR became a new NASA start, led to a postponement of the launch schedule to April 1988. This postponement provided the Soviet Union an opportunity to obtain the first SAR images of Venus. In this case, the Cold War rhetoric of the White House did not have its equivalent in the Space Race. The Space Race was dead. Starting in the early 1970s, as the United States withdrew from the war in Vietnam and the Apollo program's objective had been met several times over, a period of détente started. The U.S. and U.S.S.R. signed an accord in 1972 to allow the exchange of scholars between the two countries. A decade later, however, the United States let the accord lapse in protest over the Soviet imposition of martial law in Poland. Nonetheless, many U.S. and Soviet scientists sought to collaborate, not compete, and they did so with the tacit approval of their governments. Cold War rivalry and competition no longer held sway.[29]

The justification for canceling VOIR was its high cost. The project, conflated into an exploration of the surface, interior, atmosphere, and ionosphere of Venus, carried a total price tag of $680 million. NASA and JPL sought ways to slash that price tag to $200 to $300 million.[30] In the opinion of Gordon Pettengill, the project "was climbing a cliff. The project people at NASA Headquarters were told that if they could cut the cost in half, they could have their project. In other words, they had to do it for $300 million instead of $600 million. So an ad hoc group of JPL and NASA Headquarters people was put together to study ways of cutting costs."[31]

NASA renamed the low-cost reduced mission the Venus Radar Mapper (VRM).[32] The trick was to lower the price tag, while still getting the science done. A number of approaches were suggested and taken, not all of which were technological, such as the reduction of personnel levels. Many of the cost-cutting decisions directly reduced the scientific scope of the mission. For example, one of the earliest decisions was to jettison all scientific experiments that did not use the radar. Only the altimetry and imaging experiments, which used the radar instrument, and the gravity experiment, which was carried out by the Deep Space Network, remained. Among the rejected experiments was the airglow photometer.[33] As Pettengill pointed out, however, "they saved $150 million by

28.   Pettengill 28 September 1993; Saunders, Pettengill, Arvidson, William L. Sjogren, William T. K. Johnson, and L. Pieri, "The Magellan Venus Radar Mapping Mission," *Journal of Geophysical Research* vol. 95, no. B6 (1990): 8339; Waff, *Jovian Odyssey: A History of NASA's Project Galileo*, chapter "Surviving the Reagan Revolution," pp. 8–10, Waff materials.
29.   Henry S. F. Cooper, Jr., "A Reporter at Large: Explorers," *The New Yorker* 64 (7 March 1988): 50.
30.   "VOIR, Venus Mapper New Start Plans, 3/82," and "VOIR, Venus Radar Mapper, A Proposed Planetary Program for 1988," Box 10, JPLMM.
31.   Pettengill 28 September 1993.
32.   For a brief period in 1981 and 1982, project documents used the name Venus Mapping Mission (VMM).
33.   "VOIR, Project Management, Venus Orbiting Imaging Radar, 1981-82," Box 14; "VOIR, Venus Mapper Briefing to NASA Headquarters, 1/82," Box 10; and "VOIR, Request for Proposal for VOIR Synthetic Aperture Radar, 7/81, 3/3," Box 13, JPLMM.

getting rid of the four non-radar experiments that originally were intended for the mission."[34]

Throughout various iterations of the project, the dimensions of the high and low resolution radar images vacillated. In fact, for a while, the high resolution detailed images of selected surface features disappeared entirely. In an early 1981 iteration, the VRM was to map at least 70 percent of Venus with a resolution of 600 meters and take high resolution (150-meter) data over about one percent of the planet. As described at a January 1982 briefing at NASA Headquarters, however, the VRM was to have no high resolution capability and would image only 70 percent of the planet with a resolution of better than one km. At a February 1982 conference held at JPL for the selected contractors, Hughes (SAR) and Martin Marietta (spacecraft), the SAR performance parameters called for coverage of 90 percent of the planet with a single resolution of 300 meters. By 1984, though, when the VRM became a NASA new start, the baseline performance had been raised to resolutions of 215 meters by 150 meters and 480 meters by 250 meters.[35]

The resolution, and consequently the science that the VRM would achieve, was a trade-off against the cost of the project. Only by lowering overall costs did JPL and NASA manage to put together a mission capable of high resolution. One of the key cost-reduction approaches was to "maximize inheritance," a term that meant to borrow as much technology from other projects as possible. Magellan was to be pieced together from other NASA projects.

Among the projects from which the VRM borrowed, or considered borrowing, were Viking, Voyager, Galileo, and ISPM (International Solar Polar Mission). The VRM proposed borrowing such hardware items as the Voyager 3.7-meter dish antenna for its synthetic aperture radar, Galileo's tape recorder, and Viking's S-band low-gain antenna. Also, JPL suggested using NASA standard equipment as well as various SEASAT parts, such as sun sensor and solar array drive electronics and the solar array actuators.[36]

In order to improve the VRM's data handling capabilities, JPL modified the radar guidelines in order to use the Galileo Golay code, rather than the Golay code planned by Hughes (contractor for the SAR). The Galileo Golay code and a restructuring of the radar burst header format (for more efficient handling by the Deep Space Network) resulted in a considerable saving in ground software costs.

Another key decision was the switch from a circular to an elliptical orbit. With an elliptical orbit, the parameters of the radar varied as a function of the spacecraft's altitude above the planet's surface. Mapping from an elliptical orbit eliminated the need for aerobraking. Aerobraking is a technique for trimming a spacecraft's orbit by having it pass repeatedly through a planetary atmosphere. Its use would reduce the amount of propulsion needed for initial orbit insertion. Aerobraking offered a low-cost, low-risk option that would both save fuel, and therefore mission weight, and lower mission costs.[37]

Using digital processing to simplify the electronics was a significant saver of money. Original VOIR planning centered on analog processing for the radar, but by 1981 it had become clear that using digital circuitry was the preferred technology. The parameters of an analog system could not change during flight; so, aerobraking and a circular orbit were necessities. Digital processing allowed the radar parameters to change during flight, thereby tolerating the variations of a less expensive elliptical orbit.[38]

---

34.   Pettengill 28 September 1993.

35.   "VOIR, Request for Proposal for VOIR Synthetic Aperture Radar, 7/81, 3/3," Box 13; "VOIR, Venus Mapper Briefing to NASA Headquarters, 1/82," Box 10; "VOIR, Venus Mapper Conference w/Hughes and MMC, 2/82," Box 10; and "VOIR, Project Management Report, 1984, 1/2," Box 14, JPLMM.

36.   "VOIR, Venus Mapper Briefing to NASA Headquarters, 1/82," Box 10; "VOIR, Venus Mapper New Start Plans, 3/82," Box 10; and "VOIR, Venus Radar Mapper, A Proposed Planetary Program for 1988," Box 10, JPLMM.

37.   "VOIR, Project Management Report, 1984, 1/2," Box 14, JPLMM.

38.   Pettengill 28 September 1993.

The Microsymposia

The decisions to change the orbital geometry, use digital processing, and borrow technology from other projects lowered project costs to the point where VRM became a new NASA start in 1984. As a result of the postponed launch of VRM, Soviet scientists gained an important scientific opportunity to image Venus. When it appeared that the United States would launch VOIR on schedule, Soviet scientists decided to launch their own Venus imaging mission only if the United States did not send a Venus radar mapper before 1984. Once NASA delayed launch of the VRM beyond 1984, Soviet scientists had to move forward their own Venus radar mapper very quickly in order to seize the opportunity.

In June 1983, the Soviet Union flew two spacecraft equipped with radar mappers to Venus; they arrived four months later. Venera 15 and 16 covered the same polar region of Venus (30° North to the pole), probably on the assumption that one of the spacecraft might fail. Their goal was to map that region at a resolution of one to two km in daily, 150 by 7,000 km strips 10° to the side of the orbital track, covering a total area of 115 million square kilometers by the time the main mission ended in July 1984.[39]

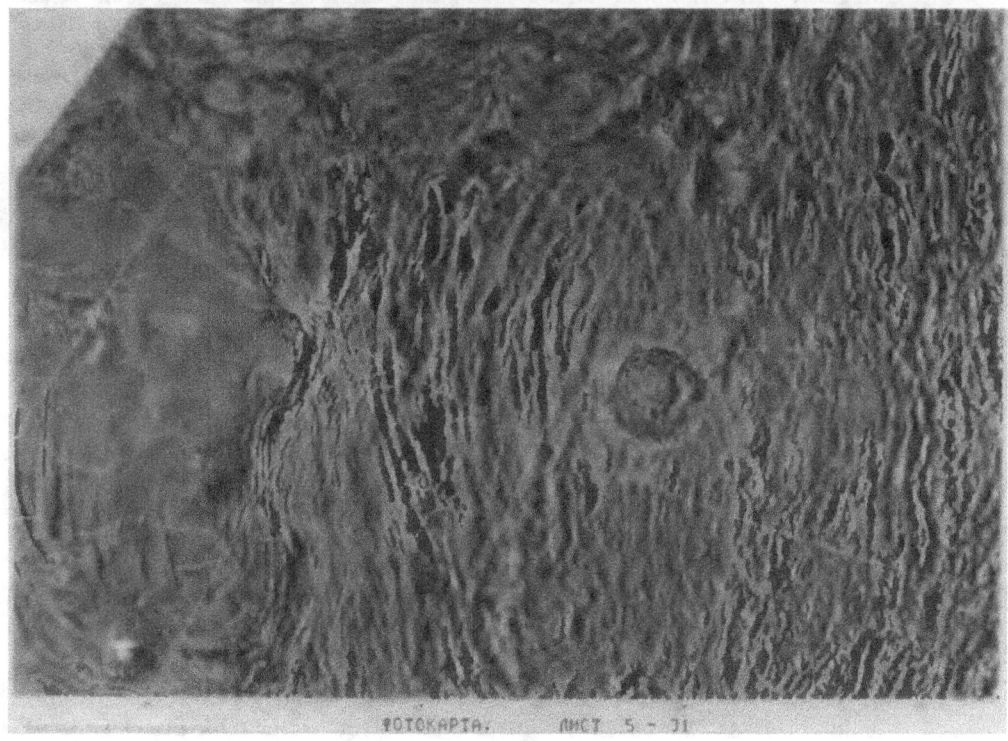

Figure 31
Radar image of Venus, near Maxwell Montes, made by Venera 15 and 16. (Courtesy of NASA, photo no. 88-H-8.)

---

39.   Andrew Wilson, *Solar System Log* (London: Jane's, 1987), pp. 112–113.

Interpreting the images from the Venera 15 and 16 mission required more information about Venus surface features than the Soviet Union had available. Previous Soviet missions had landed only in limited areas of the planet. Soviet scientists, desperately in need of information, turned to their American colleagues to exchange Venus data.

Since the Apollo era, several American scientists had made frequent trips to Moscow and to international meetings where they met Soviet planetary scientists. Of those American scientists, two of the most important ones for the Magellan mission were Jim Head of Brown University and Hal Masursky (USGS), a member of the Pioneer Venus science team. As Venera 15 and 16 data became available, its value to future American exploration of Venus, especially the VRM mission, was apparent, and a parallel American interest in collaboration developed.

On 25 March 1984, Alexander Basilevsky, a geologist and chief of the Vernadsky Institute Planetology Laboratory, and Valery L. Barsukov, director of the Vernadsky Institute (the Soviet equivalent of the USGS), presented Venera 15 and 16 results at the Lunar and Planetary Science Conference held in Houston. United States scientists appreciated that the Venera 15 and 16 SARs had yielded mosaickable images of a large part of the northern hemisphere.

COMPLEX, the Committee on Planetary and Lunar Exploration of the Space Science Board, requested that VRM scientists present an assessment of the Venera 15 and 16 accomplishments, as well as a summary of VRM capabilities and, if deemed desirable, ways of improving VRM. Gordon Pettengill presented what was known about the Soviet Venera mission, including SAR characteristics, range of resolution, and coverage, and he compared Venera results with Arecibo high resolution range-Doppler images. Having reliable images of Venus was vital to the planning of the VRM mission. Although VRM scientists already had data with which to plan the mission, the Venera 15 and 16 data would have added important information on the northern hemisphere. Only two other sources of images of Cytherean surface features were available.

One source was Pioneer Venus. Its radar altimeter measured the height of about 90 percent of the surface at roughly 75 km intervals, while the mapper mode furnished low (20 to 40 km) resolution radar images of only the equatorial region. Pioneer Venus had not covered the northern polar region, unlike Venera 15 and 16. Higher resolution imaging was available from Arecibo, the second source of Venus surface images.[40] Arecibo covered about 25 to 30 percent of the planet at resolutions around 2 to 4 km. However, Arecibo could image well only the hemisphere of Venus facing Earth at inferior conjunction.[41]

If they could be had on magnetic tape in a digital format, the Venera 15 and 16 data would have assisted VRM planning significantly. The data did become available, but not through any political maneuvering by the corresponding state departments or other high-level official channels. The exchange of scientific results between Soviet and U.S. scientists interested in the surface features of Venus came about as the result of an arrangement made among the scientists themselves and their parent institutions.

The 11 March 1985 session on Venus at the Lunar and Planetary Science Conference featured Soviet presentations of their recent interpretations of Venera 15 and 16 results by Alexander Basilevsky, Valery L. Barsukov, and two others from the Vernadsky Institute. Subsequently, on 19–20 March 1985, the first microsymposium took place at Brown University. The four Soviet scientists reviewed recent results of Venera 15 and 16, Arecibo, Pioneer Venus, as well as future Venus missions and Venus science in general. Among those attending were geologists James Head and Harold Masursky and radar astronomers Gordon Pettengill and Don Campbell of Arecibo.

---

40.   "VOIR, Project Management Report, 1984, 1/2," Box 14, JPLMM.
41.   Campbell 8 December 1993.

It was at the March 1985 microsymposium that James Head reported that the Soviets appeared to be receptive to the idea of providing some of their data. Preliminary results indicated that the Venera SAR radar parameters would not be a major obstacle to their use by American scientists. Moreover, both Soviet and American investigators had reached a preliminary agreement on the choice of a particular small feature for the definition of the Venus prime meridian. The features had appeared in both the Arecibo range-Doppler images and the Venera 15 and 16 SAR images. Establishment of a coordinate system was important to the planned VRM cartography efforts.

In November 1985, Vladimir Kotelnikov, the leader of Soviet ground-based radar astronomy research, then head of Interkosmos, delivered to Jim Head a tape with one strip of digital Venera image data with accompanying altimetry. Head distributed the tape to Saunders, Pettengill, Campbell, and Masursky for analysis. They had no difficulty in displaying the image using conventional American image processing techniques.

The Soviet-American agreement to exchange Venus data was underway. The agreement materialized as a protocol signed in 1982 between the Governor of Rhode Island (the location of Brown University) and the Soviet Academy of Sciences. Under the agreement, one microsymposium per year was to take place in each country. Traditionally, the American microsymposium has been held in March or April at Brown University; while the Soviet meeting takes place at the Vernadsky Institute in Moscow in August. James Head organized the Brown University group, and Valery Barsukov, director of the Vernadsky Institute, organized the Soviet group. The creation of the microsymposia owed much to the fact that Head was a guest investigator on Venera 15 and 16.[42]

The Soviet data delivered over the following years at subsequent microsymposia played an important role in the creation of planning maps for the VRM/Magellan mission. The microsymposia were but one forum within which geology and radar communities worked together. The VRM Radar Investigation Group (in charge of the radar science) was another forum that brought the two communities together in a common effort. The Radar Investigation Group (RADIG) was a large and multifaceted organization typical of Big Science. In order to more effectively coordinate and carry out VRM and Magellan science, Pettengill divided the group into smaller subgroups (Table 6).

The VRM (and later Magellan) Radar Investigation Group combined the former Synthetic Aperture Radar Group and the Altimetry Investigation Group of the VOIR project. Gordon Pettengill headed the Radar Investigation Group (RADIG). Three RADIG subgroups dealt with mission design, while three other subgroups concerned themselves with scientific interpretation. These last three subgroups treated cartography and geodesy, surface electrical properties, and geology and geophysics. Geology and geophysics, the largest and most complex area of scientific interpretation, consisted of even smaller groups dealing with volcanic and tectonic processes; impact processes; erosional, depositional, and chemical processes; and isostatic and convective processes.[43]

Not only did the RADIG bring together planetary radar and geology communities, but it illustrated how space flight science groups organized Little Science to function as Big Science, if only on a temporary basis within ephemeral organizations. Ordinarily, in a way characteristic of Little Science, scientists work alone at a university or technical school with a small budget and modest laboratory equipment. NASA space missions bring these individual scientists together and make them function in ways customarily associated with Big Science, mainly as part of a large group. Any given scientist works as a member of two groups, one defined by a flight instrument and the other by the scientist's discipline or

42.   Cooper, "A Reporter," p. 50; Ford 3 October 1994; Campbell 8 December 1993; "VOIR, Report Project Management, 1985," Box 14, JPLMM. The August 1991 microsymposium was delayed until November because of the putsch.
43.   *V-Gram* no. 8 (24 March 1986): 2–4.

Table 6
Members of Magellan Radar Investigation Group (RADIG)

Scientist	Institution
Raymond E. Arvidson	Washington University
Victor R. Baker	University of Arizona
Joseph H. Binsack	MIT
Donald B. Campbell	NAIC, Cornell
Merton E. Davies	Rand Corporation
Charles Elachi	JPL
John E. Guest	University of London
James W. Head, III	Brown University
William M. Kaula	UCLA
Kurt L. Lambeck	Australian National University
Franz W. Leberl	Independent Consultant
Harold C. MacDonald	University of Arkansas
Harold Masursky	US Geological Survey
Daniel P. McKenzie	Cambridge University
Barry E. Parsons	Oxford University
Gordon H. Pettengill	MIT
Roger J. Phillips	Southern Methodist University
R. Keith Raney	Canada Center for Remote Sensing
R. Stephen Saunders	JPL
Gerald G. Schaber	US Geological Survey
Gerald S. Schubert	UCLA
Laurence A. Soderblum	US Geological Survey
Sean C. Solomon	MIT
H. Ray Stanley	NASA, Wallops Island
Manik Talwani	Gulf Research and Development
G. Leonard Tyler	Stanford University
John A. Wood	Harvard-Smithsonian Astrophysical Observatory

subdiscipline. Grouped together around a common instrument, scientists jointly design the instrument that will generate their data. Grouped together around a common scientific interest, such as magnetospheres or geology, scientists jointly utilize data derived from the operation of all flight instruments. However much these scientists function within a Big Science organization, the organization itself is defined by the temporary lifetime of the project. In the end, they are once more Little Science.

Osmosis

In December 1985, NASA Headquarters notified JPL that the VRM had a new name, Magellan. The name reflected NASA's general plan of naming major planetary missions after famous scientists and explorers (Galileo, Magellan, Cassini).[44] Ferdinand Magellan had been a Portuguese navigator and explorer who led an expedition into the Pacific Ocean under the Spanish flag.

By the end of 1985, construction of the Magellan radar instrument was underway. After Hughes Aircraft Company and Goodyear Aerospace Corporation completed Phase B studies of the project in June 1980, JPL issued a Request for Proposals for the synthetic aperture radar system, including the antenna design, in April 1981. The selection of the SAR and spacecraft contractors were separate processes.[45]

---

44.   "VOIR, Report Project Management, 1985," Box 14, JPLMM.
45.   "VOIR, Venus Orbiting Imaging Radar Review, 4/80," Box 10, and "VOIR, Request for Proposal for VOIR Synthetic Aperture Radar, 7/81, 1/3," Box 13, JPLMM.

Hughes had hoped to turn its experience with the Pioneer Venus orbiter mapper into an advantage, while Goodyear had been one of the first firms to commercialize aircraft SAR systems to study the Earth. In 1983, NASA and JPL signed contracts with Hughes and Martin Marietta for the SAR and spacecraft. Hughes signed the definitive radar contract on 24 January 1984, and the contract was executed 27 January 1984. Throughout 1985 and 1986, Hughes increased the number of employees working on the Magellan radar. The project had the second highest priority within the Hughes Space and Communication Group, behind a smaller classified project.[46] Hughes' Pioneer Venus gambit had paid off.

Magellan was on schedule and under budget when the Space Shuttle *Challenger* blew up on 28 January 1986. The tragedy caused a serious delay in the Magellan launch schedule. In fact, the disaster adversely affected all Shuttle flights. The Shuttle would not fly until the cause of the *Challenger* accident was determined and corrective solutions found to prevent future repetitions of the accident. Only then would a new Shuttle flight schedule be drawn up.

In February 1986, Magellan mission personnel began to appraise probable launch dates. Realizing the uncertainties of the Shuttle launch schedule, they investigated two launch windows that followed the approved launch period in April 1988. One was between 28 October and 16 November 1989, the other between 25 May and 13 June 1991. In each case, Magellan would spend eight months in orbit performing its prime mission, and the mission would end at superior conjunction, in November 1990 or in June 1992, depending on the launch window.[47]

A delayed launch also raised the likelihood of conflicts with the Galileo launch. If Magellan held to its approved launch schedule in April 1988, and Galileo delayed 13 months, then coverage conflicts on the Deep Space Network eased considerably. Whatever launch window Magellan eventually had, conflict with the Galileo launch and scheduling of the Deep Space Network would have to be taken into consideration. Further complicating the launch schedule was the cancellation in June 1986 of the Shuttle/Centaur, which was to launch Magellan. After a study of alternate launch vehicles, in October 1986 NASA settled on a combination of the Shuttle and a launcher known as an Inertial Upper Stage (IUS) and assigned Magellan a position on the Shuttle manifest for April 1989.[48]

The change required reduction of the spacecraft mass, as well as new structural loads analyses. In order to undertake the analyses, a second spacecraft structure was needed for static load tests. The only one available was on the Voyager spacecraft hanging in the Smithsonian Air and Space Museum in Washington. NASA made arrangements to borrow the Voyager bus from the museum and conducted the tests.[49]

The *Challenger* accident also affected Magellan's use of Galileo technology. Because Magellan launched before Galileo, the extra Galileo components were not available. Ground support equipment to be borrowed from Galileo were unavailable. Now the "spare parts" Magellan was to borrow from Galileo had to be returned to Galileo and purchased new for Magellan.

The delay of Magellan also raised the cost of the project. The total dollar impact, including the cost of hardware, mission design, and mission operations, was estimated to be about $150 million. Gordon Pettengill summed up the situation: "That disaster need not have happened, but it did; it was just one of those things. Magellan would not have been as expensive, if we had launched when we were originally planned to launch."[50]

---

46.   Various documents, Box 6, and "VOIR, Report Project Management, 1986, 1/2," Box 14, JPLMM.
47.   "VOIR, Report Project Management, 1986, 1/2," Box 14, JPLMM.
48.   *V-Gram* no. 10 (January 1987): 1 and 4.
49.   *V-Gram* no. 10 (January 1987): 1.
50.   Pettengill 28 September 1993; *V-Gram* no. 10 (January 1987): 1.

JPL received unofficial notification in May 1986 from NASA Headquarters that Magellan had slipped to the October-November 1989 launch window, but no official launch date had yet been established. Nonetheless, the Magellan project proceeded on the assumption of that launch window.[51]

Meanwhile, the collection and exchange of radar data for the assembling of maps to be used in planning the mission proceeded. The Brown University-Vernadsky Institute microsymposia continued to play a vital role in the exchange of scientific information between American and Soviet scientists. In April 1986, the third international microsymposium on Venus took place at Brown University. Valery Barsukov, Alexander Basilevsky, and four other Soviet scientists presented preliminary scientific results of the Venera 15 and 16 missions and a description of the radar system.

The Soviet scientists presented the Magellan project with three Venera data tapes consisting of unpublished SAR digital data. They stipulated that the data be used strictly for planning the Magellan project; it was not for scientific publication or distribution, until the Soviet scientists had published the information. The request was reasonable; it protected their priority of discovery. In exchange, the Soviet scientists received high resolution digital data from the Viking mission to assist them in planning their Phobos mission to Mars's moon.[52]

The following year, Magellan investigators James Head, Steve Saunders, Hal Masursky, Gerald Schaber, and Don Campbell attended a microsymposium held 11 to 15 August 1986 at the Vernadsky Institute in Moscow. They and their Moscow colleagues exchanged views on the interpretation of Venus data from Venera 15 and 16 and Arecibo. The Soviet investigators presented the Magellan scientists with eight tapes of Venera 15 and 16 digital radar images and altimetry profiles for use by the Magellan project for planning purposes.[53]

At the following microsymposium held at Brown University in March 1987, scientists debated the origin and evolution of volcanic structures and deposits, domes, parquet terrain, impact craters, ridge and linear mountain belts, and plate tectonics. Only slight consensus over the interpretation of features emerged, because the resolution of features in Pioneer Venus images (25 km) and Venera 15 and 16 images (1–3 km) was sufficiently coarse to give rise to ambiguities in interpretation. Magellan's higher global resolution (about 300 meters) promised to resolve many questions of geologic interpretation. Soviet scientists provided the Magellan project with additional Venera 15 and 16 digital tapes; in return they received more high-resolution Viking imaging data of Phobos and the surface of Mars.[54]

The microsymposia demonstrated the fruitful cross-fertilization of planetary geology and radar. In order to facilitate the use of radar data by geologists, Magellan Project Manager John Gerpheide, Program Scientist Joseph Boyce, Principal Investigator Gordon Pettengill, Project Scientist Steve Saunders, and Science and Mission Design Manager Saterios Sam Dallas formulated preliminary plans in July 1986 for various radar workshops. The first, to be held in 1987, was to cover radar operation and processing, the second the interactions between radar waves and planetary surfaces, and the third interpretation of SAR images. The second and third workshops were held in 1988 and 1989, respectively. The sessions were open to Magellan scientists and to the Planetary Geology and Geophysics Program investigators. In addition, they planned one-day Venus science symposia to be held in conjunction with other project meetings for each year between 1987 and 1989.

51.   "VOIR, Report Project Management, 1986, 1/2" and "VOIR, Report Project Management, 1986, 2/2," Box 14, JPLMM.
52.   "VOIR, Report Project Management, 1986, 1/2," Box 14, JPLMM.
53.   "VOIR, Report Project Management, 1986, 2/2," Box 14, JPLMM.
54.   V-Gram no. 12 (July 1987): 1.

In 1987, 32 scientists and project personnel participated in the field trip to various sites in the Mojave Desert and Death Valley.[55] The goal was to compare a variety of geologic features with SAR images of the areas. Steve Wall, Magellan Radar Experiment representative, organized the field trip, which Tom G. Farr of JPL's Geology and Planetology Section led. Gerald Schaber of the USGS contributed to the technical presentation by sharing his knowledge of Death Valley.[56]

In May 1988, the USGS Flagstaff hosted another field trip, which was incorporated as part of the quarterly meeting of Magellan scientists and project staff. The major objective was to familiarize participants with specific radar geology targets in a semi-arid, relatively vegetation-free environment. The trip also entailed comparing geologic features with X-band and L-band SAR images. The field exercise was planned and led by Gerald Schaber, Richard Kozak, and George Billingsley, all three with the USGS Flagstaff.[57]

These field trips helped to introduce geologists to the interpretation of radar data. Geologists learn from "hands-on" experience, but that kind of experience is impossible when dealing with the geology of Venus. Radar images, moreover, are not created by the reflection of light, but by the scattering and reflection of electromagnetic waves. They cannot be read like photographs, and radar maps cannot be read like ordinary geological maps.

In order to fill in that gap, data to create a series of S-band radar images of the lunar surface were collected at the Arecibo Observatory between 1982 and 1992. The images were made at various angles of incidence at a number of known lunar locations, such as the Apollo 15 and 17 landing sites, Mare Imbrium, and craters Copernicus and Tycho, in order to provide experience in interpreting surface geology in radar images. Don Campbell, assisted by Peter Ford of MIT and later by Cornell graduate student Nick Stacy, made the observations and images in collaboration with Jim Head of Brown University. While initial image resolutions ranged from 200 to 300 meters, Nick Stacy brought image resolution down to 25 meters beginning in 1990. Elaborate data processing techniques attempted to replicate the synthetic aperture radar techniques used from spacecraft and aircraft.[58]

As Gordon Pettengill pointed out, the workshops were not the main path for geologists to learn about radar. "The people who attended those made up a small fraction of the overall community. That route is an exception to what I would call the more general experience. Generally, people become part of a team, and they work with radar people, like myself, who then, by a process I would call osmosis, pass along the mystique of what is going on, when you see these structures on a radar image, how to interpret them, and what to look out for, so you don't make errors."

This process of osmosis, Pettengill explained, "is the best way to go. A formal course is difficult. They call them workshops. They are useful. But you need both. You need the workshop as well as years of working with other people and growing used to what you are seeing."[59]

That process of osmosis was most evident at the Arecibo Observatory, where Don Campbell and his graduate students Barbara Burns and Nick Stacy and Research Associate John K. Harmon, collaborated with Jim Head and other geologists at Brown University through an informal accord between the NAIC and Brown University beginning around 1980. The heart of the accord was a cooperative effort to analyze Arecibo Venus imagery.

---

55.   "VOIR, Report Project Management, 1986, 2/2," Box 14, JPLMM.
56.   *V-Gram* no. 11 (April 1987): 1.
57.   *V-Gram* no. 15 (January 1989): 15.
58.   Ford 3 October 1994; Campbell 8 December 1993; Nicholas John Sholto Stacy, "High-Resolution Synthetic Aperture Radar Observations of the Moon," Ph.D. diss, Cornell University, May 1993; NAIC QR Q1/1982, Q4/1986, Q2/1990, Q4/1990, and Q3/1992.
59.   Pettengill 29 September 1993.

As a result of the arrangement, a number of Brown students, such as Richard W. Vorder Brueggie and David A. Senske, became involved in the analysis of Arecibo radar range-Doppler imagery and wrote their theses from the data.

"The effort was not under any formal agreement between the NAIC and Brown University," Don Campbell explained. "We badly needed the backing of a planetary geology group. We were into geology at this point. We were down to a few kilometers of resolution, and they were extremely enthusiastic. Jim Head was very enthusiastic and had a lot of students. They were very intent on getting ready for the Magellan mission and spent a lot of effort on both the Pioneer Venus and Venera data sets."[60]

The Arecibo-Brown arrangement thus fostered the geological interpretation of Venus radar images well before Magellan began its mapping mission. A major area of interest was in identifying and explaining tectonic activity on the planet. Some of the 1979 high-resolution Arecibo radar images suggested Earth-like tectonic features, such as folds and faults, while 1983 Arecibo radar images confirmed the presence of rifting in the southern Ishtar Terra and surrounding plains and general tectonic activity in Maxwell Montes.[61] Later studies examined evidence for tectonic activity in Beta Regio, Guinevere Planitia, Sedna Planitia, and western Eistla Regio in the planet's equatorial region, as well as in the southern latitudes around Themis Regio, Lavinia Planitia, Alpha Regio, and Lada Terra.[62]

Don Campbell also collaborated with Jim Head's group in searching for evidence of volcanism. Arecibo radar images of southern Ishtar Terra and the surrounding plains revealed significant details of volcanic activity. Images made from data gathered at Arecibo during the summer of 1988 of the area extending from Beta Regio to the western Eistla Regio furnished strong evidence that the mountains in Beta and Eistla Regiones, as well as the plains in and adjacent to Guinevere Planitia, were of volcanic origin. Arecibo radar images of the southern latitudes showed additional evidence for past volcanic activity on Venus.[63]

The study of cratering on Venus started by Barbara Burns for her doctoral thesis continued at Arecibo, too. She based her initial analysis on data collected in 1977 and 1979. As of 1985, Burns was able to identify only two features that exhibited unambiguous radar characteristics that could tentatively distinguish them as either volcanic (Colette) or impact (Meitner) in origin. Don Campbell, with Jim Head and John Harmon, continued

---

60.   Campbell 8 December 1993.

61.   Campbell, Head, John K. Harmon, and Alice A. Hine, "Venus: Identification of Banded Terrain in the Mountains of Ishtar Terra," *Science* 221 (1983): 644–647; L. S. Crumpler, Head, and Campbell, "Orogenic Belts on Venus," *Geology* 14 (1986): 1031–1034; Stofan, Head, and Campbell, "Geology of the Southern Ishtar Terra/Guinevere and Sedna Planitae Region on Venus," *Earth, Moon, and Planets* 38 (1987): 183–207; R. W. Vorder Brueggie, Head, and Campbell, "Orogeny and Large-Scale Strike-Slip Faulting on Venus: Tectonic Evolution of Maxwell Montes," *Journal of Geophysical Research* vol. 95, no. B6 (1990): 8357–8381.

62.   David A. Senske, Campbell, Stofan, Paul C. Fisher, Head, Stacy, J. C. Aubele, Hine, and Harmon, "Geology and Tectonics of Beta Regio, Guinevere Planitia, Sedna Planitia, and Western Eistla Regio, Venus: Results from Arecibo Image Data," *Earth, Moon, and Planets* 55 (1991): 163–214; Bruce A. Campbell and Campbell, "Western Eistla Regio, Venus: Radar Properties of Volcanic Deposits," *Geophysical Research Letters* vol. 17, no. 9 (1990): 1353–1356; Senske, Campbell, Head, Fisher, Hine, A. de Charon, S. L. Frank, S. T. Keddie, K. M. Roberts, Stofan, Aubele, Crumpler, and Stacy, "Geology and Tectonics of the Themis Regio-Lavinia Planitia-Alpha Regio-Lada Terra Area, Venus: Results from Arecibo Image Data," *Earth, Moon, and Planets* 55 (1991): 97–161.

63.   Stofan, Head, and Campbell, "Geology of the Southern Ishtar Terra/Guinevere and Sedna Planitae Region on Venus," *Earth, Moon, and Planets* 38 (1987): 183–207; Campbell, Head, Harmon, and Hine, "Venus: Volcanism and Rift Formation in Beta Regio," *Science* 226 (1984): 167–170; Campbell, Head, Hine, Harmon, Senske, and Fisher, "Styles of Volcanism on Venus: New Arecibo High Resolution Radar Data," *Science* 246 (1989): 373–377; Campbell, Senske, Head, Hine, and Fisher, "Venus Southern Hemisphere: Geologic Character and Age of Terrains in the Themis-Alpha-Lada Region," *Science* 251 (1991): 180–183; Senske, Campbell, Head, Fisher, Hine, de Charon, S. L. Frank, S. T. Keddie, K. M. Roberts, Stofan, Aubele, Crumpler, and Stacy, "Geology and Tectonics of the Themis Regio-Lavinia Planitia-Alpha Regio-Lada Terra Area, Venus: Results from Arecibo Image Data," *Earth, Moon, and Planets* 55 (1991): 97–161.

Burns's crater studies. Images made from the data collected during the inferior conjunction of 1988 of the area from Beta Regio to western Eistla Regio revealed a low density of impact craters greater than 15 km in diameter in that region compared to the average density for the higher northern latitudes. These crater densities suggested that the plains were geologically younger than the northern regions.[64]

Campbell, with graduate student Nick Stacy and computer software manager and part-time radar astronomer Alice Hine, made a further analysis of cratering by looking at diameter-frequency distributions in the low northern latitudes and the southern hemisphere. The Arecibo investigators found that the average crater density for all craters in the northernmost quarter, using Venera 15 and 16 data, was 1.27 per million square kilometers, while the average for the southern hemisphere (as imaged by the Arecibo radar) was 0.95 per million square kilometers. The different crater densities suggested that the southern latitudes were geologically younger than the low northern latitudes imaged by Venera 15 and 16.[65]

Don Campbell also participated in the microsymposia organized by Brown University and the Vernadsky Institute. As a result, he also came to collaborate with Alexander Basilevsky and other Soviet geologists on the interpretation of Venera 15 and 16 results, and that collaboration led to co-authorship of a paper with combined Vernadsky Institute and Brown University authors.[66]

Don Campbell's osmotic infiltration of the scientific community interested in Venus typified the shifting paradigm of ground-based planetary radar astronomy toward geology. Further facilitating that shift was the availability of techniques, hardware, and software at Arecibo that yielded high-resolution range-Doppler images and topographical data. Image resolution improved to one to three km in 1983 and to 1.5 km in 1988, the last observations made before the arrival of Magellan at Venus.

Because Magellan used a frequency close to that of the Arecibo radar, there was some concern that the Arecibo radar might contaminate the Magellan data or endanger the spacecraft, so Don Campbell did not pursue Venus mapping after 1988.[67] Nonetheless, the participation of Arecibo ground-based investigators in Venus radar geology illustrated that the marriage of radar and geology was not limited to Magellan and space-based radars.

---

64.   Burns and Campbell, "Radar Evidence for Cratering on Venus," *Journal of Geophysical Research* vol. 90, no. B4 (1985): 3037–3047; Campbell, Head, Hine, Harmon, Senske, and Fisher, "Styles of Volcanism on Venus: New Arecibo High Resolution Radar Data," *Science* 246 (1989): 373–377.

65.   Campbell, Stacy, and Hine, "Venus: Crater Distributions at Low Northern Latitudes and in the Southern Hemisphere from New Arecibo Observations," *Geophysical Research Letters* vol. 17, no. 9 (1990): 1389–1392.

66.   A. T. Basilevsky, B. A. Ivanov, G. A. Burba, L. M. Chernaya, V. P. Kryuchkov, O. V. Nikolaeva, Campbell, and L. B. Ronca, "Impact Craters on Venus: A Continuation of the Analysis of Data from the Venera 15 and 16 Spacecraft," *Journal of Geophysical Research* vol. 92, no. B12 (1987): 12,869–12,901; Stofan, Head, Campbell, Zisk, A. F. Bogomolov, Rzhiga, Basilevsky, and N. Armand, "Geology of a Rift Zone on Venus: Beta Regio and Devana Chasma," *Geological Society of America Bulletin* 101 (1989): 143–156.

67.   Campbell 8 December 1993; Burns, "Cratering Analysis of the Surface of Venus," p. 1; Stofan, Head, and Campbell, "Geology of the Southern Ishtar Terra/Guinevere and Sedna Planitae Region on Venus," *Earth, Moon, and Planets* 38 (1987): 183–207; Richard W. Vorder Brueggie, Head, and Campbell, "Orogeny and Large-Scale Strike-Slip Faulting on Venus: Tectonic Evolution of Maxwell Montes," *Journal of Geophysical Research* vol. 95, no. B6 (1990): 8357–8381.

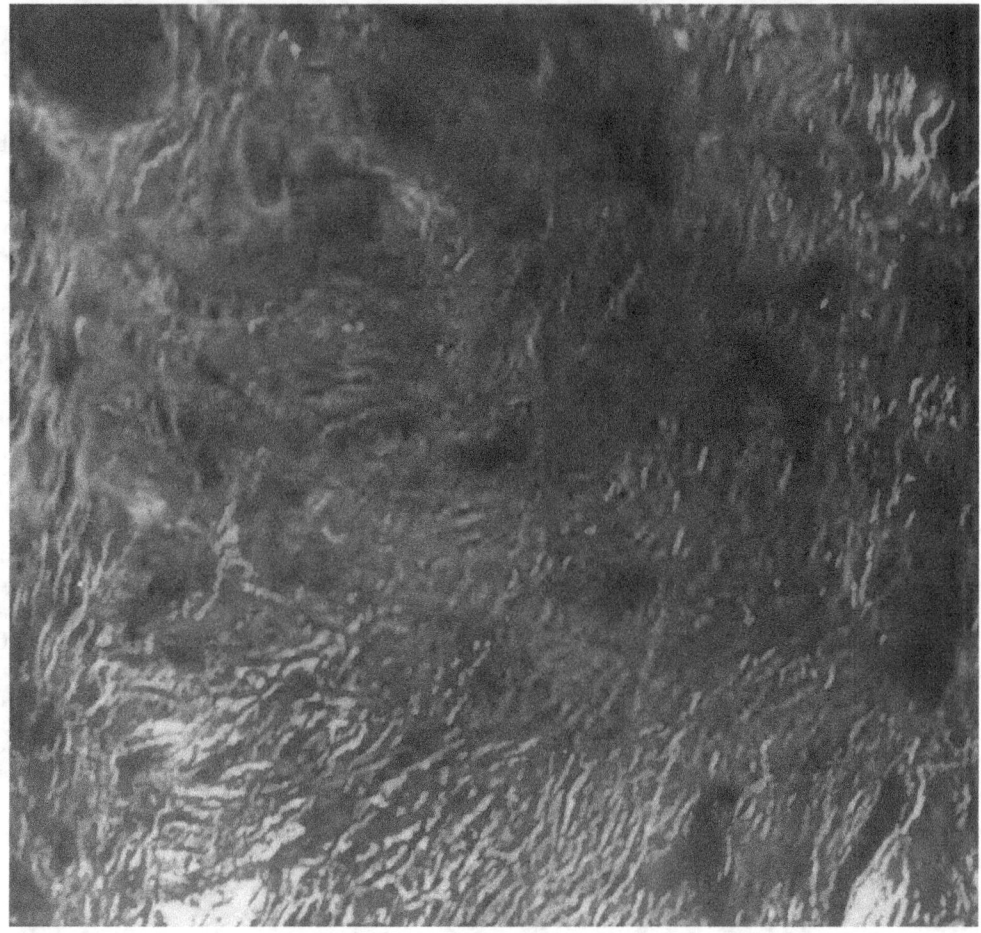

Figure 32
Radar image of the central portion of Alpha Regio, Venus, at a resolution of about 1.5 km, 1988. This, and the image in Fig. 33, illustrate the fine resolutions achieved by the ground-based Arecibo Observatory radar as Magellan began imaging Venus. (Courtesy of National Astronomy and Ionosphere Center, which is operated by Cornell University under contract with the National Science Foundation.)

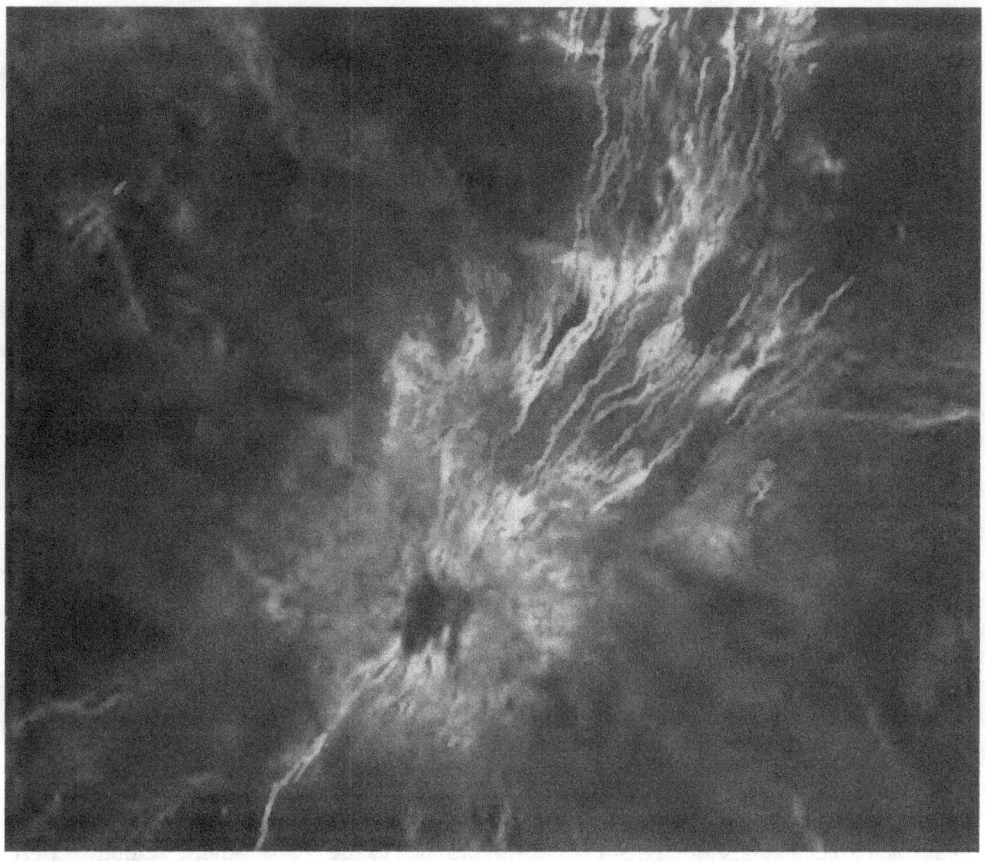

Figure 33

Radar image of Theia Mons in Beta Regio, Venus, at a resolution of 2 km made from data gathered with the Arecibo Observatory radar, 1988. (Courtesy of National Astronomy and Ionosphere Center, which is operated by Cornell University under contract with the National Science Foundation.)

Magellan

Throughout 1987 and into 1988, assembly of the Magellan spacecraft and final testing of the radar proceeded. Hardware, testing, and integration costs, coupled with an overall tight NASA budget, necessitated cutbacks and deferrals from Magellan's fiscal 1988 budget to later years. Some of the top staff transferred to other projects. Magellan Science Manager Neil Nickle, for instance, stepped down, and Thomas Thompson replaced him. Thompson had carried out lunar radar research at Arecibo and Haystack as early as the 1960s, and he was still making lunar observations with the Arecibo UHF radar as late as 1987. Also, he had been on the SEASAT radar team in the 1970s and more recently had made radar observations of Mars with the Goldstone Mars Station.[68]

68.   Thompson 29 November 1994; *V-Gram* no. 15 (January 1989): 16; *V-Gram* no. 14 (May 1988): 2; NAIC QR, Q2/1987.

In September 1988, a month ahead of schedule, the completed craft was shipped to Kennedy Space Center, where final assembly and testing took place. The Magellan launch date was moved up on the Shuttle manifest from October-November 1989 to April-May 1989 to accommodate the launch of Galileo, which needed to go to Venus for a gravity boost. The next launch window, June 1991, would have brought Magellan to Venus nearly a year later than the April–May 1989 opportunity. Launching six months earlier also meant that Magellan would have to circle the Sun one and a half times, rather than the usual one-half circuit, before encountering Venus. Although this trajectory took Magellan almost a year longer to reach Venus than the October–November 1989 opportunity, it still saved a year over the June 1991 trajectory. On 4 May 1989, after trouble with software, a hydrogen pump, and the weather, the Shuttle *Atlantis* carried Magellan aloft from Kennedy Space Center. Magellan became the first planetary mission launched by the Space Shuttle. More problems, including several losses of signal, plagued Magellan's mission.[69]

Magellan entered orbit around Venus on 10 August 1990, 15 months after launch. On 15 August, the radar sensor was turned on and powered up in preparation for the first in-orbit radar test. The next day, during the radar test, the spacecraft lost its "heartbeat" and protected itself by invoking on-board fault-protection routines. Ground control noted this immediately by the terrifying loss of signal. Communications were re-established, then lost a few days later. After a shaky start, the radar began mapping on 15 September 1990.

Mission personnel arranged the first images into mosaics. The mosaics covered about 500 km segments of 30 or more individual image strips. One of the first mosaics was centered at 27° South latitude and 339° longitude in the Lavinia region of Venus. It showed three large impact craters, with diameters ranging from 37 to 50 km. The craters showed many features typical of meteorite impact, including rough, radar-bright ejecta, terraced inner walls, and central peaks. Numerous domes of probable volcanic origin were visible in the southeastern corner of the mosaic. The domes ranged in diameter from 1 to 12 km; some had central pits typical of volcanic shields or cones.[70]

During its 243-day prime mission, Magellan amassed more imaging data than all previous U.S. planetary missions combined.[71] Magellan mapped over 90 percent of the planet's surface, covering regions from 68° South latitude to the North pole. The images were to have a resolution of about 120 meters near the equator, degrading slightly to about 190 meters near the poles because of the elliptical nature of the orbit. Although budgetary cuts had threatened to lower the resolution of Magellan radar images, the application of advanced digital electronic circuitry had restored the mission's high resolution capability.

SAR data from each orbit was to be processed to make image strips about 350 pixels wide in the across-track dimension by 220,000 pixels in the along-track direction. Some 1,852 such SAR image strips were to be generated by JPL's Multimission SAR Processing Laboratory during the primary mission. These strips were to be sufficient in number and coverage to encircle the planet, with overlap of adjacent strips even in lower latitudes. Image element widths were 75 meters to properly preserve both the along and cross-track spatial resolutions.

Each strip is called a Full-Resolution Basic Image Data Record or F-BIDR. In total, the 1,852 F-BIDR SAR image strips formed a data set in excess of 100 billion bytes. The large volume and the unwieldy width-to-length ratios for the data made them unsuitable for general use. Thus, further processing was necessary to produce mosaicked images (Mosaicked Image Data Records or MIDRs) that could be more readily used in photo-

69.   *V-Gram* no. 15 (January 1989): 1; *V-Gram* no. 16 (August 1989): 1.
70.   *V-Gram* no. 18 (October 1990): 1–2.
71.   *V-Gram* no. 13 (October 1987): 1.

interpretative studies and in comparisons with the other Magellan data. Generating full-resolution mosaics for the 90 percent of the planet covered by F-BIDRs created an enormous data set, severely taxing available processing facilities. To streamline processing and to focus efforts toward production of sets of mosaics that could be used for a variety of studies, a decision was made to compile and distribute global mosaics from compressed F-BIDR data.[72]

The USGS converted the data into a set of 62 maps in the standard 1:5,000,000 USGS planetary series. The maps showed SAR data at a resolution of about one km, and they were to contain altitude contours. In addition, a set of about 200 photomosaics were to show the entire mapped area of the planet at a resolution of 225 meters, and an additional set of about 250 photomosaics at the highest resolution, about 100 meters, were to be prepared for selected sections of the planets. Complementary data products were to include a topographic map at about 10-km surface resolution with a height accuracy of better than 50 meters, as well as special products displaying surface roughness, reflectivity, brightness temperature, and emissivity. Today, the radar data is also available in annotated digital form on CD-ROMs.[73]

Key to creating these and other Venus images was an accurate knowledge of the planet's pole position and spin vector. An analysis by Irwin Shapiro and John Chandler of 1988 Arecibo radar data supplied by Don Campbell, Alice Hine, and Nick Stacy provided a new pole position, accurate to better than 3 km, and a more accurate measurement of the planet's rotational period.[74] Such participation in NASA space missions by radar astronomers as "mission support" already had been the norm for two decades.

Don Campbell and Gordon Pettengill also worked closely with Stanford scientists Len Tyler and Dick Simpson, who participated on the science team. Tyler chaired the Surface Electrical Properties (SEP) Team, composed of Tyler, Campbell, and Gerald Schaber (USGS). Tyler, Simpson, and John Vesecky used the altimeter function of Magellan's radar to look at dielectric constants and roughness, to study the top meter of Venus's surface, and to relate its structure to its interaction with radar waves. They transferred their data to a CD, with the intention of sending copies to scientists with whom they

---

72.   *V-Gram* no. 10 (January 1987): 9–10.
73.   *V-Gram* no. 8 (24 March 1986): 2–3.
74.   *Magellan Final Science Reports*, Report D-11092 (Pasadena: JPL, 22 October 1993), p. 25; Shapiro, Chandler, Campbell, Hine, and Stacy, "The Spin Vector of Venus," *The Astronomical Journal* 100 (1990): 1363–1368. See also the analysis done at Goldstone: Slade, Zohar, and Jurgens, "Venus: Improved Spin Vector from Goldstone Radar Observations," *The Astronomical Journal* 100 (1990): 1369–1374.

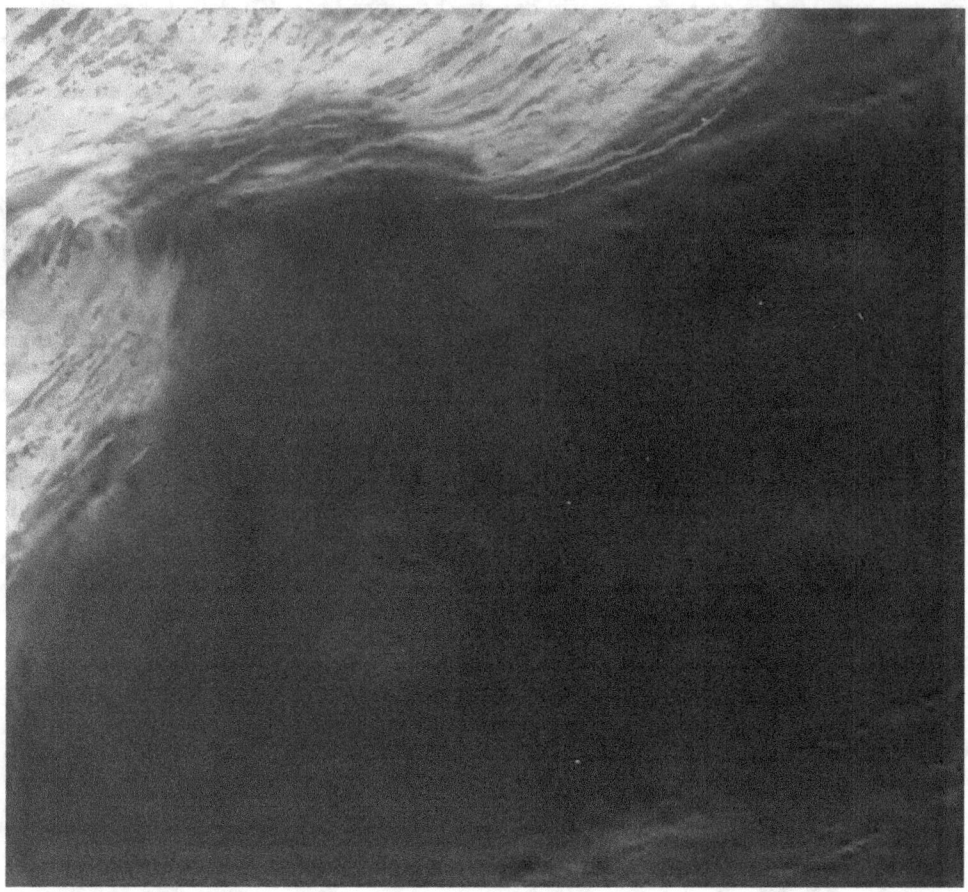

Figure 34

Radar image of Venus at 65 degrees east longitude, along the western edge of Maxwell Montes, made from Magellan observations. The sloping edge of Maxwell Montes, the highest mountain on Venus, is visible along the right hand side of the image. The imaged area is 300 km wide. (Courtesy of NASA, photo no. 90-H-752.)

collaborated, such as Don Campbell, Peter Ford, and Gordon Pettengill, as well as interested geologists.[75]

Typical of Big Science projects, Magellan thus became a meeting ground for different scientific disciplines and subdisciplines. Its broad tent covered traditional ground-based radar astronomy and Stanford bistatic radar astronomy, as well as planetary geology. Magellan accelerated cross-fertilization between planetary geology and radar that

    75.   Simpson 10 May 1994; Simpson and Tyler, "Venus Surface Properties from Magellan Radio and Radar Data," *V-Gram* 18 (October 1990): 12–18. For the results, see Tyler, Ford, Campbell, Charles Elachi, Pettengill, and Simpson, "Magellan: Electrical and Physical Properties of Venus' Surface," *Science* 252 (1991): 265–270; Tyler, Simpson, Michael J. Maurer, and Edgar Holmann, "Scattering Properties of the Venusian Surface: Preliminary Results from Magellan," *Journal of Geophysical Research* 97 (1992): 13,115–13,139. Pettengill and Ford also produced dielectric-constant and roughness maps to accompany the global topography and emissivity data they produced. The Stanford investigators used different, but complementary, algorithms that combined the altimetry and imaging SAR data to obtain estimates of surface roughness and dielectric constant. Both data sets were made available on CD-ROMs.

made radar results (mainly range-Doppler images and topography) more accessible to a larger community of investigators. As Don Campbell reflected: "We are suddenly much more respectable than we used to be! I don't want to characterize what people thought of us, but to some degree I suspect that we were regarded as a little bit of the fringe. Radar astronomy was regarded as a messy and expensive occupation. We came up with good stuff, but how we did it was not all clear!"[76]

As radar astronomers grew closer to planetary geology, they sought out their new audience in new scientific settings. Radar astronomers still discussed their findings at meetings of the IAU, the AAS Division for Planetary Science, and URSI, but also at American Geophysical Union (AGU) meetings. General science and astronomy journals, such as *Science* and *The Astronomical Journal*, and even more so the specialized planetary science journals, such as *Icarus* and *Earth, Moon, and Planets*, remained forums for publication. In addition, because they had added the planetary geology community to their audience, radar astronomers now published in the *Journal of Geophysical Research* and *Geophysical Research Letters*.

The new audience also shaped radar astronomy funding, although less so at the Arecibo Observatory, where the NSF-NASA agreement assured an annual budget for radar astronomy research. Researchers elsewhere seeking NASA money for planetary surface studies faced the demands of the NASA planetary geology program. When Dick Simpson or Len Tyler, for instance, applied for geology program funds to study planetary surfaces, geologists reviewed their proposals. One of the frequent comments by those reviewers was that the proposal should include a geologist on the science team. As a result, Dick Simpson approached USGS Menlo Park geologist Henry Moore to collaborate with him.[77] Through their role as proposal reviewers, then, planetary geologists began to shape radar astronomy research proposals.

Throughout the 1970s, as planning for Magellan and the flight of Pioneer Venus took place, the field of radar astronomy, measured in terms of active practitioners and telescopes, grew smaller. In 1980, the Arecibo Observatory was essentially the sole active telescope; it supported four active investigators. In contrast to this Little Science reality stood the Big Science of Magellan. Around a single radar instrument, the big-budget, multi-year mission organized individual scientists into groups that crossed turf boundaries (radar astronomy versus Stanford "space exploration") and that fostered common interests among fields (planetary radar and geology scientists).

Although the exploration of planetary surfaces with space-based radars seemed to invigorate radar astronomy, the space-based approach has its limits in an era of budgetary limits. Cassini probably will be the last mission to carry a radar experiment into space. As currently conceived, Cassini will explore Saturn's cloud-covered moon, Titan, with a SAR. No other solar system bodies have impenetrable atmospheres that lend themselves to radar investigation. The problem of transmitting data back to Earth at distances beyond the orbit of Saturn is a major, though not insurmountable obstacle (as Voyager has shown). The use of laser rather than radar altimeters on future missions means that modifying the altimeter to carry out imaging, as was done on Pioneer Venus, has reached its technological limit (although military research may well yield a laser altimeter capable of imaging).

However, the most formidable barrier to any future mission is the shrinking space and national budgets. The Voyager, Galileo, and Magellan spacecraft were expensive, costing $2–3 billion, huge, standing seven meters high, as tall as most homes, and heavy, weighing several tons. Galileo, for example, weighed three tons. In order to accommodate a future of smaller budgets, NASA has initiated the Discovery program, in which low-cost ($150 million limit) small, lightweight spacecraft with limited scientific objectives carry

---

76.   Campbell 9 December 1993.
77.   Simpson 10 May 1994.

out solar system exploration. One problem with this approach is that missions to Jupiter and Saturn or beyond simply cost too much to fit the budgetary limits set for Discovery missions.[78] Such is the price of practicing science on a large scale.

Magellan also effectively ended ground-based radar observations of Venus. Although a few experiments were still possible, for example, the detection of rain on Venus with an X-band radar or polarization studies of surface scattering properties,[79] they likely will not achieve prominence. Indeed, Don Campbell, who has spent his scientific career doing radar studies of Venus, volunteered to Nick Renzetti of JPL at the Lunar and Planetary Conference at Houston in 1985 that he was not likely to do any more Venus observations; instead, he planned to concentrate on asteroid and comet experiments.[80]

Campbell typified the new direction that planetary radar astronomy began taking after 1975, when the Arecibo and Goldstone upgraded radars became available. Technology still drove planetary radar astronomy. New and better instruments and innovative techniques allowed radar astronomers to solve problems previously unsolvable and to detect and study solar system objects never before explorable with radar. The exploration of those objects in turn presented unusual radar characteristics that led radar astronomers to solve new scientific problems. The dynamic resonance between radar techniques (epistemological issues) and problem solving (scientific questions) thus remained at the heart of planetary radar astronomy. Nonetheless, despite a short spurt of growth following the inauguration of the upgraded Arecibo and Goldstone radars, by 1980 the planetary radar literature had reached a plateau of activity; the field had reached the limits to its growth.

78.   Richard A. Kerr, "Scaling Down Planetary Science," *Science* 264 (1994): 1244–1246.
79.   Goldstein 14 September 1993; Pettengill 4 May 1994. Bill Smith tried to look for rain in Venus' atmosphere at the Haystack Observatory in the 1960s. Smith 29 September 1993.
80.   GSSR Min. 28 March 1985.

Chapter Eight

The Outer Limits

Planetary radar astronomy was a problem-solving activity, an algorithm in search of a problem. Its fundamental driving force was the dynamic interaction between radar techniques and the kinds of problems radar astronomy solved. Improvements in radar hardware and innovative radar techniques, such as range-Doppler mapping, allowed radar astronomy to solve scientific problems of interest to astronomers and geologists. Conversely, problem-solving could bring attention to radar techniques and properties previously neglected or little used, such as the polarization of echoes.

The institutional and financial linking of radar astronomy to NASA at Arecibo and JPL gave the field a mission-oriented cast. The justification for funding was the field's utility to NASA space missions, and access to Goldstone antenna time required specific mission support. Beginning with Viking, participation in NASA missions also brought ground-based radar astronomers into closer collaboration with the radar scientists at the Stanford Center for Radar Astronomy. The distinction made in the 1960s between ground-based planetary radar astronomy and Stanford's "space exploration" held less and less meaning.

Planetary radar astronomy after about 1975 also remained above all else a science driven by technology, namely, access to radars with the transmitter power and antenna and receiver sensitivity to explore the planets. Without those radars, radar astronomy could not exist. The decline of radar astronomy at JPL followed directly from the deteriorating state of the Goldstone radar. Improvements in radar hardware, on the other hand, drove planetary radar forward.

Additional transmitter power and receiver sensitivity meant access to previously unexplored targets. The orbit of Mars defined the outer reaches of planetary radar astronomy until 1975, when both the Arecibo and Goldstone radars underwent upgrades that significantly enhanced their value as research tools, as discussed in Chapter Four. For the first time, the Galilean satellites of Jupiter, the rings of Saturn, cometary nuclei, and a number of both Earth-approaching and mainbelt asteroids came within reach of those planetary radars. Those targets represent considerable radar distances; the round-trip radar time to the moons of Jupiter is about 1 hour and 12 minutes and to Saturn's rings around 2 hours and 15 minutes.

Meanwhile, the planetary radar astronomy community remained small, and Arecibo and Goldstone were the only active research facilities. Arecibo was still a major NSF-funded center for radio astronomy and ionospheric research. On the other hand, funded by NASA, not the NSF, and associated with exploration of the solar system, radar astronomy there occupied a small, peculiar niche, a niche that, nonetheless, furnished a research facility for both Cornell and MIT graduate students to be trained as future radar astronomers.

In contrast, Goldstone did not train graduate students. The radar astronomers at JPL did not hold the kind of appointment at Caltech that permitted them to train graduate students as future radar astronomers, and no Caltech professor was interested in training radar astronomers. A similar situation had existed at Lincoln Laboratory during the 1960s until Pettengill's appointments at Arecibo and his subsequent teaching position at MIT changed that situation and provided the institutional matrix for the training of graduate students as future radar astronomers. In short, the teacher-disciple pattern that prevailed at Arecibo was lacking at JPL, where radar astronomers propagated through job hiring. Planetary radar astronomy at JPL remained unofficial and invisible. Between 1978 and 1986, furthermore, essentially no radar astronomy work took place at Goldstone, because investigators lacked a reliable research instrument.

The Galilean Moons of Jupiter

Among the new radar targets brought into range by the Goldstone X-band and Arecibo S-band upgrades were Ganymede, Europa, Callisto, and Io, named the Galilean moons of Jupiter after their discoverer, Galileo Galilei. The radar exploration of those moons illustrated the interactions between radar astronomers and geologists, as well as the increasing collaboration with Stanford researchers that came to typify ground-based planetary radar. Those moons also puzzled radar astronomers. Never before had they encountered such peculiar radar characteristics among the terrestrial planets. An explanation for the bizarre radar readings came from Earth and from leading edge research in the physics of light.

The first, though unsuccessful, attempt at the Galilean moons took place in 1970. Dick Goldstein (at Goldstone) and Dick Ingalls and Irwin Shapiro (at Haystack) tried to detect echoes from Callisto using the bistatic Goldstack radar, in which the Haystack 300-kilowatt telescope transmitted and Goldstone received.[1] The experiment did not work, however, because of a misunderstanding over polarization.

After unsuccessfully attempting Venus with the Goldstack radar, Ingalls and Goldstein pointed the radar at the Moon and received "the weakest of signals." Goldstein, trained as an electrical engineer, realized what was wrong. Bistatic radars require investigators to agree on the polarization of the wave. Physicists, like Shapiro, use one definition for left-handed polarization, defining handedness from the view of a person looking in the direction that the wave is travelling, while electrical engineers use the opposite convention, defining handedness from the view of the receiving antenna, so left and right are reversed. Goldstack eventually searched for Ganymede and Callisto in late May and early June 1970.[2] The polarization of radar echoes was about to become a key radar technique for studying the Galilean moons and other solar system bodies.

Dick Goldstein and George A. Morris succeeded in detecting Ganymede with the 400 kilowatts of Goldstone S-band radar power on six nights in late August 1974. Those echoes set a record for the longest time of flight to a radar target, one hour and seven minutes. The echoes, though, were very weak, well below the noise level. From those weak echoes, Goldstein and Morris drew conclusions about the surface of Ganymede.

From the total signal power returned and the width of the spectrum, they concluded that Ganymede "must have a considerable degree of roughness." Their data did not agree with accepted theory, derived from infrared spectra and polarization studies, that Ganymede's surface consisted mostly of ice.[3] Goldstein and Morris ventured that the most

---

1.   Referred to in Campbell, Chandler, Pettengill, and Shapiro, "Galilean Satellites of Jupiter: 12.6-Centimeter Radar Observations," *Science* 196 (1977): 650.

2.   Shapiro 1 October 1993; "Funding Proposal, 'Plan for NEROC Operation of the Haystack Research Facility as a National Radio/Radar Observatory,' NSF, 7/1/71–6/30/73," 26/2/AC 135, and Sebring to Hurlburt, 27 March 1970, 18/2/AC 135, MITA; NEROC, Proposal to the National Science Foundation for Programs in Radio and Radar Astronomy at the Haystack Observatory, 8 May 1970, pp. III.8- III.10, LLLA; JPL 1970 Annual Report, p. 14, JPLA.

3.   See Joseph Veverka, "Polarization Measurements of the Galilean Satellites of Jupiter," *Icarus* 14 (1971): 355–359; John S. Lewis, "Low Temperature Condensation from the Solar Nebula," *Icarus* 16 (1972): 241–252. Although Io, Ganymede, and Europa were believed covered with frost, Callisto was believed to be different, more like the Moon, though with some frost possibly present.

likely possibility was for the surface to consist of rocky or metallic material from meteoric bombardment embedded in a matrix of ice.[4]

Soon after the Arecibo S-band upgrade reached completion, Don Campbell (NAIC Research Associate) and Gordon Pettengill (MIT) made the first radar detections of Callisto and Europa on 28 September and 5 October 1975, respectively, and detected Ganymede on 30 September. Pettengill and Campbell noticed that the satellites had an unusual radar signature. The three moons were almost uniformly radar bright; they lacked the bright specular return from the subradar point, the area on the target closest to the Earth, that all terrestrial planets exhibit. The uniformity of brightness suggested that the satellite surfaces were probably extremely rough on scales comparable to or larger than the wavelength of 12 cm.

Io remained an elusive radar target. The innermost of the Galilean moons, Io is inside Jupiter's magnetosphere, which may have interfered with the radar waves aimed at Io. Campbell and Pettengill unsuccessfully attempted the satellite twice in 1975, and their attempt to detect Io in January 1976 yielded only a weak echo that indicated an error in the ephemeris large enough to explain the previous failed attempt. Not until 1987, when improved hardware was available, did radar astronomers begin to receive good echoes from Io.

After reducing their January 1976 data on the four Galilean moons, Campbell and Pettengill found surprisingly large radar cross sections for Europa and Ganymede, approximately 1.5 and 0.9 times the geometric cross section, respectively, while those for Callisto and Io were around 0.4 and 0.2, respectively. The radar cross section is a measure of target brightness. Although the values for Callisto and Io were low and typical of the terrestrial planets, the radar cross sections for Europa and Ganymede were abnormally high.[5]

When Pettengill and Campbell resumed their observations of Jupiter's moons in October 1976, the Arecibo radar had a dual polarized circular feed paid for with NASA S-band operations funds. The feed increased total system sensitivity over that available in 1975 and displayed the peculiar radar polarization properties of the Galilean satellites.

Previously, all observations of the Galilean moons had been made with linear feeds in both orthogonal linear polarizations. The transmitter sent out signals with one sense of polarization, and the antenna received both the same linear and orthogonal linear polarizations. The same linear echoes are much stronger than the orthogonal linear echoes for all targets detected by radar. Although the switch from linear to circular polarization did not alter the general character of the spectra for Callisto, Ganymede, and Europa, the circular polarization ratios of the echoes were totally unanticipated.

When radar astronomers transmit a right-handed circularly polarized signal, they expect the echo to return mostly left-handed circularly polarized, the opposite handedness. This type of polarization return is called variously the "expected," "polarized," or "opposite circular" (OC). The echo power returned right-handed circularly polarized is said to have "unexpected," "depolarized," or "same circular" (SC) polarization. The SC-to-OC ratio is known as the circular polarization ratio.

The terminology "expected" and "unexpected" is out of place today. The "unexpected" polarization returns from the Galilean moons and other icy targets are no longer considered unusual or "unexpected." The terms, however, reflected the surprise of radar astronomers in the past, as they discovered polarization returns that differed markedly from those of the terrestrial planets. For the sake of preserving that historical flavor of discovery, and to avoid using terms likely unfamiliar and perhaps confusing to the reader (such as "polarized" and "depolarized"), the terminology "expected" and "unexpected," or OC and SC, will be used throughout.

---

4.   Goldstein and Morris, "Ganymede: Observations by Radar," *Science* 188 (1975): 1211–1212.
5.   Campbell 8 December 1993; NAIC QR Q3/1975, 4–5; NAIC QR Q4/1975, 5; NAIC QR Q1/1976, 6.

In radar observations of the terrestrial planets and the Moon, more power normally returns in the expected than in the unexpected mode. The circular polarization ratio for these targets is about 0.1; for Venus and the Moon, it is only about 0.05. In the case of Jupiter's moons, however, more power returned in the unexpected mode, a phenomenon called circular polarization inversion. For Europa, Ganymede, and Callisto, the average circular polarization ratios were 1.61 ± 0.20, 1.48 ± 0.27, and 1.24 ± 0.19, respectively. They were the first solar system objects for which circular polarization inversion was observed.[6]

The dominance of unexpected polarization from the Galilean satellites was enigmatic and even unbelievable. "That was a bit of a puzzle," Don Campbell recalled. "There was a lot of skepticism, frankly, about the results....That was a really significant puzzle to everybody."[7] The phenomenon was also a puzzle to Steve Ostro, then a graduate student at MIT working under Gordon Pettengill. Ostro was looking for a dissertation topic. He joined Pettengill and Campbell in observing the Galilean satellites at Arecibo in late 1976. "The anticipation," Ostro explained, "was that working on those observations, as well as on the data reduction and interpretation, would evolve into a good thesis topic."[8]

When the bizarre circular polarization inversion first appeared during the 26 October through 7 December 1976 observations, Ostro recalled, "We tested to the point of grasping at straws. Maybe we had crossed the cables. Or maybe somebody had screwed up in the data acquisition program. We checked everything. We couldn't believe it, just couldn't believe it." A test on Venus returned normal echoes. Then they pointed the telescope at Europa, and the circular polarization ratio was about one and a half. At that point, Ostro remembers watching Pettengill reflecting then saying, "Well, now I have to believe it." Then he turned to Ostro and said, "If you can explain this, it would be a good thesis topic."[9]

In order to investigate systematically the unusual radar cross sections and polarization ratios of the Galilean moons, Ostro, Campbell, and Pettengill undertook a new series of 20 observation sessions in November and early December 1977 and obtained results similar to those found the previous year.[10]

Ostro, Pettengill, and Campbell continued their campaign on the Galilean satellites in February 1979 and March–April 1980, when the satellites were in different phases. Also, in order to determine whether the strange polarization ratios were a function of frequency, Don Campbell undertook a separate series of observations with the old 430-MHz (70-cm) radar and obtained a weak detection of Europa, but not of Ganymede.[11] Jupiter then left the declination window of the Arecibo Observatory until 1987.

In order to account for the unusual radar signatures of Europa, Ganymede, and Callisto, Steve Ostro developed a model, published in 1978. The model postulated a thick surface layer of ice saturated with nearly hemispherical surface craters. Hemispherical craters would favor double reflection of radar waves at a 45° angle at each reflection, so that most of the signal would return with the same handedness of polarization. The same craters could be made to explain the high radar cross sections, as well.[12]

----

6.   Campbell, Chandler, Pettengill, and Shapiro, "Galilean Satellites of Jupiter: 12.6-Centimeter Radar Observations," *Science* 196 (1977): 650–653; Campbell, Chandler, Steven J. Ostro, Pettengill, and Shapiro, "Galilean Satellites: 1976 Radar Results," *Icarus* 34 (1978): 254–267; NAIC QR Q1/1976, 17; Ostro, "Radar Properties of Europa, Ganymede, and Callisto," in David Morrison, ed., *Satellites of Jupiter* (Tucson: University of Arizona Press, 1982), p. 213.

7.   Campbell 8 December 1993.

8.   Ostro 18 May 1994; NAIC QR Q1/1976, 6.

9.   Ostro 18 May 1994.

10.   Campbell, Chandler, Ostro, Pettengill, and Shapiro, "Galilean Satellites: 1976 Radar Results," *Icarus* 34 (1978): 254–267; Ostro, "The Structure of Saturn's Rings and the Surfaces of the Galilean Satellites as Inferred from Radar Observations," Ph.D. dissertation, MIT, 1978; NAIC QR Q4/1977, 5–6; NAIC QR Q1/1978, 6.

11.   Ostro, Campbell, Pettengill, and Shapiro, "Radar Observations of Europa, Ganymede, and Callisto," *Icarus* 44 (1980): 431–440; NAIC QR Q1/1979, 10; NAIC QR Q2/1980, 11.

12.   Ostro and Pettengill, "Icy Craters on the Galilean Satellites?" *Icarus* 34 (1978): 268–279.

Dick Goldstein and Richard R. Green at JPL proposed a different model based on their own observations of the Galilean satellites. After the pioneering observations of 1974 at S-band, Goldstein took additional data on Ganymede during six nights in December 1977 with the Goldstone X-band radar and received alternately right-handed and left-handed circular polarization, in order to compare the expected and unexpected echo strengths. Despite the high transmitter power (343 kilowatts) and low system noise temperature (23 K), the Ganymede echoes were noisy. Nonetheless, the Goldstone data confirmed the Arecibo results, which had been the subject of great incredulity. As Don Campbell recalled, "That confirmation started a significant discussion about the phenomenon. Why were we getting these odd reflections?"[13]

From the spectral data, Goldstein and Green measured the radar cross section and polarization ratios and posited a model of Ganymede's surface. They assumed that the upper few meters of its surface consisted of ice "crazed and fissured and covered by jagged ice boulders." The critical part of the model was a large number of interfaces between ice and vacuum where, depending on the angle of incidence above or below a certain limit (called the critical angle), the sense of polarization was largely preserved and most of the power remained in the original polarization sense. In a 1982 review article, Steve Ostro concluded that "many questions remain about interpretation of the radar results, but we seem to be pointed in a sensible direction."[14]

Voyager 1 had begun sending back pictures of the Jupiter system in early 1979. Geologic activity on Ganymede appeared varied, while Callisto's entire surface was densely cratered. Europa probably was covered completely by ice.[15] More information than ever was available about the surfaces of the Galilean satellites, yet none of it resolved the questions raised by planetary radar astronomers, who, in the meantime, attempted to explain the strange radar characteristics of the Galilean satellites based on reflection geometries and radar scattering rules, not the geology of those worlds as revealed by Voyager imagery.

Among those offering explanations for the high cross section and circular polarization inversion was Tor Hagfors. He proposed that the satellites' unusual radar signatures were due not to reflections at the interfaces of ice and vacuum, as Goldstein and Green had suggested, but rather to the bending of the incident wave around continuous gradients in refractive index.[16] Von Eshleman developed an argument around refraction scattering from imperfect spheroidal lenses. Then he modified his argument and incorporated Ostro's notion of hemispheroidal impact craters, as well as elements from the Goldstein-Green model.[17]

The Ostro, Goldstein-Green, Hagfors, and Eshleman models all rested on radar geometries and scattering mechanisms. Not a single model linked surface or subsurface structure realistically to the radar signatures, nor did the models explain the origins of those structures. Positing the existence of hemispherical craters was one thing; finding geologic evidence for them was another. Not surprisingly, Voyager revealed no hemispherical craters on any of the Galilean satellites. Ostro now sought an explanation for the radar signatures of the Galilean moons in collaboration with USGS planetary geologist Eugene Shoemaker.

13.   Campbell 8 December 1993.

14.   Goldstein and R. Green, "Ganymede: Radar Surface Characteristics," *Science* 207 (1980): 179–180; Ostro, "Radar Properties of Europa, Ganymede, and Callisto," in Morrison, *Satellites of Jupiter*, pp. 225–233, and quote p. 235.

15.   Morrison and Jane Samz, *Voyage to Jupiter*, NASA SP-439 (Washington: NASA, 1980), pp. 58, 60 and 142.

16.   Hagfors, Gold, and M. Ierkic, "Refraction Scattering as Origins of the Anomalous Radar Returns of Jupiter's Satellites," *Nature* 315 (1985): 637– 640.

17.   Eshleman, "Mode Decoupling during Retrorefraction as an Explanation for Bizarre Radar Echoes from Icy Moons," *Nature* 319 (1986): 755–757; Eshleman, "Radar Glory from Buried Craters on Icy Moons," *Science* 234 (1986): 587–590.

Shoemaker had a rather simple and elegant geologic solution to the problem. In developing his solution, Shoemaker drew upon his knowledge of the lunar regolith and Voyager data. He assumed that the surfaces of the Galilean moons were exactly like that of the Moon. From statistics of craters observed in Voyager images of Ganymede and Callisto, Shoemaker inferred that the surfaces of those moons had a history of meteor bombardment similar to that of the Moon. He concluded that they were probably blanketed with fragmental debris produced by prolonged meteoroid bombardment. The only difference, then, between the Moon and Jupiter's moons was that the rocks on the Galilean satellites were made of ice, and the ice, given the extremely low ambient temperatures, would behave like a silicate rock. Ice is highly transparent to radar waves, so the icy surfaces of the Galilean moons would permit radar waves to penetrate those surfaces to a far greater extent than if they were made of silicate rock. The combination of the greater penetrating depth and the greater number of scattering events could provide an explanation for the peculiar radar signatures of the Galilean satellites.[18]

The primary contribution of the Ostro-Shoemaker model was its geological perspective. Nonetheless, the model only partially explained the radar results; a satisfactory understanding of the detailed scattering mechanism that gave rise to the odd radar signatures still remained beyond reach. Meanwhile, Steve Ostro and Don Campbell had begun a new series of radar observations of the Galilean satellites at Arecibo in 1987. Unlike the previous campaign, Stanford researchers under the leadership of Von Eshleman participated. Dick Simpson took data at Arecibo, while a graduate student, Eric Gurrola, was charged with the analysis. Tor Hagfors, who also was interested in experimenting on the Galilean satellites for reasons similar to those of the Stanford researchers, joined their group.

This new series of S-band observations was to provide thorough phase coverage for all three icy satellites (Ganymede, Callisto, and Europa). Started in November 1987, the campaign continued into 1988, then November–December 1989, January 1990, and February–March 1991, when Ostro observed the satellites at rotational and orbital phases chosen to fill in gaps in the 1987–1990 phase coverage.[19] Then Jupiter left the Arecibo declination window.

At the same time, Arecibo obtained the first good echoes from Io. Its radar properties were unlike those of the other Galilean satellites. Data collected in 1976 already had shown that Io's surface was significantly rougher on average than the terrestrial planets, but much smoother than the other Galilean moons. Its radar cross section and polarization ratio were more typical of the inner planets, however, and argued strongly against the presence of significant quantities of surface ice.[20]

In parallel with the 2,380-MHz (12.6-cm) observations, Don Campbell studied the Galilean moons with the 430-MHz (70-cm) radar beginning in November 1988, the first time in 25 years that the UHF radar had been used in the continuous-wave mode. He detected Ganymede and Callisto, then in November–December 1989, made the first UHF detection of Europa. The purpose of the experiment was to compare the polarization properties of the Galilean satellites at both S-band and UHF. Campbell discovered that the echoes from Ganymede at UHF were reminiscent of those at S-band. Additional UHF measurements made in January 1990 apparently confirmed that the peculiar polarization ratios of the Galilean moons were independent of frequency.[21]

18.   Shoemaker 30 June 1994; Ostro and Eugene M. Shoemaker, "The Extraordinary Radar Echoes from Europa, Ganymede, and Callisto: A Geological Perspective," *Icarus* 85 (1990): 335–345.

19.   E-mail, Simpson to author, 9 November 1994; NAIC QR Q2/1987, 7; Q3/1987, 8-9; Q4/1987, 9; Q2/1988, 9; Q4/1988, 8; Q4/1989, 7; Q1/1990, 7; Q1/1991, 7; Q1/1992, 8.

20.   Campbell, Chandler, Ostro, Pettengill, and Shapiro, "Galilean Satellites: 1976 Radar Results," *Icarus* 34 (1978): 254–267; NAIC QR Q1/1976, 6; Q4/1977, 5-6; Q2/1987, 7; Q3/1987, 8–9; Q4/1987, 9.

21.   NAIC QR Q4/1989, 7; Q1/1990, 7.

Steve Ostro, who now had a position at JPL, also observed the Galilean satellites with the Goldstone X-band radar between 1987 and 1991 and measured polarization ratios and radar cross sections. The combined X-band, S-band, and UHF radar data taken over a long period of time documented the degree to which the satellites' radar properties depended on target, rotational phase, and frequency.[22] They provided a considerable base upon which to explain the bizarre radar signatures of the Galilean moons, and a reasonable explanation soon was in hand.

Toward the end of the Arecibo and Goldstone campaign on the Galilean satellites, Bruce Hapke, an optical astronomer and scattering expert, drew attention to a growing body of literature on laboratory and theoretical investigations of a phenomenon called alternatively "coherent-backscatter effect" or "weak localization." The effect has potential application in a new class of semiconductors in which photons, rather than electrons, perform circuitry functions. Weak localization of light takes place at the microscopic level and arises from a combination of coherent multiple scattering and interference. Backscattered intensity is enhanced, and the forward diffusion through the low-loss medium reduced, by constructive interference between fields propagating along identical but time-reversed paths.[23]

At the suggestion of Steve Ostro, Kenneth J. Peters of Caltech did calculations that demonstrated that coherent backscattering from forward scatterers could explain the high reflectivity and polarization ratios of the Galilean satellites.[24] Coherent backscattering now appeared to explain adequately the high radar cross sections and circular polarization ratios of the icy satellites, and it was consistent with the geologic picture of those moons painted by Gene Shoemaker. The scattering might arise less from individual pieces of ejecta, but more likely from uncoordinated changes in porosity (and hence refractive index) that occur randomly throughout "smoothly heterogeneous" regoliths, argued Ostro and Shoemaker.[25]

Additional data on the radar properties of icy surfaces came from observations of the Earth. In June 1991, the NASA/JPL airborne synthetic aperture radar (AIR-SAR) flew over a vast portion of the Greenland ice sheet called the percolation zone, where summer melting generates water that percolates down through the cold, porous dry snow then refreezes in place to form massive layers and pipes of solid ice. The AIR-SAR radar observed the Greenland ice sheet at several wavelengths (5.6-, 24-, and 68-cm) and obtained values for the circular polarization ratio greater than one.[26]

The riddle of the strange radar signatures of the Galilean satellites focused radar astronomers' attention on epistemological questions, the fundamental need to understand and interpret radar echoes and their relationship to the target. Such questions, though, were of interest only to radar astronomers; their solutions contributed to an

22.   Ostro, Campbell, Simpson, R. Scott Hudson, Chandler, Keith D. Rosema, Shapiro, Standish, R. Winkler, Donald K. Yeoman, Ray Vélez, and Goldstein, "Europa, Ganymede, and Callisto: New Radar Results from Arecibo and Goldstone," *Journal of Geophysical Research* 97 (1992): 18,227–18,244. The Goldstone observations were made 10–11, 13, 15–16, 22, 26, and 29–30 November 1988; 5 and 8 December 1988; 13, 14, 15, 18, 19, 20, 22, 24, 27, and 29 December 1989; 13, 18, 22 and 27 December 1990.

23.   Ostro 18 May 1994; Bruce Hapke, "Coherent Backscatter and the Radar Characteristics of Outer Planet Satellites," *Icarus* 88 (1990): 407–417; Hapke and David Blewett, "Coherent Backscatter Model for the Unusual Radar Reflectivity of Icy Satellites," *Nature* 352 (1991) 46–47; Sajeev John, "Localization of Light," *Physics Today* 44 (May 1991): 32–40.

24.   Kenneth J. Peters, "Coherent-Backscatter Effect: A Vector Formulation Accounting for Polarization and Absorption Effects and Small or Large Scatterers," *Physical Review* B 46 (1992): 801–812; John, "Localization of Light," *Physics Today* 44 (May 1991): 32–40; Ostro 18 May 1994.

25.   Ostro 18 May 1994; Ostro and Shoemaker, "The Extraordinary Radar Echoes from Europa, Ganymede, and Callisto: A Geological Perspective," *Icarus* 85 (1990): 335–345.

26.   Eric J. Rignot, Ostro, Jakob J. Van Zyl, and K. C. Jezek, "Unusual Radar Echoes from the Greenland Ice Sheet," *Science* 261 (24 September 1993): 1710–1711.

understanding of the radar characteristics of planetary surfaces, but not to the more general scientific questions posed by non-radar planetary astronomers. However, if radar astronomers were going to contribute to our knowledge of the Jupiter and Saturn systems, they first had to resolve such basic epistemological issues relating to the radar properties of those planetary systems.

Although the central focus of radar research on the Galilean satellites had been the solution of the satellites' strange radar signatures, the data also has served to correct their ephemerides as part of the Planetary Ephemeris Program of Irwin Shapiro and John Chandler of the Harvard-Smithsonian Center for Astrophysics. The radar data uncovered errors in the ephemerides as early as 1976. A round of Callisto observations carried out beginning in 1987, though, were intended mainly for orbital ephemeris refinement in support of the Galileo mission.[27]

Sensitized to the needs of planetary geologists, Ostro also attempted to relate radar data collected at Arecibo and Goldstone between 1987 and 1991 to surface features on the Galilean moons. The most prominent features tentatively identified in the echo spectra were Ganymede's Galileo Regio and Callisto's Valhalla Basin.[28] Using a new radar coding technique, John Harmon and Steve Ostro observed Ganymede and Callisto at Arecibo from February to March 1992 and obtained the first range-Doppler images of the moons. These observations also constituted the first successful ranging measurements to the Galilean satellites and the farthest radar distance measurements ever reported.[29]

The exploration of the Galilean moons of Jupiter illustrated the increasing complexity of the planetary radar paradigm. Hardware improvements, coding techniques, and even discoveries made in optics laboratories shaped the science done by radar astronomers. Moreover, despite the shift toward geology, planetary radar remained oriented toward astronomical questions and NASA missions, such as Galileo.

The Outer Limits

The rings of Saturn, like the Galilean moons of Jupiter, presented radar astronomers with a target very different from the terrestrial planets. The rings of Saturn were believed to be icy and until the 1970s, were thought to consist of tiny, micron-sized particles. Radar astronomy upset that conception of the rings. In doing so, radar astronomy also set a distance record: the round-trip light time to the rings was about 2 hours and 15 minutes.

After an unsuccessful try in 1967, Haystack researchers successfully bounced X-band radar waves off the rings in 1973.[30] Earlier, however, in December 1972 and January 1973, Richard Goldstein and George A. Morris, Jr., at JPL detected the rings with the S-band Goldstone Mars Station. Making the observation was not easy. The orientation of the rings is optimum for radar observations only twice during each 29-year orbit of Saturn, when the rings are most tilted to the line of sight and present the largest projected area. At the same time, the Doppler spreading and consequent dilution of the signals in the noise is the least.

---

27.   NAIC QR Q1/1976, 7; Q4/1977, 5–6; Q3/1987, 8–9; Q2/1988, 9; Q1/1992, 8; Campbell, Chandler, Pettengill, and Shapiro, "Galilean Satellites of Jupiter: 12.6-Centimeter Radar Observations," *Science* 196 (1977): 651; Ostro, Campbell, Simpson, Hudson, Chandler, Rosema, Shapiro, Standish, Winkler, Yeoman, Vélez, and Goldstein, "Europa, Ganymede, and Callisto: New Radar Results from Arecibo and Goldstone," *Journal of Geophysical Research* 97 (1992): 18,227-18,244.

28.   Ostro, Campbell, Simpson, Hudson, Chandler, Rosema, Shapiro, Standish, Winkler, Yeoman, Vélez, and Goldstein, "Europa, Ganymede, and Callisto: New Radar Results from Arecibo and Goldstone," *Journal of Geophysical Research* 97 (1992): 18,227–18,244; NAIC QR Q1/1991, 7.

29.   Ostro, Pettengill, Campbell, Goldstein, *Icarus* 49 (1982): 367.

30.   NEROC, *Final Progress Report Radar Studies of the Planets*, 29 August 1974, pp. 1, 3, 6 and 8–9; Log Book, Haystack Planetary Radar, HR-73-1, 27 June 1973 to 26 November 1973, SEBRING; and Goldstein, R. Green, Pettengill, and Campbell, "The Rings of Saturn: Two-Frequency Radar Observations," *Icarus* 30 (1977): 105.

The echoes Goldstein and Morris found were unexpectedly strong. The rings were inclined at an angle about 26° with respect to the line of sight, and the amount of power returned from the rings was about 10 times that for Mercury and five times that for Venus. Moreover, wrote Goldstein and Morris: "Particles of any material that are much smaller than our wavelength [12.6 cm] are ruled out by our data....Large (compared to the wavelength), irregular, rough particles could produce the observed echoes."[31]

Shortly thereafter, on 31 July and 1 August 1973, JPL organized a workshop on Saturn's rings at the request of S. Ichtiaque Rasool of the Planetary Programs Office, NASA Headquarters. Gordon Pettengill organized the scientific program. The workshop responded to an upsurge in interest in the Saturn system, and the outer systems in general, in anticipation of the 1977 Mariner Jupiter/Saturn mission, later known as Voyager.

The interpretation of the JPL radar experiment on Saturn's rings surprised astronomers[32] and caused rethinking about the ring particles and models published by radio astronomers. The amazingly large particle size also raised questions about the safety of a spacecraft near the rings and gave rise to NASA and JPL interest in the radar results, which George Morris discussed at the workshop. Excited by the Goldstone radar findings, astronomers during the general discussion expressed an interest in obtaining more radar data on the rings.[33]

The JPL results also surprised radar astronomers. For example, Gordon Pettengill (MIT) and Tor Hagfors (then at the Department of Electrical Engineering of the Norges Tekniske Hogskole, Trondheim, Norway), based on their own radar experience with the terrestrial planets and the asteroids Icarus and Toro, felt that the radar cross section observed by Goldstein and Morris, 0.62 ± 0.15, was unreasonably high. "Even by assuming the particulate matter in the rings to have linear dimensions comparable to or larger than the radar wavelength," they wrote, "we are left with the need to explain a radar scattering mechanism more efficient by a factor of about 10 than that of the inner planets, unless we wish to postulate an unreasonable ring particle density or composition."[34]

Astonished, too, were radio astronomers. The high radar return had to be reconciled with the rings's low radio emission, as well as with optical and infrared results.[35] As the enigma of Saturn's rings continued to puzzle astronomers, the Arecibo S-band upgrade reached completion. It seemed only natural, as Don Campbell explained, that the first radar experiment with the upgraded telescope should be an attempt to detect echoes from the rings of Saturn: "When Arecibo first came on line in 1974, the very first thing we did to test the transmitting system, apart from trying to communicate with a star system 25,000 light years away, was to run a bistatic radar measurement on the rings of Saturn

31.   Goldstein and Morris, "Radar Observations of the Rings of Saturn," *Icarus* 20 (1973): 260–262; Morris, "Distribution and Size of Elements of Saturn's Rings as Inferred from 12-cm Radar Observations," in Frank Don Palluconi and Pettengill, eds., *The Rings of Saturn*, SP-343 (Washington: NASA, 1974), p. 73.

32.   Campbell 8 December 1993. See, for example, Allan F. Cook, Fred A. Franklin, and F. D. Palluconi, "Saturn's Rings: A Survey," *Icarus* 19 (1973): 317–337 and Pollack, "The Rings of Saturn," *American Scientist* 66 (1978): 30–37.

33.   Rasool, "Foreword," in Palluconi and Pettengill, pp. v-vi; ibid., pp. 192–195; and Morris, "Distribution and Size of Elements of Saturn's Rings as Inferred from 12-cm Radar Observations," pp. 73–82. Interestingly, when a subsequent workshop on Saturn's rings was held at the Reston International Conference Center, Reston, Virginia, 9–11 February 1978, and sponsored by the NASA Office of Space Science, no radar presentations were made. The purpose of the workshop was more tightly defined than the 1973 workshop; the 1978 workshop strictly prepared for the Voyager mission.

34.   Pettengill and Hagfors, "Comment on Radar Scattering from Saturn's Rings," *Icarus* 21 (1974): 188–190, esp. 188.

35.   Jeffrey N. Cuzzi and David Van Blerkom, "Microwave Brightness of Saturn's Rings," *Icarus* 22 (1974): 149–158; Pollack, A. L. Summers, and B. Baldwin, "Estimates of the Size of the Particles in the Rings of Saturn and their Cosmogonic Implications," *Icarus* 20 (1973): 263–279; Morrison and D. P. Cruikshank, "Physical Properties of the Natural Satellites," *Space Science Review* 15.(1974): 722–732; Pollack, "The Rings of Saturn," *Space Science Review* 18 (1975): 3–97.

with Goldstone. At that time, we had transmitting capability, but we had not yet installed the receivers. The dedication of the upgraded telescope had been in November 1974, and this was in December, when we were trying to get the transmitter really working properly."[36]

Despite equipment difficulties at Arecibo, Goldstone received echoes from Arecibo by way of Saturn.[37] In addition to the bistatic Arecibo-Goldstone radar test on Saturn's rings in December 1974, Arecibo and Goldstone performed dual-polarization experiments on two nights in January 1975. These bistatic linear polarization experiments established that echoes from the rings of Saturn were highly depolarized, that is, more power appeared in the unexpected than in the expected polarization.

Goldstein also conducted monostatic dual-polarization observations with the Goldstone X-band radar on five nights in December 1974 and January 1975 and measured a high circular polarization ratio. Goldstone and Arecibo investigators now knew that Saturn's rings exhibited high linear and circular polarization ratios and that the phenomenon was independent of frequency. Moreover, they confirmed at both X-band and S-band that the rings had high radar cross sections.[38]

The high radar cross sections and polarization ratios of Saturn's rings were puzzling. Campbell and Goldstein considered several possible explanations for those radar properties. Two models appeared plausible. One model hypothesized a thick cloud of irregular water-ice chunks a few centimeters or larger in radius. The other posited a monolayer of multimeter-sized water-frost-coated metallic chunks. Voyager data later rejected the metallic composition of the rings.[39] In summing up the state of knowledge on Saturn's rings in 1975, Allan F. Cook and Fred A. Franklin of the Smithsonian Astrophysical Observatory speculated that the ring particles consisted of water ice, clathrated hydrates of methane, and ammonia hydrates,[40] in agreement with one of the radar models.

Meanwhile, James Pollack and other astronomers proposed that the ring system was diffuse and many particles thick. In order to determine whether the rings of Saturn consisted of one or several layers, and in general to test various models of the thickness and composition of the rings, Gordon Pettengill, Don Campbell, and Steve Ostro undertook further radar observations in 1977, 1978, and 1979 on a total of 13 nights. Like those on the Galilean satellites of Jupiter, the observations became part of Ostro's thesis.[41] In March 1977, also, Gordon Pettengill and Dick Goldstein resumed bistatic observations of Saturn's rings with the Arecibo and Goldstone S-band radars.[42]

The key to the radar observations made in 1977, 1978, and 1979 was the differing tilt angles of the rings during the 13 total nights of observations. The tilt angle of the rings relative to the line of sight declined over those three years from 18.2° to 11.7°, then to 5.6°. The astronomers also received in both senses of circular polarization in order to measure the polarization ratio as a function of tilt angle. Their results, when combined

36.   Campbell 7 December 1993.

37.   NAIC QR Q1/1975, 4.

38.   Goldstein, R. Green, Pettengill, and Campbell, "The Rings of Saturn: Two-Frequency Radar Observations," *Icarus* 30 (1977): 104–110.

39.   L. W. Esposito, Cuzzi, J. B. Holberg, E. A. Marouf, Tyler, and C. C. Porco, "Saturn's Rings: Structure, Dynamics, and Particle Properties," in ̹Tom Gehrels and Mildred Shapley Matthews, eds., *Saturn* (Tucson: University of Arizona Press, 1984), p. 466.

40.   Allan F. Cook and Fred A. Franklin, "Saturn's Rings: A New Survey," in Joseph A. Burns, ed., *Planetary Satellites* (Tucson: University of Arizona Press, 1977), pp. 412–419. See also Cuzzi and Pollack, "Saturn's Rings: Particle Composition and Size Distribution as Constrained by Microwave Observations." *Icarus* 33 (1978): 233–262.

41.   Campbell 8 December 1993; Esposito, Cuzzi, Holberg, Marouf, Tyler, and Porco, "Saturn's Rings: Structure, Dynamics, and Particle Properties," in Gehrels and Matthews, *Saturn*, p. 467; NAIC QR Q1/1978, 7; NAIC QR Q1/1979, 9; Ostro, "The Structure of Saturn's Rings," pp. 105–157.

42.   NAIC QR Q1/1977, 7.

with earlier radar data and the theoretical calculations of Jeffrey N. Cuzzi and James Pollack,[43] provided significant constraints on ring structure.

The observations confirmed that the radar reflectivity of the rings was quite high and that depolarization was also high. The polarization ratio for the Galilean satellites, a mystery not yet solved, however, was higher. The data ruled out all large-particle monolayer models. On the other hand, the polarization and radar cross section results favored ring models of several layers. The radar data also appeared to support particle composition of ice or metal, but not silicate rock.[44]

Ostro, Pettengill, and Campbell also concluded that the A and B rings (the outermost rings) were responsible for most, if not all, of the S-band radar echoes, and that the radar reflectivity of the A ring was nearly as great as the B-ring radar reflectivity. The radar reflectivity of the C ring was notably less than that of the B ring. Also, they found no evidence for radar echoes from beyond the A ring or from the planet itself.[45]

The case of Saturn's rings resulted in radar astronomers contributing to planetary science, in contrast to their studies of the Galilean moons. Those studies for a long time had been limited to epistemological issues, namely, what caused the Galilean moons' strange radar signatures? Radar contributed to Saturn science, on the other hand, by focusing less on such questions of radar technique and more on scientific questions, such as the size of the ring particles and the number and thickness of the ring layers. Although the solution of technical problems was a prerequisite for any radar astronomy problem solving, the lack of obvious relevance to planetary science was a serious matter; the ability to solve scientific problems, especially those relating to NASA space missions, was the basis on which scientists judged the value of radar astronomy and on which funding decisions were made.

Cometary Nuclei

The nuclei of comets provided radar astronomers additional icy research subjects. Comets are believed to represent samples of the most primitive material of the solar nebula and to hold clues to the origin of the solar system.[46] They make challenging radar targets, because close approaches are rare. The relatively small size of comets dictates that they be studied by radar only when they approach Earth at distances of a fraction of an astronomical unit. Also, ephemerides derived from optical data lack the accuracy demanded for radar observations. Only the S-band and X-band upgrades of the Arecibo and Goldstone antennas made radar studies of comets possible.

---

43.   Cuzzi and Pollack, "Saturn's Rings: Particle Composition and Size Distribution as Constrained by Microwave Observations." *Icarus* 33 (1978): 233–262.

44.   Ostro, Pettengill, and Campbell, "Radar Observations of Saturn's Rings at Intermediate Tilt Angles," *Icarus* 41 (1980): 381–388.

45.   Ostro, Pettengill, Campbell, and Goldstein, "Delay-Doppler Radar Observations of Saturn's Rings," *Icarus* 49 (1982): 367–381. See also Ostro and Pettengill, "A Review of Radar Observations of Saturn's Rings," in A. Brahic, ed., *Planetary Rings 1982* (Toulouse: CEPADUES Editions, 1982), pp. 49–55.

Later radar data collected at Goldstone by Goldstein and Jurgens and at Arecibo by Ostro, Pettengill, and Campbell in 1981, when the rings were at a 6° tilt angle, confirmed that the ring particles were large, irregular, and jagged in shape and made of ice; the researchers finally abandoned the notion that they might be metallic. Moreover, they affirmed the conclusion that the A and B rings reflected most, if not all, of the radar echo from Saturn's rings. Goldstein and Jurgens, "Radar Observations of the Rings of Saturn," *Journal of Geophysical Research* submitted for publication; Ostro, Pettengill, Campbell, and Goldstein, "Delay-Doppler Radar Observations of Saturn's Rings," *Icarus* 49 (1982): 367–381; Ostro, Pettengill, and Campbell, "Radar Observations of Saturn's Rings at Intermediate Tilt Angles," *Icarus* 41 (1980): 381–388. This research is summarized in: Ostro and Pettengill, "A Review of Radar Observations of Saturn's Rings," pp. 49–55.

46.   Whipple, "Comets," in J. A. M. McDonnell, ed., *Cosmic Dust* (New York: John Wiley & Sons, 1978), pp. 1–73.

Early attempts all ended in failure. For example, after an attempt in January 1971 on Comet Kohoutek stymied by rain and snow, the Haystack telescope again failed to detect that comet in January 1974. Although Irwin Shapiro had prepared an accurate ephemeris in advance, neither the bandwidth nor the center frequency of the radar echo was known precisely, so they had to search for the echo.[47]

It took the S-band upgrade of the Arecibo Observatory to make the first comet detections possible. Paul G. D. Kamoun, a French student of Gordon Pettengill at MIT, built his dissertation research around those detections. The main objective of his dissertation was to use cometary radar data to discriminate between two different models of cometary nuclei.[48] One model was that proposed by Fred Whipple, who served on Kamoun's dissertation committee, and supported by Zdenek Sekanina, an established expert on comets.

In the Whipple model, the cometary nucleus was like a rotating "dirty snowball," an icy matrix of water ammonia, methane, carbon dioxide, or carbon monoxide, combined with rock, dust and other meteoric debris. A popular model for the nucleus in the early 20th century predicated a "dust swarm" or swarm of solid particles of unknown sizes, each particle carrying with it an envelope of gas, mostly hydrocarbons. However, that model had a number of difficulties, and by the 1970s Whipple's "dirty snowball" model prevailed.[49] Consequently, Kamoun's dissertation did not contribute meaningfully to the comet debate.

Kamoun's research on comets turned around the unsuccessful cometary research begun at Arecibo by Gordon Pettengill, Brian Marsden (Harvard-Smithsonian Astrophysical Observatory), and Irwin Shapiro (who prepared the ephemerides). In late July 1976, they attempted to detect echoes from Comets d'Arrest and Grigg-Skjellerup during three observing sessions. Both attempts failed, although Comet d'Arrest came within 0.15 astronomical units of Earth.[50]

The first comet detected by radar was Comet Encke. As Don Campbell explained, "It was a historic first. We had never actually seen a comet before."[51] French and German astronomers had observed Encke earlier; its name came from the German mathematician and physicist Johann Encke, who initially suggested an elliptical orbit with a period of 12.2 years, then correctly recalculated an elliptical orbit of 3.3 years, the shortest period of any known comet.[52] Comet Encke was due back in November–December 1980. Although Encke had a relatively stable and therefore predictable orbit, optical observations were neither sufficiently numerous nor sufficiently accurate to formulate a satisfactory ephemeris for the radar. Irwin Shapiro and Antonia Forni (Lincoln Laboratory) based the radar ephemerides on optical data from both past appearances and new observations associated with the 1980 appearance supplied by Brian Marsden. The ephemeris difficulties resolved, Kamoun, Campbell, and Ostro observed Encke for 12 hours on seven consecutive days, 2–8 November 1980, about 30 days before the comet reached perihelion and at a distance of slightly more than 0.3 astronomical units from Earth. They found distinct, but very weak, echoes during each observing session.[53]

47.   Log book, Haystack Planetary Radar, HR-73-2, 9 December 1970 to 11 August 1971, SEBRING; Shapiro 1 December 1993; Eric J. Chaisson, Ingalls, Rogers, and Shapiro, "Upper Limit on the Radar Cross Section of the Comet Kohoutek," *Icarus* 24 (1975): 188–189.

48.   Paul Gaston David Kamoun, "Radar Observations of Cometary Nuclei," Ph.D. diss., MIT, May 1983.

49.   Whipple, "A Comet Model. I. The Acceleration of Comet Encke," *Astrophysical Journal* 111 (1950): 375–394; Whipple, "A Comet Model. II. Physical Relations for Comets and Meteors," ibid., 113 (1951): 464–474.

50.   Kamoun, p. 31; NAIC QR Q3/1976, 6-7.

51.   Campbell 9 December 1993.

52.   John E. Bortle, "Comet Digest," *Sky and Telescope* 60 (1980): 290; Kamoun, pp. 37–38.

53.   Kamoun, p. 51; Kamoun, Campbell, Ostro, Pettengill, and Shapiro, "Comet Encke: Radar Detection of Nucleus," *Science* 216 (1982): 293–295; NAIC QR Q4/1980, 8–9.

Next, Kamoun attempted radar observations of the Comet Grigg-Skjellerup, which was discovered in 1902 by Grigg in New Zealand, then re-discovered as a new comet in 1922 by Skjellerup in South Africa. Grigg-Skjellerup has an orbital period of 5.1 years, making it the second shortest periodic comet after Encke. The time of perihelion passage was 15 May 1982, at a perihelion distance of nearly one astronomical unit (0.989).

Compared to other cometary experiments, Kamoun spent an unprecedented and never repeated 49 hours observing the comet between 20 May and 2 June 1982, about a week after it passed perihelion, while the comet was about 0.33 astronomical units from Earth. He received echoes in both senses of circular polarization, but technical problems prevented the acquisition of data on five days. An interesting feature was the very narrow (less than one Hz) Doppler bandwidth of the echo, which indicated either a very specular echo, a slow rotation rate, or collinearity of the polar axis with the line-of-sight.[54]

Comet Austin came next. Unlike Encke and Grigg-Skjellerup, Comet Austin had only been discovered on the morning of 19 June 1982 by Rodney Austin in New Zealand. Alan Gilmore, of Mount John University Observatory, New Zealand, confirmed the discovery. The comet was first reported on 21 June 1982 in IAU circular 3705 of the Central Bureau for Astronomical Telegrams by Brian Marsden, who also computed and made public a set of orbital elements showing that the comet was moving on a parabolic orbit. From the Marsden ephemeris, it appeared that Comet Austin would pass close enough to Earth to detect it with the Arecibo radar.

Following receipt of IAU circular 3706 containing the improved elements of the comet's orbit, Kamoun undertook the task of obtaining telescope time. He attempted to observe the comet on the mornings of 8–12 August 1982. Despite equipment problems that plagued observations on 8 and 9 August, the last three days yielded normal performance. On the last day, 12 August, the analyzing bandwidth was doubled from 380 to 760 hz, with a corresponding increase in the frequency resolution, in order to widen the search window. They computed an ephemeris after the experiment, using all the astrometric observations available for Comet Austin between June 1982 and November 1982. That ephemeris turned out to be substantially different from the ephemeris used during the actual radar observations. Despite correcting for this, and despite the distance from Earth being very similar to that of Comets Encke and Grigg-Skjellerup, five days of observations in August 1982 did not result in a successful detection.[55]

Radar detections of comets were obviously fairly difficult to make, even with the best radar telescope then available. Another opportunity to attempt a newly-discovered comet came later that year. Comet Churyumov-Gerasimenko was discovered on a photograph taken on 11 September 1969 at the Alma-Ata observatory in the Soviet Union by K. I. Churyumov and S. I. Gerasimenko. At the time of Kamoun's radar observations in November 1982, Comet Churyumov-Gerasimenko was 0.39 astronomical units from Earth. It ought to have been detectable by the Arecibo radar. Kamoun attempted Comet Churyumov-Gerasimenko for 33 hours between 7 and 16 November 1982. Serious technical problems on 7 and 16 November prevented acquisition of data. Further difficulties on 8 and 11 November caused loss of some data. In the end, the attempt on Comet Churyumov-Gerasimenko was not successful.[56]

From his successful and unsuccessful observations of comets, Kamoun estimated the radii of their nuclei, which were 0.4–3.6 km for Encke, 0.4–2.2 km for Grigg-Skjellerup,

54. Kamoun, pp. 90 and 85; NAIC QR Q2/1982, 7–8.
55. Kamoun, pp. 108–110; NAIC QR Q3/1982, 8.
56. Kamoun, pp. 21 and 122; NAIC QR Q4/1982, 7–8; K. I. Churyumov and S. I. Gerasimenko, "Physical Observations of the Short-Period Comet 1969 IV," in G. A. Chebotarev, E. I. Kazimirchak-Polonskaya, and Brian G. Marsden, eds., *The Motion, Evolution of Orbits, and Origin of Comets* IAU Symposium 45 (New York: Springer-Verlag, 1972), pp. 27–34. Both Churyumov and Gerasimenko were in the Department of Astronomy, University of Kiev.

less then 1.5 km for Austin, and less than 2 km for Churyumov-Gerasimenko. He also placed upper limits on the number of millimeter and centimeter-sized particles in the coma of the four comets (Table 7).[57]

Table 7
Upper Limits on the Number of Grains in the Coma of Four Comets

Comet	Grain-Size	Assumed Ice	Grain Olivine	Composition Magnetite Iron Sulfide
Encke	mm	4.5×10^{17}	1.5×10^{17}	7.5×10^{16}
	cm	1.5×10^{11}	6×10^{10}	3×10^{10}
Grigg-Skjellerup and Austin	mm	3×10^{17}	10^{16}	10^{16}
	cm	10^{11}	4.5×10^{10}	2×10^{10}
Churyumov-Gerasimenko	mm	6×10^{17}	2×10^{17}	10^{17}
	cm	2×10^{11}	9×10^{10}	4×10^{10}

In setting forth a program of future cometary radar studies, Kamoun noted that the comets attempted in his dissertation could not be observed again during the next 10 years. Despite the scheduled reappearances of Encke in 1984 and 1987, of Grigg-Skjellerup in 1987, and of Churyumov-Gerasimenko in 1989, none of the comets would approach close enough for radar observation. On the other hand, he calculated, even if no improvement in radar sensitivity occurred, other comets would be accessible, particularly Comets Haneda-Campos (1984), Giacobini-Zinner (1985); Borelly and Denning-Fujikawa (1987), and Brorsen-Metcalf and Dubiago (1989).[58] None of those comets, however, was ever observed by radar.

Instead, opportunities, in fact far better opportunities, came from comets never before seen. In early May 1983, as Paul Kamoun was writing his dissertation, preparations were underway at Arecibo to observe Comet IRAS-Araki-Alcock. On 25 April 1983, the Infrared Astronomical Satellite (IRAS) discovered Comet IRAS-Araki-Alcock. Initially, scientists believed it was an asteroid. In either case, it was sure to approach near the Earth. Astronomers calculated that the object would pass Earth at a distance of only 0.03 astronomical units (450,000 km), that is, about 10 times closer than any other comet that Kamoun had observed for his dissertation. In fact, such a close approach for a comet had not been known to have occurred in more than two hundred years. Although Kamoun had pioneered cometary radar, he would miss the most spectacular cometary opportunity. After writing up his thesis, he returned to France and took a position with a French aerospace firm.[59]

But observing Comet IRAS-Araki-Alcock was not going to be easy. Its orbit was highly inclined relative to the Earth's equator, and to make observation at Arecibo that much harder, as Don Campbell explained, "It was moving in declination so rapidly, that it actually went through the entire sky coverage of Arecibo in one day. We had a two-and-a-half-hour observing window, and that was it!"[60]

---

57.   Kamoun, p. 230.
58.   Kamoun, p. 237.
59.   Campbell 9 December 1993; Kamoun, p. 238; Jurgens, "Seeing Comet IRAS," p. 221; information supplied by Pettengill.
60.   Campbell 9 December 1993.

The ability to get good data on Comet IRAS-Araki-Alcock depended heavily on having an accurate ephemeris. That was the job of Brian Marsden and Irwin Shapiro, who had just become Director of the Harvard-Smithsonian Astrophysical Observatory in January 1983, four months before the comet's discovery. "Taking over this place was an all-consuming job," he recalled. "I worked day and night. But for a few days, I dropped this job like a ton of bricks, literally, to develop the ephemeris needed to observe IRAS-Araki-Alcock at Arecibo and Goldstone."[61]

Working closely with Brian Marsden, Shapiro generated an ephemeris for the comet. "It was a big mess," Shapiro explained. "I was up until 2:30 in the morning every night. The difficulty was due to there being very few comet observations, mostly bad. We had to try numerous combinations to sort the good from the bad." Then Shapiro turned to the task of preparing an ephemeris for the radar. "The radar ephemeris was prepared at Lincoln Laboratory; the radar observations were to be made at Arecibo. It was a logistical nightmare, because of the incredible time pressure," Shapiro explained. "As the time of close approach of the comet to Arecibo neared, we sent the ephemeris electronically. It arrived an hour before the comet was to make its one and only pass over head. It worked brilliantly."[62]

Don Campbell took high quality data on IRAS-Araki-Alcock for about three hours during the single observation evening of 11 May when the comet was in the telescope's declination window. Campbell recalled: "We got extremely nice data. You could actually see the echo on the oscilloscope right there in the control room. It was all over the place. A nice sine wave popping in and out. It was all very exciting. We measured only spectra and obtained a lot of very interesting data on IRAS-Araki-Alcock in just that two-hour period."[63]

More surprising than a powerful echo from a relatively large nucleus, the spectra showed a broad low-level skirt distinct from the nucleus echo. The skirt suggested the possible existence of a cloud of unexpectedly large, centimeter-sized ejected particles from the comet. The IRAS-Araki-Alcock skirt spectrum appeared to be consistent with a model in which large grains were ejected from the nucleus by the same gas-drag mechanism used to explain the ejection of the smaller particles making up the dust coma and tail.[64] "This was the first time that such particles had ever been discovered," Campbell explained. "It made the whole experiment much more interesting."[65]

At the same time, Dick Goldstein and Ray Jurgens, in collaboration with JPL comet specialist Zdenek Sekanina, prepared to look at IRAS-Araki-Alcock with the Goldstone radar. Previously, they had made failed attempts at Comets d'Arrest (1976), Kohoutek (1974), and Bradfield (1974).[66] IRAS-Araki-Alcock would be their first successful cometary detection. Their chief obstacle was the resuscitation the Goldstone radar. As Jurgens wrote: "As luck would have it, the JPL radar system had been shut down following

61. Shapiro 1 October 1993.
62. Shapiro 1 October 1993.
63. Campbell 9 December 1993; Harmon, Campbell, Hine, Shapiro, and Marsden, "Radar Observations of Comet IRAS-Araki-Alcock 1983d," *The Astrophysical Journal* 338 (1989): 1071; Harmon, Campbell, Hine, Shapiro, and Marsden, *Radar Observations of Comet IRAS-Araki-Alcock (1983d)* Report 245 (Ithaca: NAIC, September 1988), Pettengill materials.
64. Campbell 9 December 1993; NAIC QR Q2/1983, 7; Harmon, Campbell, Hine, Shapiro, and Marsden, "Radar Observations of Comet IRAS-Araki-Alcock 1983d," *The Astrophysical Journal* 338 (1989): 1071-1093; Campbell, Harmon, Hine, Shapiro, Marsden, and Pettengill, "Arecibo Radar Observations of Comets IRAS-Araki-Alcock and Sugano-Saigusa-Fujikawa," *Bulletin of the American Astronomical Society* 15 (1983): 800; Goldstein, Jurgens, and Zdenek Sekanina, "A Radar Study of Comet IRAS-Araki-Alcock 1983d," *The Astronomical Journal* 89 (1984): 1745–1754; and Shapiro, Marsden, Whipple, Campbell, Harmon, and Hine, "Interpretations of Radar Observations of Comets," *Bulletin of the American Astronomical Society* 15 (1983): 800.
65. Campbell 9 December 1993.
66. Jurgens, "Seeing Comet IRAS," p. 221; Goldstein, Jurgens, and Sekanina, pp. 1745–1754.

the unsuccessful tracks of asteroid 4 Vesta on 28 May 1982. Since the radar system has seen only sporadic usage over the past few years, the X-band transmitter, the 20 year old computer and the data acquisition equipment were unreliable. We were in the midst of a major rebuilding project that would not be put into operation until March 1985. Fortunately, we had not removed the old equipment."[67]

Jurgens and a team of JPL engineers refurbished the radar equipment, while Mike Keesey prepared a radar ephemeris based on orbital elements supplied by Brian Marsden and Irwin Shapiro, who also had supplied the Arecibo ephemeris. The Goldstone observations took place on 11 and 14 May 1982 at both S-band and X-band. On a few runs, echoes were received in the same circular polarization.[68] Goldstein, Jurgens, and Sekanina concluded that the nucleus of Comet IRAS-Araki-Alcock was very rough on a scale larger than the radar wavelength. They did not believe that the predominant backscattering mechanism was similar to that observed from the icy surfaces of the Galilean satellites, but instead consisted of single reflections from very rough surfaces. They posited, furthermore, that the shape of the nucleus appeared to be irregular. Jurgens believed that the nucleus's shape could be represented fairly well by a triaxial ellipsoid having equatorial radii in a ratio of two to one. The JPL radar astronomers estimated its radius to be between three and six km (larger than any comet observed by Kamoun) and its rotational period to be from one to two days.

Because of Jurgens' interest in asteroids, he and his JPL colleagues compared the comet to known asteroids. "The observed spectral shapes are typical of those measured for small Earth-crossing asteroids except for the broadband skirt," they noted. "Due to distance and sensitivity limitations, such a skirt would not have been detected on any asteroid observed so far even if it existed."[69] However, they did not carry out a detailed analysis of the skirt.

Within weeks after Comet IRAS-Araki-Alcock, another new comet, Sugano-Saigusa-Fujikawa, passed the Earth. The two comets coming so closely together created a "once in a lifetime" opportunity. Comet Sugano-Saigusa-Fujikawa came within 0.06 astronomical units of Earth in early June 1983. Don Campbell attempted Sugano-Saigusa-Fujikawa on the one day it was within the Arecibo telescope's declination window, while Jurgens and Goldstein tried during four full days of observations, which delayed the renovation of the Mars Station antenna for one month. "Night after night," Jurgens wrote, "we searched the sky in the area of the comet with no indication of an echo."[70] Arecibo, on the other hand, did find echoes; however, Sugano-Saigusa-Fujikawa was about three times further away than IRAS-Araki-Alcock had been, and it was a smaller comet, so that it was a less interesting and "somewhat disappointing" target.[71]

Despite the many unsuccessful and disappointing attempts to detect comets, until the passing of Comet Halley, only the radar observations of IRAS-Araki-Alcock made at Arecibo and Goldstone contributed to the vast amount of data collected by comet scientists at optical, radio, infrared, and ultraviolet wavelengths.[72] Comet Halley returns every 76 years. Its reappearance prompted a global effort, the International Halley Watch, to coordinate ground and space observations. Unlike previous comets, Halley was investigated from a number of spacecraft sent by Japan (Suisei and Sakigake), the Soviet Union (Vega 1 and 2), and the European Space Agency (Giotto). The radar results, however, did

---

67.   Jurgens, "Seeing Comet IRAS," p. 222.
68.   Jurgens, "Seeing Comet IRAS," p. 222; Goldstein, Jurgens, and Sekanina, pp. 1745–1747.
69.   Goldstein, Jurgens, and Sekanina, p. 1754.
70.   Jurgens, "Seeing Comet IRAS," p. 224.
71.   Campbell 9 December 1993; NAIC QR Q2/1983, 7.
72.   Sekanina, "Nucleus of Comet IRAS-Araki-Alcock (1983 VII)," *The Astronomical Journal* 95 (1988): 1876–1894.

not play a part in the international effort.[73] Radar was still a marginal tool for cometary research.

Comet Halley was to make two close approaches to Earth during its appearance in 1985–1986. At its closest approach in November 1985, it was to be 0.61 astronomical units from Earth, and during its second approach, even closer, 0.41 astronomical units, to Earth in April 1986. At the November 1985 approach, Halley would be visible at both Arecibo and Goldstone, though far below likely detectability at the latter site. Moreover, Halley was not within the Arecibo telescope's limited declination coverage during its closer approach to Earth in April 1986.[74] The chances for viewing Halley thus were small; the best chance was in November and December 1985, when Halley was to be 0.62 astronomical units distant from Earth, not a good distance for observing comets.

At Arecibo, John Harmon observed Halley on 24, 28, 29 November and 1 and 2 December 1985 during its inbound Earth approach and detected a weak echo from Halley at a distance of 0.62 to 0.64 astronomical units, the most distant comet yet detected with radar. With the exception of IRAS-Araki-Alcock, comets observed earlier generally had been about 0.3 astronomical units away. A broadband feature with a high radar cross section and a large Doppler bandwidth dominated the echo spectrum, properties that were inconsistent with an echo from the nucleus. Halley, then, became the second comet to yield a radar detection of grains larger than two cm in radius ejected from the nucleus. Comet Halley also was the first radar bright comet observed; it had the largest radar cross section to date of any comet detected by radar. "If our interpretation of the echoes is correct," Don Campbell explained, "Halley is the first comet to give a stronger echo from particles than from the nucleus itself."[75]

The Arecibo attempt on Halley in 1985 was the last successful radar detection of a comet. In 1990, John Harmon attempted Comet Austin in cooperation with Steve Ostro, who tried to obtain echoes with the Goldstone X-band radar. Harmon also attempted Comet Honda-Mrkos-Pajddusakova in 1990, but again without success.[76] These failures only served to highlight the extreme difficulty of doing radar research on comets and, as a result, the lack of major radar contributions to cometary science.

A Vision of Things to Come

Asteroids did not make easy radar targets, either. Their small size and distance from Earth placed them at the limits of planetary radar capabilities. Also, the known population of asteroids outside the mainbelt between Mars and Jupiter, that is, the known number of asteroids that might approach Earth close enough for radar study, was far smaller than the quantity we know today. After the detection of Icarus at Haystack and Goldstone in June 1968, only six more asteroids came under radar investigation between then and July 1980: five near-Earth asteroids (1566 Icarus, 1685 Toro, 433 Eros, 1580 Betulia, and Phocaea) and two mainbelt asteroids (1 Ceres and 4 Vesta).

73.   E. Grün, ed., "Halley and Giacobini-Zinner," *Advances in Space Research* vol. 5, no. 12 (1985): 1–344; J. W. Mason, ed., *Comet Halley: Investigations, Results, Interpretations*, 2 vols. (New York: Ellis Horwood, 1990); R. Reinhard and B. Battrick, eds., *The Giotto Mission: Its Scientific Investigations* (Noordwijk: ESTEC, European Space Agency, 1986); M. Grewing, F. Praderie, and R. Reinhard, eds., *Exploration of Halley's Comet* (New York: Springer-Verlag, 1986).

74.   Kamoun, pp. 239–240; Campbell, Harmon, and Shapiro, "Radar Observations of Comet Halley," *The Astrophysical Journal* 338 (1989): 1094–1105; Campbell, Harmon, and Shapiro, *Radar Observations of Comet Halley* Report 246 (Ithaca: NAIC, September 1988), Pettengill materials.

75.   Campbell 9 December 1993; Campbell, Harmon, and Shapiro, "Comet Halley," pp. 1094 and 1103; NAIC QR Q4/1985, 8.

76.   NAIC QR Q2/1990, 6.

Interest in asteroids was growing among astronomers during that 12-year period. Tom Gehrels, University of Arizona at Tucson, was the most vocal advocate of asteroid research. During the 1970s, he organized three asteroid conferences at Tucson which provided much of the impetus for the modern investigation of asteroids. He also initiated a program of asteroid detection called Spacewatch. Spacewatch, a survey telescope located on Kitt Peak to discover new asteroids, started operating in May 1963. Tom Gehrels also led an effort to use a modern CCD scanning camera on a specially designed telescope beginning in 1979. In its first two years, Spacewatch discovered 69 new asteroids. The rapid discovery rate of asteroids that started in the 1970s was due largely, however, to the Palomar Planet-Crossing Asteroid Survey (PCAS), begun in 1973 by Eleanor Helin and Eugene Shoemaker. The Survey initially used a 46-cm Schmidt camera to detect asteroids on the four to five nights each month around the new Moon. The exposed photographic plates were subjected to stereoscopic examination the same night they were taken, in case a new asteroid was recorded on the film. If an object were discovered, positional data was relayed by telephone to Brian Marsden at the Harvard-Smithsonian Astrophysical Center, where he headed a center for data on minor planets starting in 1978. Marsden then computed the orbit and ephemerides for further observations.

As a result of the Spacewatch and PCAS programs, the asteroid literature, as measured by citations of asteroid papers, underwent the kind of swift growth that is typical of Big Science.[77] Although radar astronomers at first simply attempted to detect asteroids, both Arecibo and Goldstone investigators initiated systematic programs of asteroid detection and research in the mid-seventies. The focus was on measuring radii, surface roughness, and composition, and on improving orbits. In addition, Ray Jurgens pioneered the modeling of asteroid shapes.

Dick Goldstein, using the Goldstone Mars Station, obtained echoes from 1685 Toro, the first asteroid detected after Icarus, in 1972. After he combined the radar and optical data, Goldstein inferred that the asteroid had an irregular rocky surface slightly smoothed by a mantle of loose material.[78] The following asteroid opportunity, 433 Eros, arrived in January 1975. The experiment carried out on Eros at Goldstone was, in the words of Steve Ostro, "The most important asteroid experiment before 1980," because data was taken at two frequencies (X-band and S-band) and in both senses of circular polarization. "As a result," according to Ostro, "they achieved the best characterization of an asteroid's centimeter-to-decimeter scale surface properties until the late 1980s. By then, all work was dual polarization. Jurgens and Goldstein were well ahead of their time."[79]

The data Goldstein and Jurgens collected indicated that the surface of Eros was much rougher than the Moon or any of the terrestrial planets. They described a surface completely covered with sharp edges, pits, subsurface holes, or embedded chunks. They also estimated the asteroid to have equatorial dimensions of 18.6 and 7.9 km.[80] In order to better describe the shape of Eros and other asteroids, Ray Jurgens developed a triaxial ellipsoid model. His work represented an important first step toward modeling asteroids with radar data. Optical observations often provide the spin rate and pole, prerequisite parameters for determining the shape of an asteroid from radar data.[81]

77.	Clifford J. Cunningham, *Introduction to Asteroids: The Next Frontier* (Richmond: Willmann-Bell, 1988), pp. 2 and 97–101; Tom Gehrels, "The Asteroids: History, Surveys, Techniques, and Future Work," in Gehrels and Matthews, eds., *Asteroids* (Tucson: University of Arizona Press, 1979), pp. 4–5 and 13–14.

78.	Goldstein, D. B. Holdridge, and J. H. Lieske, "Minor Planets and Related Objects: 12. Radar Observations of (1685) Toro," *The Astronomical Journal* 78 (1973): 508–509.

79.	Ostro 25 May 1994.

80.	Goldstein and Jurgens, "Radar Observations at 3.5 and 12.6 cm Wavelength of Asteroid 433 Eros," *Icarus* 28 (1976): 1–15.

81.	Jurgens, "Radar Backscattering from a Rough Rotating Triaxial Ellipsoid with Applications to the Geodesy of Small Asteroids," *Icarus* 49 (1982): 97–108.

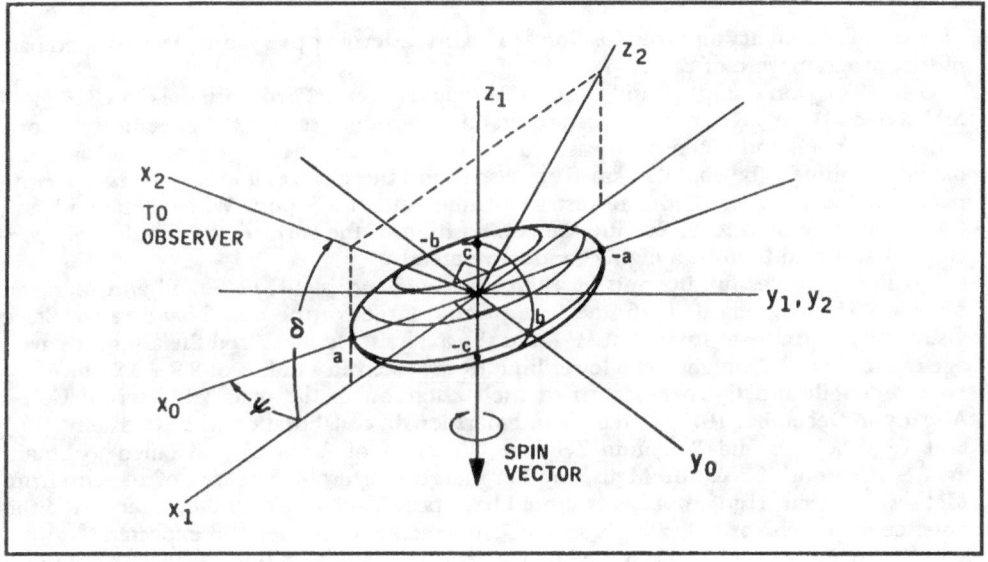

Figure 35

In order to model the nonspherical shape and numerous axes of rotation of asteroids, Ray Jurgens designed a coordinate frame to describe asteroids as rotating triaxial ellipsoids. This was the first attempt to model asteroid shapes with radar data. (Courtesy of Jet Propulsion Laboratory.)

Continuing his pursuit of radar asteroid research, Ray Jurgens outlined an ambitious 10-year program of asteroid opportunities in 1977. The program laid out the kinds of measurements that ground-based radars (both Goldstone and Arecibo) could make with currently available transmitter power and receiver sensitivity. Jurgens estimated that the number of detectable asteroids available for study over the following 10 years was 60. It took a few years longer to reach that number, however, for a variety of reasons. Jurgens also pointed out that astronomers could use the radar data in many cases to calculate the radius, average surface roughness, rotational rate, and polar axis direction, and in some cases the radar albedos and orbital parameters, of asteroids.[82]

In a memorandum to NASA Headquarters, Jurgens described the kinds of asteroid opportunities that would become available upon the upgrading of the Goldstone radar and argued for the scientific value of determining object size, rotation period, shape, and surface properties from range and Doppler measurements, in the hopes of funding asteroid radar research at JPL.[83] Jurgens had foreseen and mapped out the kind of radar asteroid research program that only a few years later would materialize, but at Arecibo. Jurgens' asteroid research program did not take root at JPL; the Goldstone radar was shut down after some unsuccessful tracks on Vesta on 28 May 1982.[84] In contrast, radar asteroid studies at Arecibo were far more energetic. There, Brian Marsden worked with Irwin Shapiro, the guru of the Planetary Ephemeris Program, to undertake a systematic study

82.   Jurgens and D. F. Bender, "Radar Detectability of Asteroids: A Survey of Opportunities for 1977 through 1987," *Icarus* 31 (1977): 483–497. The asteroid research program grew out of a larger work Jurgens did while at JPL, namely, Jurgens, *A Survey of Ground-based Radar Astronomical Capability Employing 64 and 128 Meter Diameter Antenna Systems at S and X Band*, Report 890–44 (Pasadena: JPL, March 1977). See also Pettengill and Jurgens, "Radar Observations of Asteroids," in Gehrels and Matthews, *Asteroids*, pp. 206–211.

83.   Jurgens to Geoffrey A. Briggs, 12 August 1982, Jurgens materials.

84.   Jurgens, "Seeing Comet IRAS," p. 222.

of asteroid astrometry and composition.[85] The first asteroid observations that formed part of that program were of Eros.

In 1975, Don Campbell and Gordon Pettengill observed Eros with the old UHF (430-MHz; 70-cm) transmitter. That was the first asteroid detected by the Arecibo telescope radar. Campbell and Pettengill measured the radar cross section of the asteroid and estimated its radius to be about 16 km. They also found the surface of Eros to be rough compared to the surfaces of the terrestrial planets and the Moon. When Pettengill and Campbell attempted to determine the composition of the surface, they could only conclude that it could not be a highly conductive metal.[86]

After unsuccessful attempts at asteroids Ceres and Metis, Pettengill and Marsden observed 1580 Betulia in 1976, the first asteroid target of the new S-band radar. Steve Ostro, then a graduate student at MIT, did the analysis. He measured the asteroid's average radar cross section and set a lower limit to the asteroid's radius of 2.9 ± 0.2 km.[87]

Pettengill and Ostro next turned their attention to the mainbelt asteroid Ceres. Already, in December 1975, Pettengill and Marsden, in collaboration with Goldstein (JPL) and Tom Gehrels and Benjamin Zellner (University of Arizona) had failed to obtain echoes from both Ceres and Metis, another mainbelt asteroid. The lack of an echo from Metis was not surprising, but Ceres should have been easy to detect; they interpreted the absence of an echo as indicating a smaller cross section than they had expected.[88]

Ostro, Pettengill, and Campbell finally detected Ceres with the Arecibo S-band radar in March and April 1977. This was the first mainbelt asteroid detected by radar. The greater sensitivity of the S-band instrument made it possible for the radar to reach into the mainbelt of asteroids and detect such a small body. The opportunity of March 1977 was slightly more favorable than that of 1975, thanks to the installation of a more sensitive line feed in 1976. Ceres was found to have a low radar cross section, less than that for the Moon, the terrestrial planets, and even Eros. On the other hand, the asteroid appeared to have a very rough surface at some scale comparable to, or larger than, the 12.6-cm wavelength of the radar, that is, rougher than the Moon and terrestrial planets, but smoother than the Galilean satellites of Jupiter.[89]

Noisy data taken on mainbelt asteroid Vesta during three nights of observations in November 1979 returned only a weak detection.[90] Each asteroid detection seemed to bring a new revelation; no pattern emerged. Unlike the terrestrial planets, asteroids presented not a few bodies to study but an entire population, a population, moreover, that the growing discovery rate kept increasing. Although the systems of Jupiter and Saturn defined the outer limits of planetary radar astronomy after 1975, the asteroids defined its future. They were on their way to deposing Venus from its position as the favored target of radar astronomers.

---

85.   NAIC QR Q2/1976, 6–7.

86.   Campbell, Pettengill, and Shapiro, "70-cm Radar Observations of 433 Eros," *Icarus* 28 (1976): 17–20; NAIC QR Q1/1975, 4.

87.   Pettengill, Ostro, Shapiro, Marsden, and Campbell, "Radar Observations of Asteroid 1580 Betulia," *Icarus* 40 (1979): 350–354.

88.   NAIC QR Q4/1975, 5.

89.   Ostro, Pettengill, Shapiro, Campbell, and R. Green, "Radar Observations of Asteroid 1 Ceres," *Icarus* 40 (1979): 355–358; NAIC QR Q1/1977, 6–7.

90.   Ostro, Campbell, Pettengill, and Shapiro, "Radar Detection of Vesta," *Icarus* 43 (1980): 169–171; Ostro 25/5/94; NAIC QR Q4/1979, 7.

Chapter Nine

One Step Beyond

Just as the Arecibo S-band and the Goldstone X-band upgrades had propelled radar astronomy into new directions, in 1986 a second upgrade planned for the Arecibo radar and the restoration and upgrading of the Goldstone radar stimulated new shifts in the planetary radar paradigm. Instruments and hardware continued to drive the field. Fresh techniques, either developed by radar astronomers or borrowed from other fields, namely ionospheric and radio astronomy research, allowed radar astronomers to solve new problems on the terrestrial planets. Also, the bizarre radar signatures of the icy Galilean satellites appeared once again, though closer to home on the terrestrial planets, and suggested new problems to solve.

The Goldstone Solar System Radar

Planetary radar astronomy survived at JPL in a tenuous state as a testbed for DSN technology and as a mission-oriented activity. That state depended largely on support from specific upper-management individuals, Eb Rechtin and Walt Victor. After Rechtin left JPL and Victor transferred out of the Deep Space Network to the JPL Office of Planning and Review in December 1978, radar astronomy became vulnerable to extinction. The DSN Advisory Group, headed by Rechtin, had judged that radar astronomy was no longer the testbed of DSN technology.

The Goldstone radar was in desperate need of repair, and the old equipment had become very hard to maintain; few people knew how to work with it. By 1980, much of the equipment was old and not functioning properly. During experiments, for instance, entire runs of data would be flawed or lost as a result of computer malfunctions. "You simply had to bite the bullet and rebuild the whole damned thing, particularly the data acquisition systems," Ray Jurgens explained.[1] However, nobody wanted to pay the cost of the needed repairs and upgrades.

Reviving the Goldstone radar so that planetary radar astronomy could once again prosper at JPL required that the activity have a new rationale. The initial arguments for funding needed new equipment focused on the value of radar to NASA flight missions and to planetary geology. In 1979, E. Myles Standish, Jr., who was in charge of the JPL planetary ephemeris program, wrote a memo to Richard R. Green, who had recently been promoted from the radar group to Advanced Systems, to explain that if no radar experiments were conducted, then the accuracy of the ephemerides for the terrestrial planets would suffer, and JPL would not be able to meet its ephemeris commitments to either the Galileo or Magellan missions.[2]

Obtaining mission approval had been a requisite for acquiring antenna time for radar experiments. As George Downs explained, "Dick Goldstein always wanted to

---

1. Jurgens 23 May 1994. In addition to the computer and other hardware problems, small cracks appeared in the pedestal of the Goldstone Mars Station. The repair involved raising the 3,000-ton structure and replacing a large portion of the pedestal concrete. The antenna did not return to service until June 1984, after being down a year for repairs. JPL Annual Report, 1983, p. 26, and ibid., 1984, p. 26, JPLA.
2. Memorandum, Standish to R. Green, 10 May 1979, Jurgens materials.

connect us with a project. I believe he felt that if we tried to get constituency from geologists alone, we wouldn't make it. Well, he was right."[3] In 1979, George Downs asked USGS
geologist Henry J. Moore to write a letter in support of the Goldstone radar; Moore wrote
to Arden L. Albee, Caltech professor of geology and the new JPL Chief Scientist.

Albee was sympathetic and met with members of the JPL radar group, Ray Jurgens,
George Downs, Stan Butman, and Rick Green, on 24 January 1980. As a result of the meeting, Albee wrote to the NASA Office of Space Science recommending a line item for radar
astronomy in the fiscal 1982 budget. Thomas Mutch, associate administrator, NASA Office
of Space Science, replied that he could not raise the annual allocation; the funding level
would have to remain level.[4]

Some support for resurrection of the Goldstone radar could be counted on coming
from the Planetary Radar Working Group, which consisted largely of geologists in the
USGS and academia plus smaller numbers of individuals representing SAR remote sensing and NASA Headquarters, as well as radar astronomers Pettengill, Campbell,
Goldstein, and Len Tyler. The Planetary Radar Working Group met in conjunction with
the AAS Division for Planetary Sciences and the Lunar and Planetary Science Conference
and discussed priorities in radar astronomy at Goldstone and Arecibo. Of course, the fate
of the VOIR mission, not the Goldstone radar, was the focal point of discussions.[5]

With support from the Planetary Radar Working Group, Ray Jurgens and George
Downs wrote a proposal requesting about $1.8 million to purchase a VAX computer to
reduce radar data. They submitted it to the NASA planetary geology office because they
thought planetary geologists would be the prime users of the data. In retrospect, George
Downs judged that reviewers saw the proposal as a threat to their own funding and did not
give it good reviews, while those who saw the project's usefulness gave it good reviews.

Although Jurgens and Downs did not get the amount requested, the NASA Office of
Space Science and Implementation did grant them enough to buy a new VAX-700 and
about $150,000 a year to analyze radar data. Radar astronomy also achieved a modest level
of recognition in 1982. The original 1971 NASA Management Instruction governing
ground radio science, now considered obsolete, replaced the term "radio science" with
"Radio and Radar Astronomy." Such was the state of radar astronomy when Downs left in
1982.[6]

In 1983, radar astronomy acquired a new advocate, Nicholas A. Renzetti. Originally
a DSN manager responsible for the interface between the DSN and its flight customers,
starting in 1975 with the Viking and Voyager launches, Renzetti gave less attention to
flight projects and more attention to applications of radio technology to non-flight projects, such as geodynamics, the Search for Extraterrestrial Intelligence, radio astronomy,
and starting in 1983, radar astronomy. Renzetti took on the task of convincing the Office
of Space Science and other NASA Headquarters departments that it was in NASA's interest to support the Goldstone radar as a scientific instrument.[7]

The new rationale for funding the Goldstone radar, as defined by Renzetti, would be
its use as a scientific instrument. In his campaign to garner support for the Goldstone
radar, Renzetti was assisted by Steve Ostro, who took a position at JPL in late 1984, after
leaving Cornell. They negotiated a new task in December 1987, in which the Goldstone
radar would be treated as if it were a facility, not as a science task, with an annual budget

3.   Downs 4 October 1994.
4.   Henry J. Moore to Arden L. Albee, 2 July 1979, Jurgens materials; Memorandum, William H. Bayley
to Murray, 4 February 1980, 91/7/89-13, JPLA; Various documents in "NASA Correspondence, 1980–1981,"
JPLPLC.
5.   Jurgens 23 May 1994; Planetary Radar Working Group mailing list, Jurgens materials;
Memorandum, Carl W. Johnson to Murray, 27 October 1980, 99/8/89-13, JPLA.
6.   Jurgens 23 May 1994; Downs 4 October 1994; C. H. Terhune, Jr., to B. I. Edelson and R. E. Smylie,
20 September 1982, "Chron 1982, #2," JPLPLC.
7.   Renzetti 16 April 1992; Renzetti 17 April 1992.

of about $200,000 for hardware improvements. The NASA task underwrote the interface between the DSN and the radar astronomers. The objective of the new task, called the Goldstone Solar System Radar (GSSR), was to support planning, experiment design, and coordination of data acquisition and engineering activities for all Goldstone planetary radar astronomy. As Steve Ostro explained, "This has been the financial backbone for the Goldstone radar, and it is separate from the DSN."[8]

At the same time, Renzetti created a part-time position, the Friend of the Radar. The holder of that position was to carry out a number of duties, including NASA flight project science and liaisons with Arecibo Observatory, but most importantly interfacing with the scientific community. Tommy Thompson performed those duties until he became Magellan Science Manager in 1988, when Martin A. Slade replaced him. Slade had been a graduate student of Irwin Shapiro at MIT and had had some exposure to radar astronomy during summer jobs at Haystack. His main previous research interests, however, lay elsewhere.[9]

The creation of the GSSR task and the Friend of the Radar were only first steps in addressing the core issue of funding the Goldstone radar on the basis of its use as a scientific instrument. Renzetti took tentative, unsuccessful steps to open up the Goldstone radar to outside researchers in order to operate it as a national research facility. He approached Von Eshleman and two others from outside JPL to propose radar experiments at Goldstone. Renzetti also proposed to Tor Hagfors, NAIC director, that a single peer review panel assess radar experiment proposals for both the GSSR (as the Goldstone Mars Station or DSS-14 now came to be called) and Arecibo. Moreover, hoping to acquire a facility budget for GSSR on a level with that of Arecibo, Renzetti proposed to Hagfors that Arecibo and GSSR present a common front to NASA, rather than appear as competing facilities.[10]

But it did not make sense to pursue the common budget, Renzetti reasoned, as long as the GSSR was not a national facility. The annual amount requested from NASA to make the GSSR a "first-class scientific instrument," $500,000, was not well received at NASA Headquarters. In comparison, the NASA budget for the Arecibo radar was only $362,000 in 1986.[11] Nonetheless, Renzetti, who felt there was a built-in bias in favor of Arecibo at high-level NASA meetings, submitted a formal proposal to make the GSSR a national facility, but it never got off the ground.[12]

A chief critic of the proposal to turn the Goldstone radar into a national research center was Dewey Muhleman of Caltech. He called parts of the proposal "ludicrous" and declared that it would do "nothing for Science, the Nation, NASA nor, in the long run, JPL." Moreover, he pointed out, the heavy scheduling of the antenna for spacecraft work militated against the plan. "I strongly favor," he wrote, "the idea of getting Radar Astronomy at JPL out of the closet of component development and into the light of pure science."[13]

Gradually, that was starting to take place. During a JPL administrative reorganization in the fall of 1987, the Office of Space Science and Instruments (OSSI) was created with Charles Elachi as its head. Elachi was a seasoned radar engineer with decades of SAR experience. After he obtained a modest level of funding, $150,000, from NASA Headquarters, Elachi named Steve Ostro manager of Planetary Radar Science and authorized him to allocate the funding.[14]

---

8. Ostro 18 May 1994; GSSR Min. 6 December 1984.
9. Thompson 29 November 1994; Slade 24 May 1994; GSSR Min. 6 December 1984 and 31 March 1988.
10. GSSR Min. 29 December 1986.
11. GSSR Min. 22 January 1987 and 26 February 1987; NAIC QR Q1/1986, 19.
12. GSSR Min. 26 February 1987.
13. Memorandum, Muhleman to Edward C. Posner, 28 October 1986, Ostro materials.
14. GSSR Min. 3 December 1987, 14 January 1988, 18 February 1988, and 28 April 1988.

"At that point," Ostro explained, "I had a little bit of authority. I had the program office backing me. I acted as somewhat of a filter on proposals and papers, when I could, and I acted as the voice of science for radar." Ostro agreed with Muhleman's perception that JPL placed too much emphasis on hardware and not enough on doing science. The science community in general, he pointed out, viewed the GSSR as state-of-the-art electronics, but saw Arecibo as producing state-of-the-art planetary radar data. The objective, Ostro declared in 1988, "is, a year from now, to have a sparkling list of GSSR radar articles that have appeared in high-quality journals."[15] Despite such sterling intentions on the part of Ostro and Renzetti, keeping JPL, DSN, and NASA management aware of the Goldstone radar's scientific achievements and potential has been a Sisyphean task. In contrast, the value of radar astronomy was established from the outset at the Arecibo Observatory.

Here was an important difference between the two facilities that had a profound impact on the development of radar astronomy at each site. Even more important, however, was the fact that Arecibo had acknowledged and formalized the existence of radar astronomy from the start; whereas JPL purposely had denied radar astronomy any formal existence. The difference has had long-term implications that has favored radar astronomy science at Arecibo, while holding it back at JPL.

New hardware and fresh leadership enabled radar astronomers to make new discoveries about Mars, Mercury, and the asteroids with the Goldstone radar. The major hardware upgrade did not arise from a concerted campaign on the part of Renzetti and Ostro to improve the state of radar astronomy at JPL, but rather, in a fashion typical of the history of planetary radar astronomy, came from outside radar astronomy, namely, the Voyager mission to the outer planets.

The Voyager upgrade of the main GSSR antenna, known within the Deep Space Network as DSS-14, involved enlarging the dish diameter from 64 (210 ft) to 70 meters (230 ft), increasing the surface accuracy, and improving the receiving system. These measures increased the sensitivity of the DSS-14 significantly. Tracking and acquiring data from the Voyager spacecraft, as they encountered Uranus and Neptune, stretched the capacity of the Deep Space Network. During the Neptune encounter, the Voyager X-band radio signal would be less than one-tenth as strong as during the Jupiter encounter in 1979 and less than one-half as strong as during the Uranus encounter in 1986.

A study to enlarge all the DSN 64-meter antennas to 70 meters already had been undertaken as early as 1973 in preparation for Voyager when it was still called Mariner Jupiter/Saturn. After completion of design work in 1984, the upgrade of the DSS-14 began in October 1987 and concluded in May 1988.[16] When Steve Ostro arrived at JPL in 1984, the DSS-14 lacked the threshold of sensitivity to do meaningful asteroid research. Upon completion of the initial upgrade phase, however, Ostro made his first successful asteroid observations with the DSS-14 in May 1986, when he detected echoes from 1986 JK, an asteroid only just then discovered by Eugene and Carolyn Shoemaker.

The Voyager upgrade had a profound impact on the practice of radar astronomy at JPL; it provided the GSSR the sensitivity needed to carry out research on a whole new set of targets (and to begin solving new sets of problems). Not only did the GSSR gain the ability to undertake significant asteroid research, but when linked to the Very Large Array in New Mexico, as we shall see later, it became a new radar research tool.

Despite these major upgrades, the GSSR had serious problems as a scientific instrument. The site lacked dormitory and cooking facilities for visiting or even JPL scientists, and the drive to Barstow 50 miles away on winding roads after a night of observations was dangerous. These deficiencies and dangers persist today. Furthermore, the radar itself was far from user-friendly. "It was just impossible to work," Ostro explained. "For example, the

---

15.   Ostro 18 May 1994; GSSR Min. 14 January 1988 and 18 February 1988.
16.   JPL Annual Report, 1973–1974, p. 15; ibid., 1984, p. 13; ibid., 1987, p. 41; and ibid., 1988, p. 28, JPLA.

VAX that is used for data acquisition at Goldstone is not good for radar astronomy for various technical reasons. It has been improved a lot since the mid-1980s, but even now it is difficult, for example, to stamp your data with a high-precision UTC [from the French for Coordinated Universal Time] time tag. For this kind of work, the first thing you need on your data is a UTC time tag."[17]

The Arecibo Observatory stood in sharp contrast to the Goldstone radar. It had proper quarters for visiting scientists, and the radar was far more user-friendly. Moreover, in 1986, the Arecibo Observatory proposed a major upgrade of the radar that would benefit both ionospheric and radio astronomy research and planetary radar astronomy. The Arecibo upgrade stirred Renzetti to seek funding for a Goldstone radar upgrade.

Renzetti lobbied the DSN and NASA hierarchy for funding for a Goldstone one-megawatt transmitter, which JPL engineers initially estimated would cost $12 million. A good argument for DSN use of the radar would not fly; the rationale had to be its use for scientific research, Renzetti realized. The radar upgrade, to be completed in fiscal year 1993 and costing $10 million over two years, appeared in the DSN budget for fiscal 1989. JPL viewed the price tag as "pared to the bone." In the end, Congress approved the expenditure not as a specific radar upgrade but as an ambiguous improvement of the DSN. This ambiguity freed DSN management to use the radar transmitter money to purchase low-noise supercooled masers to improve antenna sensitivity for Galileo's encounter with Io.[18]

It was not clear, moreover, that radar science at JPL needed the one-megawatt transmitter. The estimated cost of the transmitter now stood at $16 million. Steve Ostro believed that if the cost were reduced below $8 million, the improved science capacity would justify the expense. The high cost reflected JPL administrative and DSN operational support requirements that added several million dollars to the cost.[19]

Ostro favored upgrading the GSSR antenna's subreflector to improve its ability to make asteroid observations. The long time needed to rotate the subreflector, which was never designed to act as a transmit/receive switch, compromised short round-trip-time asteroid observations. "The most powerful scientific rationale for the *longterm* support of GSSR," Ostro argued, was work on near-Earth asteroids. The estimated price tag for the transmit/receive upgrade, which involved turning the transmit-only horn into a horn capable of switching quickly back and forth between transmit and receive, was $485,000.[20]

Instead of going directly through the DSN hierarchy for a radar upgrade, Renzetti changed his strategy. The one-megawatt transmitter and two other radar improvements (construction of a transmit/receive horn and modernization of the data acquisition system) were submitted to a panel of outside scientists for review. Gordon Pettengill chaired the Goldstone Planetary Radar Science Review Committee, as the panel was called. It included planetary astronomers and geologists, as well as Don Campbell and Tor Hagfors from the Arecibo Observatory.[21]

17. Ostro 18 May 1994.

18. GSSR Min. 22 January 1987, 18 June 1987, 23 July 1987, 24 September 1987, 18 February 1988, 31 March 1988, and 26 April 1990; Renzetti, Thompson, and Slade, "Relative Planetary Radar Sensitivities: Arecibo and Goldstone," TDA Progress Report 42–94 (Pasadena: JPL, April–June 1988), pp. 287–293; Arthur J. Freiley, Bruce L. Conroy, Daniel J. Hoppe, and Alaudin M. Bhanji, "Design Concepts of a 1-MW CW X-Band Transmit/Receive System for Planetary Radar," *IEEE Transactions on Microwave Theory and Techniques* 40 (1992): 1047–1055.

19. GSSR Min. 6 February 1992.

20. Memorandum, Ostro to Elachi, 29 August 1990; Memorandum, David Hills to Dick Mathison, 1 October 1990; Memorandum, Ostro to Larry N. Dumas, 15 October 1990, Ostro materials.

21. Pettengill to Elachi, 22 August 1991, and attachments, Ostro materials. The members of the Goldstone Planetary Radar Science Review Committee were Gordon H. Pettengill, MIT; Michael J. S. Belton, Kitt Peak National Observatory; Donald B. Campbell, NAIC; Clark R. Chapman, Planetary Science Institute; Tor Hagfors, NAIC; Bruce W. Hapke, University of Pittsburgh; Randolph L. Kirk, USGS; David Morrison, NASA Ames Research Center; and F. Peter Schloerb, University of Massachusetts.

Although invited to join the Committee, Muhleman declined. He "took the attitude, well, this is one more panel, it can't be that important. How about if I don't come? Let me know how it comes out. That was a terrible mistake. It really was....The JPL viewpoint was not represented."[22] More importantly for Muhleman, *his* viewpoint was not represented, and he paid the price. The Committee met on 8 August 1991 and presented its conclusions later that month. The Committee applauded "the efforts currently underway by JPL management to broaden the usage of the Goldstone facilities (including observations jointly with the VLA) to include members of the larger North American and global planetary communities."

Of the three improvements, the committee gave the highest priority to the single-horn, fast-transmit/receive-switchover system. That improvement would serve asteroid work only. "At a lower, but still high, priority," the committee endorsed the modernization of the data acquisition system and recommended that the output protocols and formats of the new system be coordinated with those of the Arecibo planetary radar. Each of these two improvements had a modest cost of about $500,000 spread over one to two years.

The one-megawatt transmitter, the committee judged, "seems less attractive as an upgrading option than the first two presented." The cost was too high for the amount of sensitivity gained. The value of the transmitter upgrade, the committee decided, lay in observing Titan, "but we do not find the scientific argument compelling for what appears to be a fairly narrowly focused study of a single object. We note also that the improved transmitter is unlikely to be available in time to provide data that materially assist in the design of the Cassini Mission."[23]

Titan, however, was of the highest research interest to Dewey Muhleman. "In my absence," he complained, "this panel frankly wrote a silly report. It just really made me sick to read it. It said that the only advantage of going to a megawatt on the Goldstone antenna was to be able to do Titan better with the VLA. Nothing else was really important. *That* is ridiculous. For everything we do, our integration time would be cut down by a factor of four by doubling our power to a megawatt. We would be able to do much more on each one of these objects and quite frankly continue to rival Arecibo after the upgrade."[24]

The Arecibo Upgrade

The struggle at JPL to gain recognition for the GSSR as a scientific instrument stood in stark contrast to the effort to upgrade the Arecibo telescope. Both NASA and the NSF already recognized Arecibo as a national research center, and the rationale for any upgrade would be on the basis of scientific need. Furthermore, the Arecibo upgrade stood to benefit all research at the facility, radio and radar astronomy and ionospheric research, not just planetary radar astronomy. Other factors eased the process of garnering support for the Arecibo upgrade, including the method of funding what was, in relative terms, a low-cost project.

The Arecibo upgrade was a package of five interrelated improvements: 1) installation of a ground screen to virtually eliminate noise from the surrounding earth; 2) adjustment of the reflector surface to enhance antenna gain; 3) correction of the pointing system; 4) replacement of the accumulation of radio astronomy line feeds with a single reflector feed possessing large bandwidth, low loss, high gain, and continuous frequency coverage from 300 MHz (1 meter) to 8 GHz (3.75 cm); and 5) doubling the S-band transmitter power to one megawatt. The total effect of these changes was to

22.   Muhleman 27 May 1994.
23.   Pettengill to Elachi, 22 August 1991, and attachments, Ostro materials.
24.   Muhleman 27 May 1994.

increase radar sensitivity by a factor of 10 to 50 (about 20 times on average), to double its range or to detect objects 10 times smaller than previously possible.[25]

The principal objective of the upgrade, however, was to solve a problem that had plagued the telescope since its creation—the problem of spherical aberration. Unlike parabolic dishes, the Arecibo spherical antenna did not focus waves in a single point. The antenna feed system designed by the Air Force did not work efficiently, and though later feeds improved the telescope's performance, they did not perform up to the level of a Gregorian reflector, the solution recognized as early as the 1960s. Named for the astronomer John Gregory, a Gregorian reflector is concave and placed above the prime focus of a telescope. In a Cassegrain system, the type used, for example, on the Goldstone DSS-14, the reflector is convex mounted below the prime focus.[26]

Designing the Gregorian optics was a daunting task. A Cornell graduate student had considered the use of Gregorian optics, an option also studied by the AFCRL's Antenna Laboratory.[27] Frank Drake, director of the NAIC from 1971 to 1981, nurtured the Gregorian reflector idea and attempted unsuccessfully to gain financial support to gather together the necessary antenna expertise to submit a formal proposal to the NSF.[28]

Design of the Gregorian reflector did not begin until 1984, after Tor Hagfors became director of the NAIC in late 1982. After serving earlier as director of the Arecibo Observatory following the departure of Gordon Pettengill, Hagfors spent a number of years in Scandinavia building the EISCAT facility, before returning to Cornell to head the NAIC.[29]

EISCAT (European Incoherent Scatter Association) is a European consortium headquartered at Kiruna, Sweden. Inaugurated by the King of Sweden in August 1981, the EISCAT facility is a high-power radar installed at sites in Norway and Finland for the study of the Earth's ionosphere, upper atmosphere, and magnetosphere at high latitudes. Germany, France, and the United Kingdom bore the greatest share of its construction costs (25 percent each), while Sweden (10 percent), Norway (10 percent), and Finland (5 percent) contributed the rest.[30]

Under the direction of Tor Hagfors, the NAIC initiated systematic studies of several major antenna upgrading projects in 1984. As part of the upgrading project, the NAIC concluded consulting agreements with a number of antenna experts. Among them were Alan Love, who had designed the telescope's first circular feed, and Sebastian von Hoerner. Morton S. Roberts, director of the National Radio Astronomy Observatory (NRAO), and a member of the Arecibo Advisory Board, suggested that the NAIC hire as a consultant von Hoerner, a well known antenna expert working for the NRAO. The project appealed to von Hoerner's imagination, and he set to work designing the Gregorian optics and laying out the initial description of the shape and size of the reflector. He also realized the need for a tertiary reflector.[31]

25.   Hagfors, "The Arecibo Gregorian Upgrading," in Joseph H. Taylor and Michael M. Davis, eds., *Scientific Benefits of an Upgraded Arecibo Telescope* (Arecibo: NAIC, 1987), p. 4, and Ostro, "Benefits of an Upgraded Arecibo Observatory for Radar Observations of Asteroids and Natural Satellites," in ibid., p. 233.
26.   Campbell 9 December 1993.
27.   Kay, *A Line Source Feed*, passim, and Pierluissi, *A Theoretical Study of Gregorian Radio Telescopes*, passim.
28.   Hagfors, "The Arecibo Gregorian Upgrading," p. 3; Per-Simon Kildal, Lynn A. Baker, and Hagfors, "The Arecibo Upgrading: Electrical Design and Expected Performance of the Dual-Reflector Feed System," *Proceedings of the IEEE* 82 (1994): 714.
29.   NAIC QR Q3/1982, 19; Campbell 7 December 1993; Campbell 9 December 1993.
30.   Lovell, *The Jodrell Bank Telescope*, pp. 270–271.
31.   Campbell 9 December 1993; Hagfors, "The Arecibo Gregorian Upgrading," p. 3.

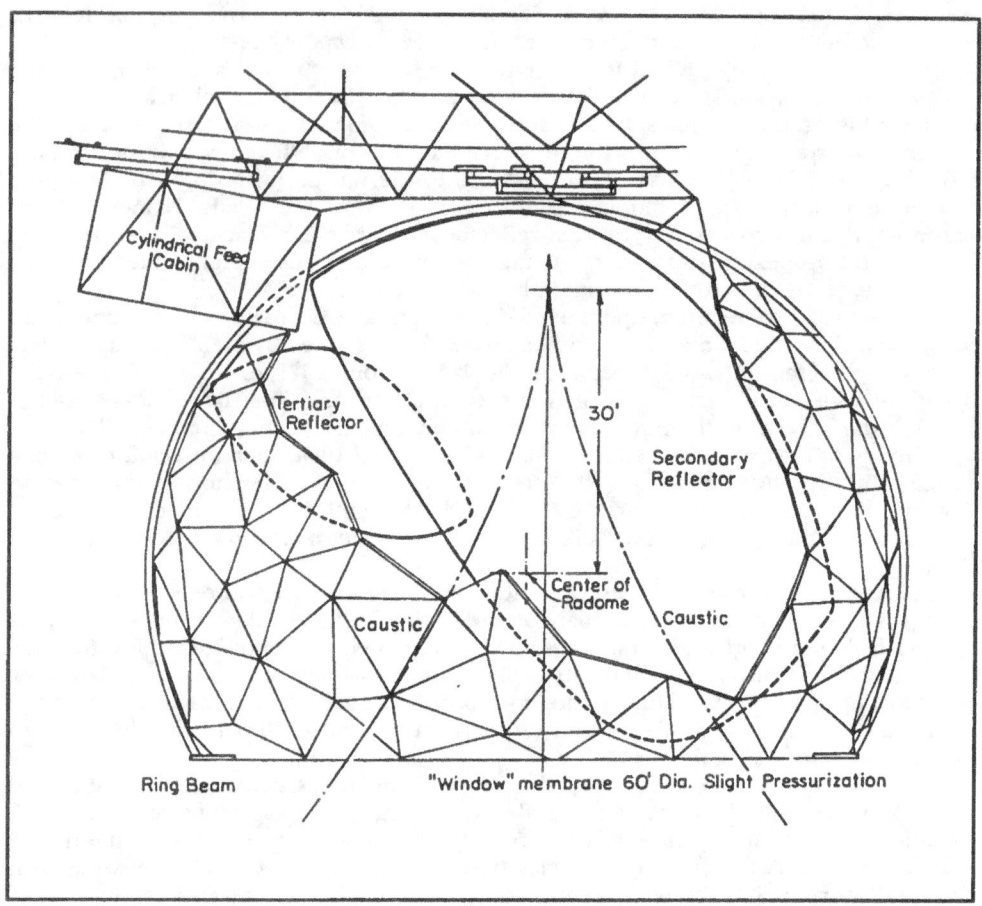

Figure 36

Diagram illustrating Gregorian optics of the Arecibo upgrade subreflector. Unlike the Lincoln Laboratory radomes, this one is not designed to allow radio signals to penetrate the radome shell. (Courtesy of National Astronomy and Ionosphere Center, which is operated by Cornell University under contract with the National Science Foundation.)

In addition, Hagfors brought in Per-Simon Kildal, a professor at Chalmers University of Technology, Gothenburg, Sweden. Kildal was an expert in the design of feed horns and antenna diffraction effects and a former student of Hagfors. He had performed some of the design work on the EISCAT antennas for his doctoral thesis. When Kildal worked for the NAIC for two months during the summer of 1984, he joined NAIC line feed designer Lynn A. Baker. Baker and Kildal devised a practical Gregorian design to correctly illuminate the primary reflector.[32]

32.   Campbell 9 December 1993; NAIC QR Q2/1984, 14; Hagfors, "The Arecibo Gregorian Upgrading," p. 3; Kildal, Baker, and Hagfors, p. 714.

Designing and installing the Gregorian reflector also changed the mechanical stress on the suspended platform. In order to work on the mechanical engineering aspects of the project, Hagfors asked Paul Stetson, an antenna builder formerly with Lincoln Laboratory, to come out of retirement. Stetson joined the NAIC in February 1984.[33]

As a test of the Gregorian feed concept, the NAIC at its own expense constructed and installed a so-called mini-gregorian antenna which was to illuminate a 107-meter (350-ft) diameter area of the reflector. Also, the ground screen underwent preliminary design, and another study determined that the dish surface could be adjusted to be operational up to 8 GHz (3.75 cm).[34]

In 1984, as these design studies were underway, the NAIC submitted a preliminary proposal to the National Science Foundation for Phase 1, the ground screen. The NAIC submitted the Phase 2 preliminary proposal in 1985 for the Gregorian reflector system, the new radar transmitter, ancillary receivers, and data processing equipment. The NAIC then entered into negotiations with both the NSF and NASA, the two NAIC funding agencies. The House subcommittee that handled NSF appropriations was well aware of the upgrade project. Jerome Bob Traxler (D-Mich.), the chairperson of the House subcommittee, Harry Block, the NSF director, and Dick Mallow, the subcommittee's chief of staff, visited Arecibo several times.[35]

The key to selling the project to the scientific community, which ultimately reviewed all NSF proposals, was the building of consensus, a standard strategy among American scientists. The NSF proposals were supposed to stand on their own merit. Whether those reviews were good or bad was critical to the success of the upgrade project. The keystone of consensus-building was a workshop held at Cornell University 13–15 October 1986. The NSF proposal for Phase 1 was already under review, when the workshop took place. Talks highlighted the kinds of scientific experiments one could do with the upgraded telescope, whether in atmospheric research or in radio astronomy. Steve Ostro, Don Campbell, and Irwin Shapiro pitched the possibilities for radar astronomy.

---

33.   NAIC QR Q2/1984, 14; Campbell 9 December 1993; Hagfors, "The Arecibo Gregorian Upgrading," p. 3.

34.   NAIC QR Q2/1984, 14, and Q3/1984, 15; Hagfors, "The Arecibo Gregorian Upgrading," p. 4; Kildal, Baker, and Hagfors, pp. 717–718 and 722.

35.   Campbell 9 December 1993; Dickman 2 December 1992.

Figure 37

View of the Arecibo Observatory dish. The completed ground screen is visible in the background. (Courtesy of National Astronomy and Ionosphere Center, which is operated by Cornell University under contract with the National Science Foundation.)

Ostro largely proposed research on mainbelt and near-Earth approaching asteroids, passing quickly over other solar system objects, such as the moons of Mars, Jupiter, and Saturn. Don Campbell emphasized exploration of the terrestrial planets and comets. The major impact of the upgrading, he and Shapiro acknowledged, would be on the observation of asteroids.[36] The scientific repercussion of the Arecibo upgrade for radar astronomy would be to sustain the observatory as the major research instrument and to make asteroid studies the predominant area of research.

The NSF sent the NAIC upgrade proposals out for review. The reviews aided the NSF in prioritizing its spending. Where the project stood within the NSF's own priority list of projects also was subject to input from the Division of Astronomy, primarily, and from the

---

36. Ostro, "Benefits of an Upgraded Arecibo," pp. 233–239; Campbell, "Prospects for Radar Observations of Comets and the Terrestrial Planets," in Taylor and Davis, pp. 243–248; Shapiro, "Radar Tests of Gravitational Theories and Other Exotica," in ibid., pp. 225–232.

Division of Atmospheric Sciences. Within NASA, the planetary program decided funding priorities. In 1988, following the Cornell workshop, the NAIC submitted the main proposal for the Gregorian system and radar transmitter. Numerous discussions, presentations, committee meetings, and reviews followed. Also providing input was the Bahcall Committee, the successor to the Whitford Panel.[37]

The Bahcall Committee, named for its chair John N. Bahcall, Princeton Institute for Advanced Study, and formally known as the Astronomy and Astrophysics Survey Committee, was a group of 15 astronomers and astrophysicists commissioned in 1989 by the National Academy of Sciences to survey their fields and to recommend new ground and space programs for the coming decade. To carry out the actual work, the Committee established 15 advisory panels to represent different subdisciplines, and those panels submitted their reports in June and July 1990.[38]

Radar astronomy came under the general umbrella of the Planetary Astronomy Panel, chaired by David Morrison, NASA Ames Research Center, chair, and Donald Hunten, University of Arizona, vice chair. Among the 22 planetary scientists constituting the panel was one radar astronomer, Steve Ostro. The Planetary Astronomy Panel recommended several facilities as "critically important" for planetary astronomy in the 1990s. Prioritized according to their cost (small, medium, large) within the categories "space-based" and "ground-based," the most important *small* ground facility for planetary astronomy was the Arecibo upgrade.[39]

The upgrade was never regarded as a huge project. The total estimated price tag of the upgrade, around $23 million spread out over four years, placed it in the "small" category; even the medium-sized proposed facilities cost substantially more. The relatively small total amount underwent further diminution in such a way that the project was never big enough to be a separate line item within the budget of the Office of Management and Budget. Both NASA and the NSF split the total cost, which underwent further division within each agency, so that the total amount per year was never a huge sum for each agency or for each agency program.

Geoff Briggs, director of the Division of Solar System Exploration within the NASA Office of Space Science, chaired discussions about the project with the NAIC, NASA, and the NSF. According to Don Campbell, "Briggs somewhat arbitrarily just took it on himself to break up who was going to pay for what right there."[40]

The allocation of the costs of what was already considered a small, low-cost project was a strategy in tune with the budgetary times. NASA would pay 100 percent of the ground screen and the one-megawatt radar transmitter costs, but the money came from the budgets of three different divisions. The Division of Solar System Exploration paid for the ground screen; the Office of Space Communications paid for the transmitter; and the Division of Biological Sciences, the source of SETI (Search for Extra-Terrestrial Intelligence) funding, contributed partially to the Gregorian reflector. The NSF paid for the remainder, with the Division of Astronomical Sciences paying for some specific equipment. The distribution of individual program contributions split the cost evenly between the two agencies and became the basis for the memorandum of understanding between NASA and the NSF that covered the upgrade.[41]

37. Campbell 9 December 1993; Kildal, Baker, and Hagfors, p. 715.
38. John Bahcall, "Preface," in National Research Council, *The Decade of Discovery in Astronomy and Astrophysics* (Washington: National Academy Press, 1991), pp. ix–xi.
39. National Research Council, *Working Papers: Astronomy and Astrophysics Panel Reports* (Washington: National Academy Press, 1991), pp. X- 1–X-20.
40. Campbell 9 December 1993.
41. Dickman 2 December 1992; Campbell 9 December 1993.

Titan

The Arecibo upgrade, when completed, promises entirely new research capabilities that will open up a new set of targets to be explored and new problems to be solved. Another upgrade, though not intended to provide new radar capability, created a research instrument that never existed before. That was the Voyager upgrade. It involved improvement of the GSSR, as well as the Very Large Array (VLA), a radio telescope located in New Mexico. For the VLA upgrade, NASA installed low-noise X-band receivers on each of the 27 VLA antennas. When radar astronomers linked the Goldstone radar and the VLA in a bistatic mode, they created a radar with an extraordinary capacity for exploring the solar system.

The upgrade of the VLA for the Voyager mission originated in the need to communicate with the spacecraft at unprecedented distances. During Voyager's encounter with Neptune, its X-band radio signal would be less than one-tenth as strong as from Jupiter and less than one-half as strong as from Uranus. In addition to the enlargement of the DSN 64-meter antennas to 70 meters in diameter, the Neptune encounter required assistance from the Parkes telescope in Australia and the VLA. Through the radio astronomy technique of arraying, and the installation of low-noise receivers on each VLA dish, the echoes received from the VLA were combined with those received at the Goldstone 70-meter and 34-meter dishes to provide a data rate more than double that which would have been available with Goldstone's antennas alone.[42]

The idea of using the VLA as a receiver in a bistatic radar system was not new; Ed Lilley had suggested some two decades earlier a bistatic radar consisting of the VLA and the NEROC transmitter for carrying out planetary radar mapping.[43] Moreover, the VLA management already had thought of the possibility of a Goldstone-VLA bistatic radar years earlier, when they were looking for a broader foundation of support for a facility strictly dedicated to radio astronomy. They, therefore, were receptive to the suggestion of Nick Renzetti (JPL) that joint Goldstone-VLA radar experiments be conducted, provided the proposed experiments first would undergo the normal review process.[44]

As the Goldstone and VLA upgrades were underway, Caltech professor Dewey Muhleman became interested in the possibilities opened up by a Goldstone-VLA bistatic radar. After abandoning a career in radar astronomy in 1966 as professor of planetary science at Caltech, Muhleman switched to the study of radio emissions from the planets. Muhleman thought the Goldstone-VLA radar an excellent tool for exploring Saturn's barely explored and poorly understood moon, Titan. Scientists knew nothing about Titan's surface, because like the surface of Venus, it is hidden by an opaque cloud cover.[45]

Despite, or perhaps because of, this lack of knowledge, scientists speculated on the nature of the satellite's surface. According to conventional wisdom, Titan's surface was an ocean of ethane and methane, which would have almost no reflecting surface at radar wavelengths.[46] In 1980, Voyager 1 flew past Titan and provided fresh facts about the moon's surface temperature (about 94° Kelvin) and surface pressure (around 1,500 millibars). Voyager found an atmosphere composed mainly of nitrogen and trace amounts of

---

42.   Murray to Morton S. Roberts, 25 February 1982, "Chron 1982 #1," and Memorandum, Associate Administrator for Space Tracking and Data Systems to Deputy Director, JPL, 28 February 1983, "NASA Correspondence, 1983, pt. #1," JPLPLC; JPL Annual Report, 1984, p. 13, and ibid., 1987, p. 41, JPLA.

43.   Memorandum, Lilley to CAMROC Project Office Members, 14 June 1966, 18/1/AC 135, MITA.

44.   Renzetti 17 April 1992.

45.   Muhleman 8 April 1993.

46.   The ethane-methane ocean model of Titan was developed by Jonathan I. Lunine, David J. Stevenson, and Yuk L. Yung. See, for example, Lunine, Stevenson, and Yung, "Ethane Ocean on Titan," *Science* 222 (1983): 1229–1230.

hydrocarbons and nitriles, including ethane, methane, and acetylene. But Voyager revealed nothing about the moon's surface features.[47]

Titan's surface remained hidden from the view of radar astronomers, too. In February 1979, using the Arecibo S-band radar, Don Campbell, Gordon Pettengill, and Steve Ostro unsuccessfully attempted to detect Titan. Later, in 1987 and 1992, Dick Goldstein and Ray Jurgens also failed to receive echoes from Titan using the Goldstone Mars Station alone.[48] The bistatic Goldstone-VLA radar, however, promised an extra measure of sensitivity.

Muhleman hoped to find land masses and challenge the ethane ocean model. He already had conducted a radio study of Titan, but that research had yielded ambiguous results. Muhleman teamed up with JPL radar astronomer Marty Slade, who oversaw operation of the Goldstone half of the bistatic radar. Muhleman's graduate students, Bryan Butler and Arie Grossman, participated in the experiments, too. In order to test the system, Muhleman, Slade, and Butler attempted a known target, the rings of Saturn, in the spring of 1988. The success encouraged them to attempt Titan.[49]

Muhleman, Butler, and Slade first observed Titan on the nights of 3, 4, 5, and 6 June 1989 with the VLA in the so-called C configuration, in which the maximum separation among the 27 25-meter (82-ft) telescopes was about three km. The echoes were marginal, although those obtained on 4 June were strong, and the detection of 5 June was "quite certain." "The data," they concluded, "appear to favor a real variation in surface properties but more observations are required."[50]

The backscatter from Titan was highly diffuse, similar to that from the Galilean satellites of Jupiter. The diffuse backscatter, they believed, was a strong argument against an ethane ocean being the reflecting medium. A liquid body without floating scatterers would be a specular not a diffuse reflector. Instead, the radar echoes from Titan suggested an icy surface similar to that of Europa, Ganymede, or Callisto. The experiment, however, did not rule out entirely the existence of liquid hydrocarbons on Titan's surface that might exist in the form of small lakes.

Muhleman, Slade, and Butler attempted Titan again in August 1992 and in the summer of 1993.[51] From these fresh echoes, they concluded that Titan does not always keep the same hemisphere towards Saturn, as had previously been believed. In addition, one region very bright to the radar consistently appeared 15 hours earlier than expected, suggesting that its rotational period was 49 minutes shorter than its orbital period of 15.945 Earth days.

More importantly, variations in radar reflectivity gave the first indications of surface conditions on Titan. Results from instruments on the Voyager spacecraft in the 1980s suggested that there might be a global ocean of liquid ethane. However, Muhleman, Slade, and Butler reported that only a few patches of liquid will be found by the European-built Huygens probe scheduled to land on Titan early in the next century after a journey

47.   Muhleman, Arie W. Grossman, Bryan J. Butler, and Slade, "Radar Reflectivity of Titan," *Science* 248 (1990): 975–980.

48.   NAIC QR Q1/1979, 9; Campbell 8 December 1993; Goldstein and Jurgens, "DSN Observations of Titan," in Posner, ed. *The Telecommunications and Data Acquisition Report: Progress Report, Jan.–Mar. 1992* (Pasadena: JPL, 1992), pp. 377-379.

49.   Muhleman, G. Berge, and D. Rudy, "Microwave Emission from Titan and the Galilean Satellites," *Bulletin of the American Astronomical Society* 16 (1984): 686; JPL Annual Report, 1988, p. 29, JPLA.

50.   Muhleman, Grossman, Butler, and Slade, "Radar Reflectivity of Titan," *Science* 248 (1990): 975–980, quote on p. 979.

51.   Muhleman, Grossman, Slade, and Butler, "Titan's Radar Reflectivity and Rotation," *Bulletin of the American Astronomical Society* 25 (1993): 1099; Butler, Muhleman, and Slade, "Results from 1992 and 1993 VLA/Goldstone 3.5 cm Radar Results," ibid., p. 1040; GSSR Min. 19 February 1993.

aboard the Cassini spacecraft. The moon's surface seems to be covered mainly by icy continents, perhaps coated in tars of hydrocarbons.

The results of Muhleman's radar research on Titan were of enormous interest to Dennis L. Matson, Cassini project scientist, and others involved in the planning of the Cassini mission. In 1989, NASA was preparing the Cassini Announcement of Opportunity for release on 1 December 1989. A major experiment on Cassini, as then planned, was a radar instrument to be built by JPL. The nature of Titan's surface was a major parameter in the design of any radar system for the Cassini mission.

If an ocean of ethane and methane really covered Titan, the radar would have to be designed to anticipate the special scattering conditions that such a surface would create. The Goldstone-VLA radar data, then, would be useful in targeting the Huygens probe, and the targeting decisions had to be made before the launch of the Cassini spacecraft itself.[52] As Nick Renzetti characterized the situation: "So why put a $20 million radar on Cassini and get zilch? That really stirred the community for the last three years."[53]

The Polar Ice Caps of Mars

The radar results from Titan were revealing but puzzling. The radar study of Titan also highlighted the continuing mission-oriented nature of radar astronomy. The same was true of Mars radar research. Although Muhleman intended to use the Goldstone-VLA bistatic radar primarily to study Titan, equally startling results came from its application to Mercury and Mars. The Goldstone-VLA system allowed radar astronomers to solve problems previously unsolved or solved unsatisfactorily. The Goldstone-VLA work added to a long tradition of studying Mars topography that began, as we saw in an earlier chapter, before Viking went to Mars, and continued in support of the Viking mission. Most of the Mars radar topography work done in the 1970s, in fact, related directly to Viking.

The exploration of Martian topography and radar reflectivity from the 1970s into the 1980s had yielded some rather interesting results. The studies done for Viking had revealed high roughness (large rms slopes) and sharp roughness transitions in the area around the Tharsis volcanoes and their associated lava flows. Tharsis itself was found to have a low overall reflectivity. The most unusual and controversial development was the claim by Stan Zisk and Peter Mouginis-Mark, from their analysis of Goldstone Mars data from 1971 and 1973, that the Solis Lacus region showed seasonal variations in its radar reflectivity which might indicate the presence of near-surface liquid water.[54]

The Tharsis and Syrtis Major regions were of special radar interest. Syrtis Major was a classical radar dark spot on Mars. From topographical data, George Downs showed that the Tharsis bulge was lower than originally thought. Geologists used the radar data to show that Tharsis had been tectonically inactive since the occurrence of the last major lava flows. Topographical data for the south Tharsis region suggested that it was an ancient impact basin. Interpretation of the radar studies of Downs and Simpson (at Arecibo) of the Syrtis Major area by USGS geologist Gerry Schaber indicated that it was a low-relief

52.   Muhleman to W. E. Giberson, 10 February 1989, and Memorandum, D. L. Matson to Dumas, 8 January 1991, Renzetti materials.

53.   Renzetti 17 April 1992.

54.   Harmon, "Radar Observations of Mars and Mercury: History and Progress," paper read at the Thirtieth Anniversary Celebration of Planetary Radar Astronomy, 3 October 1991, Caltech; Zisk and P. J. Mouginis-Mark, "Anomalous Region on Mars: Implications for Near-Surface Liquid Water," *Nature* 288 (1980): 735–738. See also Aaron P. Zent, Fraser P. Fanale, and Roth, "Possible Martian Brines: Radar Observations and Models," *Journal of Geophysical Research* vol. 95, no. B9 (1990): 14,531–14,542.

shield volcano, rather than the impact basin it had always been believed to be, because it was not very heavily cratered.[55]

John Harmon arrived shortly after the installation of the Arecibo S-band radar as a Research Associate, after graduating from the University of California at San Diego with a doctoral thesis on solar winds. John Harmon began a series of studies of Mars topography and scattering, initially under the direction of Don Campbell, and drew the first topographic profile of Syrtis Major. Starting in February 1980, Harmon and Steve Ostro undertook a study of Tharsis and the surrounding area using both the Arecibo S-band and the Goldstone X-band radars and taking data in both senses of circular polarization, in order to compare polarization ratios at both S-band and X-band.

While the initial focus in 1980 had been on the Tharsis region, the 1982 observations took in a broader area and revealed correlations between maximum depolarization and the volcanic regions Tharsis and Elysium, while the heavily cratered upland terrain yielded relatively low depolarization. This led to the suggestion by Harmon and Ostro, and confirmed independently by radar astronomer Tommy Thompson and USGS Menlo Park geologist Henry J. Moore, who used Goldstone data, that most of the strong sources of diffuse and depolarized backscatter on Mars were rough-surfaced lava flows.[56]

Such was the state of radar studies of Martian topography and scattering, when Muhleman, Butler, and Slade began looking at Mars with the Goldstone-VLA bistatic radar in 1988. The proximity of Mars, in contrast to the great distance to Titan, allowed them to construct full-disk images of the planet. During the 1988 Mars opposition, moreover, the Earth and Mars were closer than they had been for 17 years.

These images were not the product of radar range-Doppler techniques, but of standard VLA radio astronomy imaging software. The array and its software avoided the problem of north-south ambiguity that typically plagued planetary range-Doppler mapping; the VLA radio imaging software, which Muhleman regularly used in his planetary radio astronomy research, created unambiguous images. In this bistatic imaging mode, the Goldstone radar illuminated the target with a continuous-wave signal whose frequency was adjusted to remove the Doppler shift. When the VLA aimed at a target, the signal came from all over the planet, as though the target were a natural emitter of radio waves. Then the powerful imaging software of the VLA processed these echoes.

Muhleman, Butler, and Slade observed Mars twice during the opposition of 1988 and three times during the opposition of 1992–1993. They obtained surface resolutions of 80 km at the subradar point. The Mars observations differed from those of Titan, because for Mars the VLA A array (36-km maximum spacings) was used. The transmitted signal to

55. Downs, Mouginis-Mark, Zisk, and Thompson, "New Radar-Derived Topography for the Northern Hemisphere of Mars," *Journal of Geophysical Research* 87 (1982): 9747–9754; Mouginis-Mark, Zisk, and Downs, "Ancient and Modern Slopes in the Tharsis Region of Mars," *Nature* 297 (1982): 546-550; Simpson, Tyler, Harmon, and Alan R. Peterfreund, "Radar Measurement of Small-Scale Surface Texture: Syrtis Major," *Icarus* 49 (1982): 258–283; Schaber, "Syrtis Major: A Low-relief Volcanic Shield," *Journal of Geophysical Research* 87 (1982): 9852–9866; Roth, Downs, Saunders, and Schubert, "Radar Altimetry of South Tharsis, Mars," *Icarus* 42 (1980): 287–316; R. A. Craddock, R. Greeley, and P. R. Christensen, "Evidence for an Ancient Impact Basin in Daedalia Planum, Mars," *Journal of Geophysical Research* 95 (1990): 10,729–10,741; Downs, R. Green, and Reichley, "Radar Studies of the Martian Surface at Centimeter Wavelengths: The 1975 Opposition," *Icarus* 33 (1978): 441–453; Roth, Saunders, Downs, and Schubert, "Radar Altimetry of Large Martian Craters," *Icarus* 79 (1989): 289–310.

56. Harmon 15 March 1994; Harmon, Campbell, and Ostro, "Dual-Polarization Radar Observations of Mars: Tharsis and Environs," *Icarus* 52 (1982): 171–187; Harmon and Ostro, "Mars: Dual-Polarization Radar Observations with Extended Coverage," *Icarus* 62 (1985): 110-128; Thompson and Henry J. Moore, "A Model for Depolarized Radar Echoes from Mars," *Proceedings of the Lunar Planetary Science Conference 19th* (1989): 409-422; Moore and Thompson, "A Radar-Echo Model of Mars," *Proceedings of the Lunar Planetary Science Conference 21* (1991): 457–472. Later radar mapping supported these observations: Muhleman, Butler, Grossman, Slade, and Jurgens, "Radar Images of Mars," *Science* 253 (1991): 1508–1513; Harmon, Michael P. Sulzer, Phillip J. Perillat, and Chandler, "Mars Radar Mapping: Strong Backscatter from the Elysium Basin and Outflow Channel," *Icarus* 95 (1992): 153–156.

Mars was circularly polarized and both opposite circular and same circular echoes were received and mapped. As anticipated, the opposite circular echoes were dominated by the so-called specular (or phase-coherent) reflections.

Muhleman and Butler found regions with anomalously high radar cross sections on Mars, particularly around the three Tharsis volcanoes and Olympus Mons. These, Muhleman recalled, "just lit up like a Christmas tree." In contrast, the region west of Tharsis, extending over 2,000 km in the East-West direction and 500 km across at its widest point, displayed no cross section distinguishable from the noise in either polarization. "We didn't believe that result. We've never seen that on any real surface," Muhleman explained.[57]

Muhleman dubbed the area "Stealth," because it was invisible to the radar. Photographs do not indicate the nature of the Stealth region. Muhleman interpreted the lack of radar echo as arising from a deposit of ash or pumice spewed from the bordering Tharsis volcanoes and carried by winds blowing off the Tharsis ridge. He estimated that the Stealth material would have a density of less than about 0.5 grams per cubic centimeter, be free of rocks larger than one centimeter across, and have a depth of at least five, if not ten, meters.

Equally surprising was the radar signature of the residual southern polar ice cap. The 1988 observations were made in the southern hemisphere around -24° latitude in late spring, so the seasonal carbon dioxide ice cap had sublimated away and exposed the residual southern polar ice cap. That area had the highest radar cross section of any other area observed on the planet in 1988. Furthermore, the residual ice cap exhibited strong circular polarization inversion. Thus, unexpectedly, part of one of the terrestrial planets displayed radar characteristics more typical of the Galilean satellites.

When Muhleman, Butler, and Slade looked at the VLA images, they "instantly saw that the brightest thing on the planet was the South pole, which turned out to be the residual South polar ice cap," Muhleman recalled. "The amazing thing to us was that this ice was so reflecting, so bright, and its size was exactly the residual polar cap."[58] Also amazing was the fact that Dick Simpson and Len Tyler had failed to notice any unusual scattering properties from the North pole in data from a bistatic radar experiment conducted from the Viking spacecraft.[59]

Butler, Muhleman, and Slade again looked at Mars with the Goldstone-VLA radar during the 1992–1993 opposition, when the planet's North pole was visible from Earth. It was early northern spring on Mars, and much of the seasonal carbon dioxide polar ice cap was present. They were anxious to study the northern polar ice cap, but the ice was invisible to the radar. In stark contrast to the southern pole, no regions with enhanced radar cross sections appeared. "We still haven't figured that out," Muhleman admitted. "It's totally a mystery why we didn't find the residual North polar ice cap."[60]

The high radar cross section and polarization inversion of the Martian South polar ice cap were confirmed by observations made at the Arecibo Observatory during the 1988 opposition by John Harmon, Marty Slade, and R. Scott Hudson. Hudson was a Caltech graduate student working on a doctoral degree in electrical engineering and had chosen aircraft radar imaging as his dissertation topic. Like those made by Harmon and Ostro in

57.   Muhleman 27 May 1994.
58.   Muhleman 27 May 1994.
59.   Muhleman 27 May 1994; Slade 24 May 1994; Muhleman, Butler, Grossman, and Slade, "Radar Images of Mars," *Science* 253 (1991): 1508–1513; Butler, "3.5-cm Radar Investigation of Mars and Mercury: Planetological Implications," Ph.D. diss., California Institute of Technology, 9 May 1994; Simpson and Tyler, "Viking Bistatic Radar Experiment: Summary of First-Order Results Emphasizing North Polar Data," *Icarus* 46 (1981): 361–389.
60.   Muhleman 27 May 1994; Butler, "3.5-cm Radar Investigation;" Butler, Muhleman, and Slade, "Martian Polar Regions: 3.5 cm Radar Images," *Lunar and Planetary Science Conference* 25 (1994): 211–212; Butler, Muhleman, and Slade, "The Polar Regions of Mars: 3.5 cm Radar Images," *Icarus* submitted in May 1994.

1980 and 1982, these were monostatic, dual-polarization continuous-wave observations made with the Arecibo S-band and the Goldstone X-band radars.

After obtaining promising results from a comparison of the 1988 data at both wavelengths, an additional set of observations were made at S-band and X-band during the 1990 opposition. Despite scheduling difficulties and the demands of competing types of radar observations (ranging observations for altimetry and mapping were also made at the two facilities), a good continuous-wave data set for S/X-band comparison was obtained in 1990.

The Arecibo data confirmed the existence of Stealth. Using an algorithm developed by Scott Hudson and the Doppler spectra taken in the unexpected sense of polarization, they produced depolarized reflectivity maps that showed clearly the anomalously high radar reflectivity and polarization inversion of the residual South polar icecap. Hudson's algorithm allowed the investigators to use only Doppler spectra, without range measurements, to create a two-dimensional map of the Martian disk largely free of north-south ambiguity.[61]

Hudson's imaging technique was necessary in order to overcome the planet's overspread nature. In comparison to Venus, Mars rotates rapidly on its axis and causes radar echoes from the limb (beyond the subradar area) to disperse broadly. The echo delay corresponding to the radius of Mars is 22.6 microseconds, which is much greater than the maximum interval of 0.725 microseconds needed to preserve complete spectral information over the band of frequencies present in the echo. As a result, when the computer samples signals, echoes from different ranges contaminate each other and become indistinguishable. Such radar targets are called "overspread."

Arecibo scientists also had a technique for overcoming the overspread problem, but they were not motivated to apply it until the Goldstone-VLA results became known. Harmon explained: "Dewey Muhleman, with his VLA experiment, spurred us on to try and do better. I really hadn't been thinking about the overspreading problem. I probably should have; I should have been trying to figure out ways to get around it."[62]

Overspreading was a problem that ionospheric scientists had been dealing with for years, because the ionosphere is an extremely overspread target. Michael P. Sulzer, an ionosphericist at the Arecibo Observatory, solved the problem for the ionosphere by using non-repeating codes. Although Don Campbell at one time had asked Sulzer to think about applying the technique to Mars, no progress had been made until Harmon told Sulzer he was interested in trying the non-repeating code technique.

Harmon then worked with Sulzer and Phil Perillat, who wrote the modified data-taking program. Normally, when a continuous-wave radar sends out a signal, the signal carries a code with a finite number of elements, and the code repeats at a regular interval. Harmon and Sulzer tested the non-repeating code, called alternately the "random code" or "coded long pulse" technique, and it worked the first time. Then Harmon wrote programs to do the data analysis.

Harmon and Sulzer made their first random-code observations on 18 nights during the Mars opposition of September-December 1990 and created range-Doppler maps. Those maps, like all range-Doppler maps, included a north-south ambiguity around the Doppler equator. However, from eyeball comparisons with maps obtained early and late in the opposition, Harmon was able to resolve much of the ambiguity.

61.   Hudson, telephone conversation, 21 November 1994; Hudson and Ostro, "Doppler-Radar Imaging of Spherical Planetary Surfaces," *Journal of Geophysical Research* 95 (1990): 10,947–10,963; Harmon, Slade, and Hudson, "Mars Radar Scattering: Arecibo/Goldstone Results at 12.6- and 3.5-cm Wavelengths," *Icarus* 98 (1992): 240–253.

62.   Harmon 15 March 1994.

The random-code maps confirmed the observations made with the Goldstone-VLA radar and revealed new information about the Elysium region, which Harmon had spent a long time studying in previous observations of Mars. Through those and subsequent observations made during the 1992–1993 opposition, he discovered strong depolarized radar echoes from the Elysium/Amazonis outflow channel complex. He interpreted the region, which was very young by Martian standards, as having lava flows that appeared to have partially filled pre-existing channels cut by flowing water.[63]

Mercury: Baked Alaska?

The strange radar signature exhibited by the southern residual polar ice cap of Mars, reminiscent of the radar characteristics of the icy Galilean satellites of Jupiter, did not prepare Muhleman, Butler, and Slade for the surprising discovery of ice on Mercury. Mercury was simply too hot to support even the smallest ice deposit. Previous radar observations of Mercury had focused on scattering and topography and had not detected ice.

Analysis of Mercury data taken between 1963 and 1965 at Goldstone, Haystack, and Arecibo showed the planet to have a radar roughness "very similar" to that of the Moon. Dick Goldstein, from radar observations of Mercury made in 1969, started characterizing Mercury's topography. His work was the most detailed radar study of Mercury's surface prior to the Mariner 10 encounters and provided the first strong evidence for the existence of craters on the surface. Dick Ingalls at Haystack and Don Campbell at Arecibo also found altitude variations on Mercury's surface from 1971 observations. Some of the earliest topographic radar studies of Mercury were carried out at Haystack by Bill Smith and Dick Ingalls.[64]

These early radar studies of Mercury were not linked to any specific NASA mission, but not because of any radar shortcomings. NASA made no meaningful effort to study Mercury until Mariner 10 flew by and photographed that planet in 1974–1975. The Mariner 10 photographs revealed a heavily cratered, lunar-like surface, as predicted by radar. Although Mariner 10 photographed over half of Mercury's surface during its flyby mission, it did not photograph any of the side not then exposed to the Sun's light and yielded only limited topographic information. Moreover, its flyby geometry prevented Mariner 10 from examining either pole directly.[65]

These gaps in Mercury coverage motivated a program of observations at Arecibo and Goldstone. John Harmon and Don Campbell, working in collaboration with Brown University geologists D. L. Bindschadler and James W. Head, carried out a campaign of

63.   Harmon 15 March 1994; NAIC QR Q2/1990, 7; Q4/1990, 7–8; and Q1/1991, 7; Q1/1993, 9; Harmon, Sulzer, and Perillat, "Mars Radar Mapping: Strong Depolarized Echoes from the Elysium/Amazonis Outflow Channel Complex," *Lunar and Planetary Science Conference* 22 (1991): 513.

64.   Muhleman, "Radar Scattering from Venus and Mercury at 12.5 cm," *Journal of Research of the National Bureau of Standards, Section D: Radio Science* 69D (1965): 1630–1631; Evans, Brockelman, Henry, Hyde, Kraft, W. A. Reid, and W. W. Smith, "Radio Echo Observations of Venus and Mercury at 23 cm Wavelength," *The Astronomical Journal* 70 (1965): 486–501; Pettengill, Dyce, and Campbell, "Radar Measurements at 70 cm of Venus and Mercury," *The Astronomical Journal* 72 (1967): 330–337; Goldstein, "Mercury: Surface Features Observed during Radar Studies," *Science* 168 (1970): 467–469; Goldstein, "Radio and Radar Studies of Venus and Mercury," *Radio Science* 5 (1970): 391–395; Goldstein, "Radar Observations of Mercury," *The Astronomical Journal* 76 (1971): 1152–1154; Goldstein, "Review of Surface and Atmosphere Studies of Venus and Mercury," *Icarus* 17 (1972): 571–575; Zohar and Goldstein, "Surface Features on Mercury," *The Astronomical Journal* 79 (1974): 85–91; Smith, Ingalls, Shapiro, and Ash, "Surface-Height Variations on Venus and Mercury," *Radio Science* 5 (1970): 411–423; Ingalls and Rainville, "Radar Measurements of Mercury: Topography and Scattering Characteristics at 3.8 cm," *The Astronomical Journal* 77 (1972): 185–190.

65.   Murray, Michael J. S. Belton, G. Edward Danielson, Merton E. Davies, Donald E. Gault, Hapke, Brian O'Leary, Robert G. Strom, Verner Suomi, and Newell Trask, "Mercury's Surface: Preliminary Description and Interpretation from Mariner 10 Pictures," *Science* 185 (1974): 169–179.

S-band radar observations of Mercury at Arecibo from 1978 to 1984. They measured
Mercury's topography over much of the equatorial zone (between 12° North and 5° South
latitude), an area not imaged by Mariner 10, and concluded that radar depths for large
craters supported previous indications from photographs that Mercury's craters were shal-
lower than lunar craters of the same size.[66] At the same time, Ray Jurgens, using the
Goldstone S-band radar, started an ongoing series of Mercury observations to study the
planet's topography and to correlate radar measurements with Mariner 10 visual images,
in collaboration with geologists Gerald G. Schaber (USGS Flagstaff) and P. E. Clark
(JPL).[67]

Such was the state of radar research on Mercury, when Muhleman, Butler, and Slade
began their observations with the Goldstone-VLA bistatic radar during the inferior conjunction of August 1991. Although they made further observations during the inferior
conjunctions of November 1992 and February 1994, the 1992 effort failed because of
transmitter problems, and the 1994 data yet remains to be reduced.[68] The key results,
then, were those from the 1991 observations. They did nothing less than revolutionize our
knowledge of Mercury in a way that radar had not done since the discovery of the planet's 59-day spin rate by radar astronomers Gordon Pettengill and Rolf Dyce in 1965.

During the first Goldstone-VLA observation of Mercury on 8 August 1991, Ray
Jurgens coordinated activities at the Goldstone X-band transmitter, while Marty Slade and
Bryan Butler awaited the echoes at the VLA, which was operating in the so-called A array,
the most widely spaced configuration. During the 10 hours of observation, the VLA
received in both senses of circular polarization. At the time of these observations, Mercury
was at inferior conjunction and presented the hemisphere not photographed by Mariner
10, roughly between 180° to 360°, to the radar. As a result, the subradar point was far
enough North to see over the North pole and into areas believed to be permanently shadowed from the Sun.

When Muhleman, Butler, and Slade looked at their results, they were astonished;
they had found ice near Mercury's North pole. What signalled the presence of ice was the
abnormal radar signature of the spot, which was unusually bright and showed a ratio of
same circular to opposite circular polarization greater than unity, that is, a circular polarization inversion. This was the same type of radar signature displayed by Jupiter's Galilean
moons. Muhleman recalled: "We instantly looked at the first image and saw this white spot
on the North pole. We said, 'My God! Are we going to find an ice cap on every planet we
look at?' This is crazy!" Marty Slade remembered looking at the bright spot and reacting:
"It's not possible that could be ice! It's too hot!"[69]

Muhleman, Butler, and Slade again observed Mercury with the Goldstone-VLA radar
two weeks later on 23 August 1991. This time, they transmitted both right-handed (RCP)
and left-handed circular (LCP) polarization, and they received in both senses of polarization for either sense, so that they could make all four correlations of the two polarizations
(RCP to LCP, RCP to RCP, LCP to RCP, and LCP to LCP). Mercury as seen from Earth had
rotated 101°. The subradar point was around 353° and the ice near the northern polar

66.   Harmon, Campbell, Bindschadler, Head, and Shapiro, "Radar Altimetry of Mercury: A Preliminary
Analysis," *Journal of Geophysical Research* 91 (1986): 385–401.
67.   See, for example, P. E. Clark, M. E. Strobell, Schaber, and Jurgens, "Some New Radar-Derived
Topographic Profiles of Mercury," *Bulletin of the American Astronomical Society* 16 (1984): 668; Clark, Jurgens, and
M. Kobrick, "Analyses of Radar-Derived Topography and Scattering Properties of Mercury's Equatorial Region,"
Bulletin of the American Astronomical Society 17 (1985): 712; and Clark, M. A. Leake, Slade, Jurgens, Robinett, and
C. Franck, "Scattering and Altimetry Measurements from Goldstone Radar Observations of Mercury in 1987,"
Bulletin of the American Astronomical Society 19 (1987): 863.
68.   Butler, "3.5-cm Radar Investigation," preface.
69.   Muhleman 27 May 1994; Slade 24 May 1994.

region still stood out brightly and exhibited polarization inversion. The researchers now knew that this was no fluke.[70]

Surprised by their own results, Muhleman, Slade, and Butler announced their results in two separate talks given on 6 November 1991 at the meeting of the AAS Division for Planetary Science, held in Palo Alto, California.[71] The scientific community greeted the news of their discovery with a fair amount of skepticism.[72] Prior to the launch of Mariner 10, few had suggested the presence of ice on Mercury, and then for the wrong reasons. Some drawings of Mercury showed a white spot visible at the northern pole, and in 1974, on the eve of Mariner 10's first reconnaissance of Mercury, an atmospheric scientist had proposed that ice could have accumulated in the small planet's polar regions, perhaps in permanently shaded regions.[73]

The evidence for the presence of ice near Mercury's northern pole was based on an analogy between the radar signatures of known icy targets, the Galilean moons of Jupiter, and those found on Mercury. But more convincing evidence was needed, because Mariner 10 had documented that planet's intense surface heat. The landscape was a parched wasteland of impact craters and volcanic plains, where midday temperatures soared to 700° K, hot enough to melt lead. At the same time, though, Mariner 10's ultraviolet spectrometer had identified traces of hydrogen and oxygen in the tenuous atmosphere of Mercury. Project scientists had considered them to be remnants of the comets and asteroids that periodically collide with the planet.[74]

While such collisions would explain the existence of water on Mercury, an explanation for the existence of a permanent water ice deposit on the planet came from a consideration of the geometry of Mercury's orbit. An impact crater could provide an area of permanent shade, provided that the geometry was just right. Mercury spins on its axis and rotates around the Sun in such a way that its equator always lies in the same plane as the Sun. As a result, neither pole ever sees more than a sliver of the Sun's disk above the horizon. On the other hand, the plane of Mercury's orbit about the Sun is inclined by seven degrees relative to that of the Earth, so that Earth-based radars can see into impact craters that are never directly illuminated by the Sun.

David A. Paige and Stephen Wood of UCLA recomputed the thermal environment for Mercury's surface and concluded that the interior slopes of impact craters within five degrees of the poles would be cold enough to keep the loss of water ice through sublimation at essentially zero. Other planetary scientists also began to argue for the existence of ice in craters on Mercury, and they suggested that craters on the Moon might also contain ice. As early as 1961, Kenneth Watson, Bruce C. Murray, and Harrison Brown had proposed that ice might exist in permanently shadowed craters near the lunar poles, but

70.   Slade, Butler, Muhleman, "Mercury Radar Imaging: Evidence for Polar Ice," *Science* 258 (23 October 1992): 635–640; Butler, Muhleman, and Slade, "Mercury: Full-Disk Radar Images and the Detection and Stability of Ice at the North Pole," *Journal of Geophysical Research* vol. 98, no. E8 (1993): 15,003–15,023.

71.   Slade, Butler, and Muhleman, "Mercury Goldstone-VLA Radar: Part I," *Bulletin of the American Astronomical Society* 23 (1991): 1197, and Butler, Muhleman, Slade, and Jurgens, "Mercury Goldstone-VLA Radar: Part II," Ibid., p. 1200.

72.   David A. Paige, "Chance for Snowballs in Hell," *Nature* 369 (1994): 182; Chapman, "Ice Right Under the Sun," *Nature* 354 (1991): 504–505; J. Kelley Beatty, "Mercury's Cool Surprise," *Sky & Telescope* 83 (January 1992): 35–36.

73.   Richard Baum, "Radar Bright, Ice Bright: V. A. Firsoff and Ice Caps on Mercury," *Journal of the British Astronomical Association* 103 (1993): 126 and 139; Firsoff, "Could Mercury have Ice Caps?" *The Observatory* 91 (1971): 85–87; and G. E. Hunt, "There is no Evidence for Ice Caps on Mercury," *The Observatory* 92 (1972): 16; Beatty, "Mercury's Cool Surprise," *Sky & Telescope* 83 (1992): 35–36; Gary E. Thomas, "Mercury: Does its Atmosphere Contain Water?" *Science* 183 (1974): 1197–1198.

74.   Beatty, p. 36; Chapman, "Ice," p. 505; Chapman, *Planets of Rock and Ice: From Mercury to the Moons of Saturn* (New York: Scribner, 1982); and Faith Vilas, Chapman, and Matthews, eds., *Mercury* (Tucson: University of Arizona Press, 1988).

to date no lunar probe, not even the Clementine orbiter, has found any ice on the Moon.[75] A radar search at Arecibo also proved unsuccessful.

Nick Stacy, a graduate student working on a thesis in radar astronomy under Don Campbell, looked for ice on the Moon with the Arecibo radar. Earlier, starting in 1982, Don Campbell and Peter Ford had carried out high-resolution range-Doppler imaging of the Moon and found no evidence of ice, but they were not looking for it. Ford and Campbell brought the resolution of their images down from 300 to 150 meters, using the Higuillales antenna in a bistatic mode with the big dish. Stacy reduced the resolution to 20 meters and aimed at the lunar poles. Unfortunately, the radar could not see far enough into the polar craters and detected no ice, though Stacy found some unusual scattering properties around a number of lunar craters.[76]

Although the discovery of lunar crater ice remained elusive, John Harmon and Marty Slade at the Arecibo Observatory confirmed the existence of ice on Mercury. They imaged Mercury using the non-repeating code technique developed by Harmon and Mike Sulzer in order to overcome overspreading on Mars. These Arecibo images, according to David Paige, left "little room for doubt" about the presence of ice on Mercury.[77]

Soon after observing Mercury on 8 August 1991 with the Goldstone-VLA radar, Marty Slade travelled to the Arecibo Observatory to collaborate with Harmon on a different set of Mercury observations. They acquired their initial data prior to 8 August 1991, on 28 separate dates during the periods 28 March to 21 April 1991, 31 July to 29 August 1991, and 14 to 29 March 1992. During the spring 1991 observations, the subradar point of the Arecibo telescope subtended an area in the southern hemisphere of Mercury, while the summer 1991 observations covered a portion of the northern hemisphere, as the Goldstone-VLA had. The March 1992 data added to that already observed in the southern hemisphere.

When Slade arrived at Arecibo, his first time at the observatory, Harmon had not yet analyzed the spring 1991 data; he had been too busy studying Mars data. Slade suggested to Harmon that they analyze the Mercury data and look for the icy radar signature near the North pole, which he, Muhleman, and Butler had just found with the Goldstone-VLA radar. According to Harmon, Slade said, "We think it's the pole; we're not sure." The Arecibo data confirmed the Goldstone-VLA discovery. There was no question of priority; Muhleman, Butler, and Slade discovered the ice on Mercury first, with the Goldstone-VLA radar.

Harmon also examined the data collected from the southern hemisphere of Mercury in March–April 1991. "I saw a feature coming from what I figured probably had to be the South pole, because the latitude was about five degrees South [sic]," Harmon related. "I was pretty convinced it was coming from the South."[78] To confirm that the South pole was the source of the icy radar signature and not an artefact of north-south ambiguity, which would have shown a portion of the northern polar echo at the South pole, Harmon and Slade observed Mercury again in March 1992, when the subradar point was again in the southern hemisphere. The polar ice feature was seen again, confirming the presence of ice at the planet's South pole.[79]

75.   Simpson 10 May 1994; Paige, Stephen E. Wood, and Ashwin R. Vaasavada, "The Thermal Stability of Water Ice at the Poles of Mercury," *Science* 258 (1992): 643–646; Andrew P. Ingersoll, Tomas Svitek, and Murray, "Stability of Polar Frosts in Spherical Bowl-Shaped Craters on the Moon, Mercury, and Mars," *Icarus* 100 (1992): 40–47; Kenneth Watson, Murray, and Harrison Brown, "The Behavior of Volatiles on the Lunar Surface," *Journal of Geophysical Research* 66 (1961): 3033–3045.

76.   Ford 3 October 1994; Campbell 10 March 1993; Campbell 8 December 1993; Stacy, "High-Resolution Synthetic Aperture Radar Observations of the Moon," Ph.D. diss., Cornell University, May 1993.

77.   Paige, "Chance for Snowballs in Hell," *Nature* 369 (1994): 182.

78.   Harmon 15 March 1994.

79.   Harmon 15 March 1994; Harmon and Slade, "Radar Mapping of Mercury: Full- Disk Images and Polar Anomalies," *Science* 258 (1992): 640–642; Harmon and Slade, "An S-band Radar Anomaly at the North Pole of Mercury," *Bulletin of the American Astronomical Society* 23 (1991): 1121.

Next, Harmon and Slade proceeded to fit the radar results to photographic data from Mariner 10. Showing a correlation between a known crater and the radar ice would be persuasive confirmation of the discovery. Matching the northern polar radar ice location with a crater was hard; no Mariner 10 photographs were available for the entire region. Furthermore, the North polar radar anomaly was too large to fit within a single crater. The image, instead, appeared to consist of a number of crater-size (15–60 km in diameter) bright spots. Harmon and Slade plotted those features on a locating map created by NASA and the USGS and assigned letter labels to those features that lay in the photographed hemisphere and to three prominent features in the unphotographed hemisphere. Many of the radar spots (8 out of 20) appeared to correspond to impact craters. Correlating the southern polar radar image with topography was simpler. The radar spot was entirely inside a crater called Chao Meng-Fu.[80]

The Goldstone-VLA and Arecibo images of Mercury once again highlighted how planetary radar astronomy often solves problems left unsolved or unsatisfactorily solved by optical techniques. The discovery of ice near Mercury's North and South poles, moreover, has inspired the European Space Agency to mount a major "keystone" mission to Mercury in search of polar ice, as well as a more modest-sized NASA Discovery flight.[81]

Radar astronomers also sought signs of anomalous radar signatures on other terrestrial planets. Muhleman, Butler, and Slade turned the Goldstone-VLA radar on Venus twice, 18 and 25 February 1990, receiving both senses of polarization in order to detect any peculiar polarization inversion, and made two maps. The maps had several striking features. Surprisingly, Alpha Regio had a high unexpected (depolarized or SC) reflectivity on both maps and contained the second highest reflectivity values after Maxwell. On the second day's map, the point of highest reflectivity was in the Aphrodite region and was not visible in the previous map. On both maps, many very small areas, only a few pixels across, also had large unexpected (depolarized or SC) reflectivities, and some of them corresponded to mapped elevated areas such as Gula Mons, Sif Mons, and Bell Regio. Muhleman, Butler, and Slade concluded that a correlation existed between unexpected (depolarized or SC) reflectivities and elevation. Further bistatic observations of Venus in the spring of 1993 furnished fuel for another Muhleman graduate student, Albert Haldeman, to begin doctoral research, while Slade and Ray Jurgens also found highly reflective areas on Venus using just the Goldstone radar.[82]

Asteroids

Throughout the 1980s and into the 1990s, the number of asteroids discovered and the number of publications dealing with asteroids grew at an unprecedented rate, at first as a result of the Palomar Planet-Crossing Asteroid Survey studies initiated in the 1970s, then as the number of asteroid researchers swelled. In 1932, an astronomer discovered the first Earth-crossing asteroid, 1862 Apollo. By 1994, about 200 Earth-crossing asteroids were known, more than half of which had been discovered in the previous seven years; yet

80.   Harmon, Slade, Vélez, Andy Crespo, M. J. Dryer, and J. M. Johnson, "Radar Mapping of Mercury's Polar Anomalies," *Nature* 369 (1994): 213–215; Harmon and Slade, "Radar Mapping of Mercury: Full-Disk Images and Polar Anomalies," *Science* 258 (1992): 640–643.

81.   Muhleman 24 May 1994; Paige, "Snowballs," p. 182.

82.   Slade 24 May 1994; K. A. Tryka, Muhleman, Butler, Berge, Slade, and Grossman, "Correlation of Multiple Reflections from the Venus Surface with Topography," *Lunar Planetary Science* 22 (1991): 1417; Jurgens, Slade, and Saunders, "Evidence for Highly Reflecting Materials on the Surface and Subsurface of Venus," *Science* 240 (1988): 1021–1023; Butler, "3.5-cm Radar Investigation," passim; Slade 24 May 1994; and information provided by Bryan J. Butler.

the undiscovered population is huge. In the decade 1975–1985 alone, the total number of catalogued asteroids rose from 2,000 to more than 3,200.[83]

The field, as measured by the expanding literature, was undergoing the kind of swift growth that is typical of Big Science. Asteroid astronomy became a new theoretical framework with problems that radar astronomers sought to solve. Radar found its niche within asteroid astronomy because it could solve problems that other observational techniques could not do, namely, the creation of more accurate and reliable ephemerides and the imaging of asteroids.

The focus of asteroid research was on near-Earth asteroids, although main belt objects remained of interest, too. Near-Earth asteroids, like meteorites, are thought to come primarily from mainbelt asteroids (Table 8). A large population of asteroids also cross the orbits of Earth and Mars. The term near-Earth asteroid usually means any asteroid that can come close to the Earth, whether or not it crosses the orbit of the Earth. Eros, for example, crosses the orbit of Mars, but it is not an Earth-crossing asteroid and does not come near the Earth. Almost all of the near-Earth asteroids detected so far by radar are Earth-crossers.

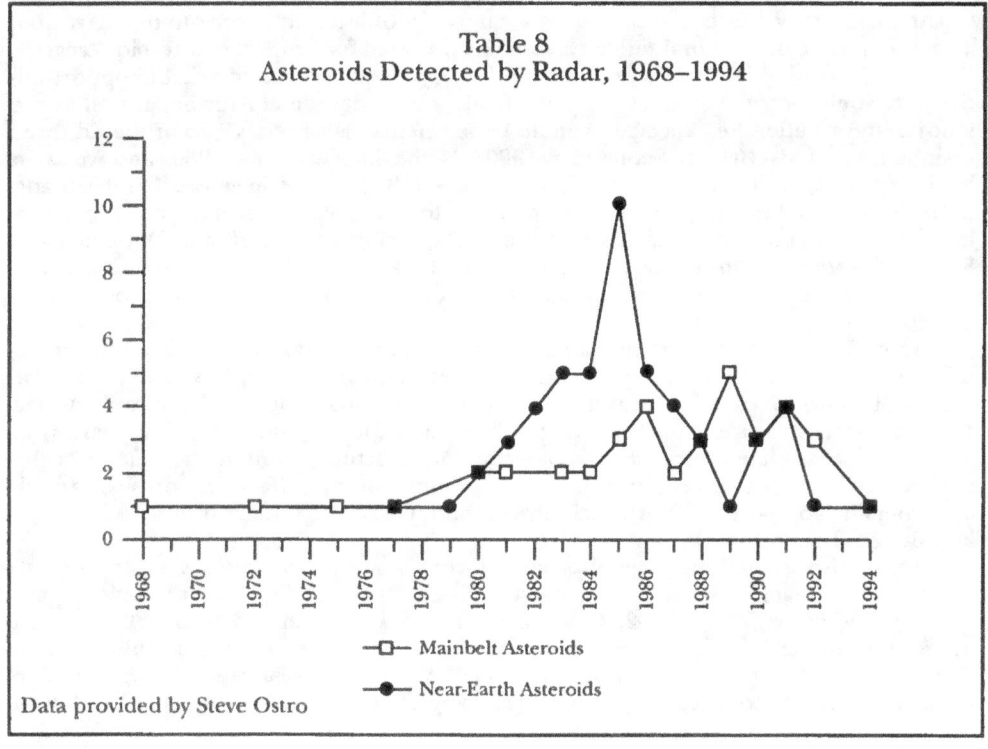

Table 8
Asteroids Detected by Radar, 1968–1994

Data provided by Steve Ostro

—□— Mainbelt Asteroids
—●— Near-Earth Asteroids

    83.   Ostro, Campbell, and Shapiro, "Mainbelt Asteroids: Dual- Polarization Radar Observations," *Science* 229 (1985): 442.

The more interesting near-Earth asteroids also were better radar targets than main belt asteroids, because now and then they come closer to the Earth. With targets as small as asteroids, some only a kilometer or two in diameter, the distance to the target is critical to radar observations. The number of asteroids observed by radar astronomers grew rapidly during the 1980s because of the availability of radars with sufficient power and sensitivity to detect and study them. Another key factor in the growth of radar asteroid studies was the decision of one radar astronomer, Steve Ostro, to begin studying asteroids almost exclusively. Quickly, his efforts dominated the asteroid study started at Arecibo and Goldstone in the 1970s.

Before beginning this intense study of asteroids, Ostro had been making radar observations of the Galilean moons and the rings of Saturn. In March 1979, about the time of Voyager's encounter with Jupiter, Ostro attended the third Tucson asteroid conference organized by Tom Gehrels. There, Ray Jurgens and Gordon Pettengill delivered a joint paper on radar observations of asteroids. The conference, especially the talks that placed the science of meteoritics and asteroid science in context with each other, gave Ostro the asteroid bug. He saw how the study of asteroids was essential to understanding the origin and evolution of the solar system. He also realized that radar was potentially the primary post-discovery technique for observing asteroids, and that asteroids, unlike planets and their moons, constitute a huge and diverse population.[84]

Later in 1979, his MIT dissertation completed, Ostro took a teaching position at Cornell University and began preparing a campaign of asteroid observations at Arecibo. The following year, he submitted his first NASA proposal for support of asteroid research. Echoing the work of Jurgens a few years earlier, Ostro laid out those asteroid opportunities that would become available over the forthcoming decade at Arecibo, as well as the kinds of information he expected from his experiments. As targets, Ostro proposed three mainbelt asteroids (Iris in September 1980, Psyche in November 1980, and Vesta in February 1981) and two Earth-crossing asteroids (1862 Apollo in November 1980 and 1915 Quetzalcoatl in March 1981). He planned to detect echoes from each target, estimate echo strength, and measure polarization, spectral bandwidth, and Doppler shift. From those four quantities, Ostro proposed to estimate asteroid size and rotation, place constraints on the composition and structure of asteroid surfaces, and improve knowledge of their orbital parameters.[85]

Over the following years, the estimation of asteroid physical properties and the determination and refinement of their orbits remained fundamental aspects of Ostro's radar studies of asteroids. He systematically took range and Doppler data on all asteroids, as well as polarization measurements (receiving in both the expected and unexpected senses) in order to best estimate their surface roughness and structure. From measurements of the surface's reflectivity came estimates of the bulk density of the surface, its porosity, and relative metallic composition. With each observation, Ostro tried to contribute to scientific knowledge about asteroids.

Ostro also studied mainbelt asteroids. "Virtually every experiment gave an interesting result, and each radar signature was different," Ostro recalled. "Every single experiment was lucrative."[86] By 1992, Ostro had observed 28 near-Earth and 36 mainbelt asteroids. Between 1980 and 1985 alone, he made dual-polarization observations of 20 mainbelt asteroids at Arecibo. These objects had low circular polarization ratios (the ratio of unexpected to expected echo power) ranging from about 0.00 to 0.40. The lowest

---

84.   Ostro 25 May 1994; Pettengill and Jurgens, "Radar Observations of Asteroids," in Gehrels and Matthews, pp. 206–211.
85.   Ostro 25 May 1994; Ostro, "Radar Investigations of Asteroids," proposal submitted to NASA in June 1980 for support 1 November 1980 through 31 October 1981, Ostro materials.
86.   Ostro 25 May 1994.

value, 0.05 ± 0.02 for the asteroid 2 Pallas, required that nearly all the echo arise from single-reflection backscattering from very smooth surface elements.

"It became clear," Ostro explained, "that the mainbelt asteroids had a dispersion of reflectivities and polarization ratios. This was evidence for diversity in surface structure and in surface bulk density."[87] The data collected helped to characterize asteroid surfaces at scales between several centimeters and several kilometers and furnished constraints on surface bulk density and metal concentration, beyond those constraints obtained by optical methods.

The metallic composition of the asteroids was an interesting question relating to possible meteoritic analogues. The radar observations suggested wide variations in metal abundance, porosity, and decimeter-scale roughness on mainbelt asteroid surfaces, underscoring the diversity of the asteroid population already evident from visible and infrared wavelength studies. Although the radar signatures of mainbelt asteroids required substantial surface roughness at some scale much larger than a meter, Ostro could not discern the precise scale of this structure, much less the actual morphologies of surface features. Similarly, the radar albedos bolstered the hypothesis that metal concentrations on asteroids run the gamut. Serious questions remain, however, about detailed mineralogies, meteoritic associations, and evolutionary histories.[88]

"Each of the near-Earth asteroids is interesting in its own way," Ostro pointed out, "and still some interesting mysteries remain."[89] Echoes from the near-Earth asteroid 1986 DA showed it to be significantly more reflective than other radar-detected asteroids. This result supported the hypothesis that 1986 DA was a piece of nickel-iron metal derived from the interior of a much larger object that melted, differentiated, and cooled, and subsequently was disrupted in a catastrophic collision. This two-kilometer-sized asteroid appeared smooth at centimeter to meter scales but extremely irregular at 10- to 100-meter scales. It might be (or have been part of) the parent body of some iron meteorites. The composition of asteroids thus bears directly on the question of their relationship to meteorites, as well as the relationship between near-Earth and mainbelt asteroids.[90]

Starting in 1983, Steve Ostro began observing echo spectra with unusual shapes, including some spectra with double peaks (called bimodal). The first asteroid to show a bimodal spectra was 2201 Oljato, observed during 12–17 June 1983 at Arecibo. Asteroid astronomers had been discussing binary asteroids and contact-binary asteroids for a long time, but no evidence of their existence was at hand. 216 Kleopatra, a large mainbelt asteroid, exhibited a strong bimodal echo spectrum. "That almost definitely is a contact binary," Ostro explained. "But almost definitely is not definitely."[91]

Proof of the existence of binary and contact-binary asteroids eventually came from radar data.[92] Finding that proof was a problem left unsolved by optical and other research techniques. To the telescope, the biggest asteroid looks like a little dot, its shape indiscernible. Radar succeeded in solving that problem through the development of new imaging and modeling techniques. The key to developing an appropriate technique, though, was to avoid simplistic models. Too, it was important that the asteroid approach Earth close enough to provide the Arecibo and Goldstone radars a sufficiently strong echo to resolve the target.

87.   Ostro 25 May 1994.
88.   Ostro, Campbell, and Shapiro, "Mainbelt Asteroids: Dual- Polarization Radar Observations," *Science* 229 (1985): 442–446.
89.   Ostro 25 May 1994.
90.   Ostro, Campbell, Chandler, Hine, Hudson, Rosema, and Shapiro, "Asteroid 1096 DA: Radar Evidence for a Metallic Composition," *Science* 252 (1991): 1399–1404.
91.   Ostro 25 May 1994.
92.   See, for example, the discussion in W. I. McLaughlin, "Radar Tracking of Asteroids," *Spaceflight* 34 (1992): 167–169.

Ray Jurgens developed the first modelling technique for describing asteroid shapes in the 1970s. He applied it to spectral data from Eros. Steve Ostro applied Jurgens' triaxial ellipsoid model to his 1980 and earlier 1972 Toro data and derived a rough description of the asteroid.[93] Similarly, when he applied Jurgens' model to the Earth-crossing asteroid 2100 Ra-Shalom in 1981, Ostro found it to have a somewhat irregular shape.[94]

With researchers at Cornell, Ostro developed a different modelling technique, one that synthesized echo spectra acquired at different rotational phases of the asteroid into a convex envelope, called the hull, which represented the asteroid's silhouette as viewed from a pole. After he fit the hull model to Jurgens' Eros data, Ostro modeled the Earth-crossing asteroids 1627 Ivar and 1986 DA, observed in July 1985 and April 1986 at Arecibo. Interestingly, the hull estimates indicated that 1986 DA's hull was "extremely irregular, highly nonconvex, and possibly bifurcated."[95]

The case of 1986 DA suggested that any asteroid model had to accommodate the possibility that the target might not be convex; both Jurgens' triaxial ellipsoid and the hull models were inadequate. Ostro also attempted to image asteroids with range-Doppler mapping techniques, beginning with 1627 Ivar in 1985.

Range-Doppler mapping revealed a bimodal distribution of echo power, suggesting that the target was not convex. All previous images had relied exclusively on Doppler spectra data; these were the first range-resolved images of an asteroid. Nonetheless, they failed to define the asteroid's global shape.[96]

The next opportunity to attempt range-Doppler imaging came in 1988, with the close approach to Earth of the small asteroid 1980 PA. The technicians at Arecibo improved the telescope's data acquisition software and hardware, in order to improve resolution of the asteroid; the resolution of the Goldstone radar on the same target was still not fine enough. The Goldstone radar did not achieve the limit needed for radar asteroid observations until 1986, when the Voyager upgrades were completed.[97]

Ostro again attempted range-Doppler images, this time of 1989 PB, later known as 4769 Castalia. On 9 August 1989, Eleanor Helin discovered the object on photographic plates taken at Palomar Observatory. Orbital calculations two days later showed that the asteroid would pass through the Arecibo Observatory's declination window during 19–22 August and that at closest approach, Castalia would be only 0.027 astronomical units from Earth. These were ideal conditions for imaging the asteroid at Arecibo, though not so at Goldstone. Communications with Voyager 2, which was making its closest approach to Neptune, occupied the Goldstone 70-meter antenna; it was unavailable for use as a radar telescope until 30 August, when some observations of Castalia took place after closest approach. "At Goldstone," Ostro recalled, "everything was a disaster. We had an eight-hour track, and we got about 20 minutes of data."[98]

Getting time on the Arecibo antenna on such short notice (10 days after detection) was not a problem; Ostro already had time to observe Victoria, a mainbelt asteroid, which had shown a hint of a double-peak spectral structure in 1982. After doing a few runs on

---

93.   Ostro 25 May 1994; Ostro, Campbell, and Shapiro, "Radar Observations of Asteroid 1685 Toro," *The Astronomical Journal* 88 (1983): 565–576.

94.   Ostro, Alan W. Harris, Campbell, Shapiro, and James W. Young, "Radar and Photoelectric Observations of Asteroid 2100 Ra-Shalom," *Icarus* 60 (1984): 391–403.

95.   Ostro, Robert Connelly, and Leila Belkora, "Asteroid Shapes from Radar Echo Spectra: A New Theoretical Approach," *Icarus* 73 (1988): 15–24; Ostro, Rosema, and Jurgens, "The Shape of Eros," *Icarus* 84 (1990): 334–351; Ostro, Campbell, Chandler, Hine, Hudson, Rosema, and Shapiro, "Asteroid 1096 DA: Radar Evidence for a Metallic Composition," *Science* 252 (1991): 1399–1404, esp. pp. 1400–1401.

96.   Ostro 25 May 1994; Ostro, Campbell, Hine, Shapiro, Chandler, C. L. Werner, and Rosema, "Radar Images of Asteroid 1627 Ivar," *The Astronomical Journal* 99 (1990): 2012–2018.

97.   Ostro 25 May 1994.

98.   Ostro 25 May 1994.

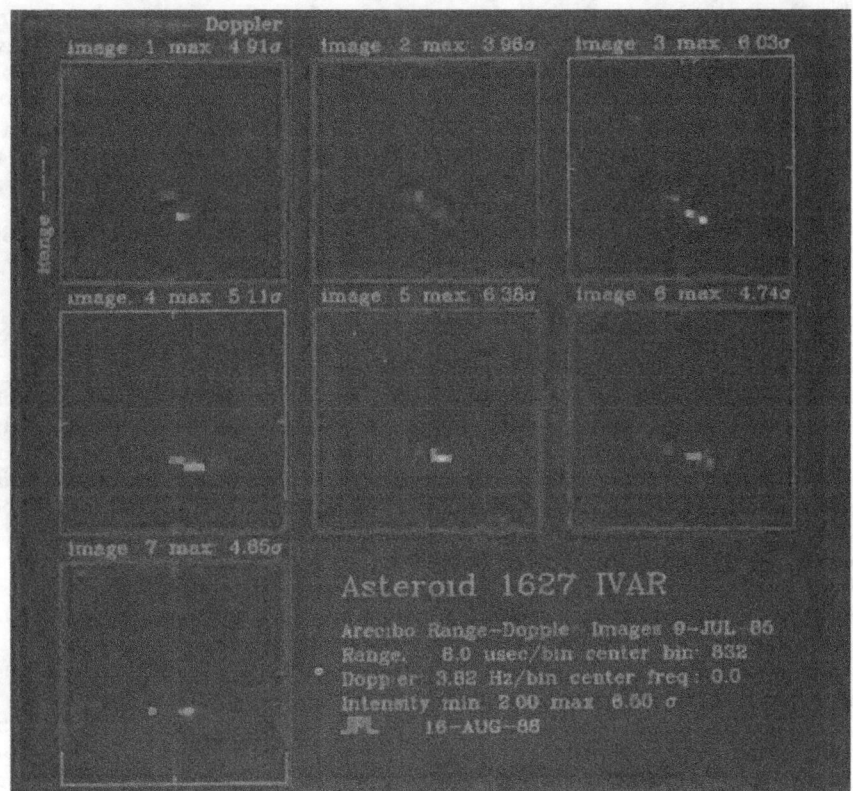

Figure 38

Range-Doppler radar images of Asteroid 1627 Ivar, 1985, made at the Arecibo Observatory by Steve Ostro. These are the first radar images made of an asteroid. (Courtesy of Jet Propulsion Laboratory.)

Victoria, Ostro spent the rest of the time on Castalia. "We saw CW [continuous-wave] echoes instantly," Ostro remembered. "A few of them from the first day looked strongly bifurcated." This was the first echo signature that said "This is a contact binary." Although he had never claimed discovery of bifurcated asteroids in print, Ostro had seen the idiosyncratic radar signatures several times before.[99]

From the Doppler and range data, Ostro created 64 images of the asteroid with an average of two dozen pixels each. Each image was bifurcated and showed a bimodal distribution of echo power. Reading the sequence of images from left to right from top to bottom, one can see the asteroid rotate.

When Ostro presented the images at the AAS Division for Planetary Science meeting a few months later, they attracted a dramatic intensity of attention; it was no less than the first time that anyone had resolved the shape of an asteroid from Earth. "This was a major breakthrough, definitely a major breakthrough," Ostro reflected.[100]

---

99.   Ostro 25 May 1994.

100.   Ostro 25 May 1994; Ostro, Chandler, Hine, Rosema, Shapiro, and Yeomans, "Radar Images of Asteroid 1989 PB," *Science* 248 (1990): 1523–1528.

About a year after the imaging of Castalia, Scott Hudson started working on a math-ematical modelling technique to reconstruct the asteroid's shape in three dimensions. While still a Caltech graduate student, Hudson had worked with Ostro in developing a technique for creating planetary Doppler images free of north-south ambiguity. Hudson devised a complex mathematical model with 169 parameters in order to capture the shape of asteroid Castalia. The resultant three-dimensional model showed indisputably that the asteroid was bifurcated into two distinct, irregular, kilometer-sized lobes.[101]

Modelling three-dimensional asteroid shapes from radar data provided further evi-dence for the existence of asteroids with exotic shapes upon the approach of 1989 AC (later known as 4179 Toutatis), discovered in January 1989. Because of its extremely close approach to Earth, 9.4 lunar distances on 8 December 1992, Toutatis showed high promise as a candidate for imaging. Ostro proposed a Toutatis experiment to the NAIC and to Nick Renzetti, explaining how extraordinary the opportunity was and urging that Goldstone also make observations.

Ostro planned to use both telescopes to take data in both senses of polarizations and to create range-Doppler images of several thousand pixels, considerably more resolution than had been achieved ever before. The high resolution was possible at Arecibo and Goldstone because of incremental improvements made in the data acquisition hardware and software over the preceding years. Ostro obtained continuous-wave echoes at Goldstone on 27 November, then range-Doppler images daily from 2–18 December 1992 and at Arecibo each day from 13–19 December. In addition to the routine monostatic observations, Ostro's team took advantage of new antennas recently made available. They observed Toutatis bistatically with the DSS-14 transmitting and a new 34-meter beam-waveguide antenna (DSS-13) 21 km away receiving, and on one day they received with both DSS-14 and DSS-13 to acquire interferometric data. On yet another day, they col-lected data with the Goldstone-VLA radar.

Preliminary analysis of the data showed Toutatis to have an unusually slow rotation rate and a maximum dimension of no less than 3.5 km. Interestingly, too, Toutatis appeared to consist of two irregularly-shaped components in close contact. The images provided a first glimpse of craters on an Earth-crossing asteroid, as well. The asteroid's roughness, as measured by the circular polarization ratio, indicated a considerable degree of general roughness at centimeter-to-decimeter scales, supporting the belief that Toutatis had undergone a complex collisional history.

---

101. Hudson, telephone conversation, 21 November 1994; Ostro 25 May 1994; Hudson and Ostro, "Shape of Asteroid 4769 Castalia (1989 PB) from Inversion of Radar Images," *Science* 263 (1994): 940–943; Hudson and Ostro, "Doppler-Radar Imaging of Spherical Planetary Surfaces," *Journal of Geophysical Research* 95 (1990): 10,947–10,963.

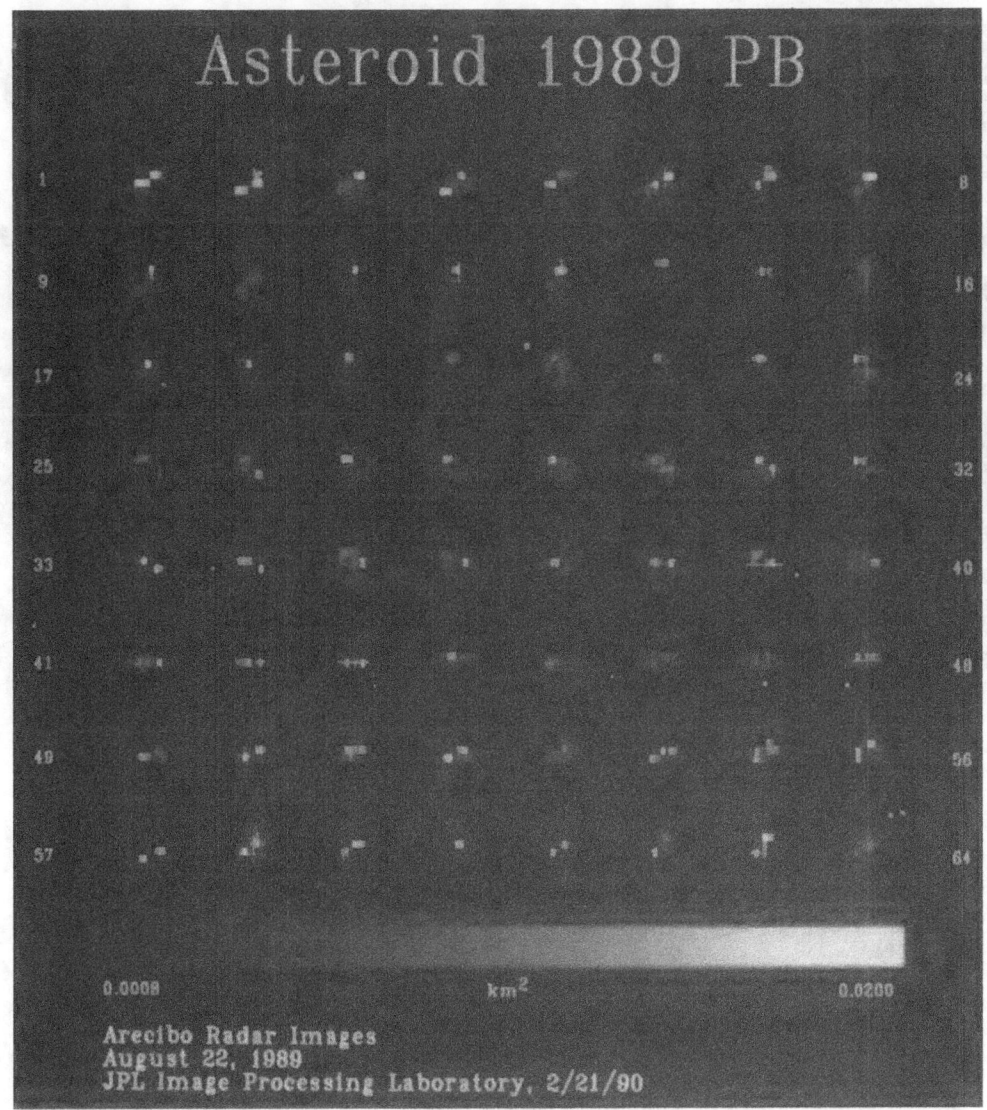

Figure 39

Radar images of Asteroid 1989 PB (later known as 4769 Castalia). The asteroid's rotation is noticeable in the 64 images. (Courtesy of Jet Propulsion Laboratory.)

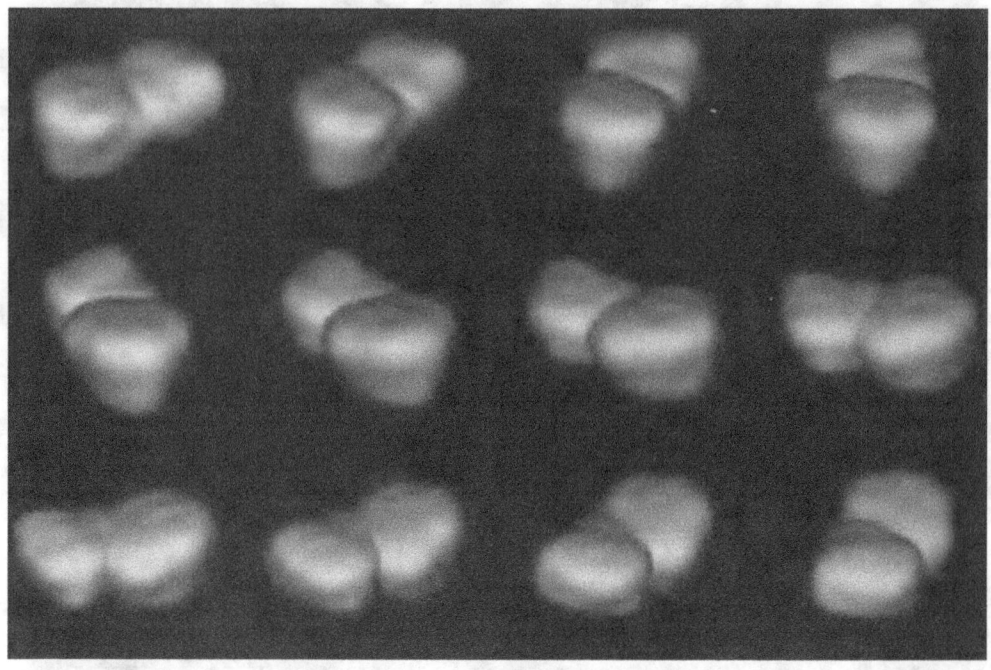

Figure 40

Model of Asteroid 4769 Castalia. It was the first three-dimensional model of an asteroid ever produced. The picture shows 16 different views of a three-dimensional model of Castalia, which is 1.8 km across at its widest. The model was created by Scott Hudson (Washington State University) and Steve Ostro (JPL) from data taken at Arecibo Observatory in 1989. (Courtesy of Jet Propulsion Laboratory, photo no. P43041A.)

Since then, Scott Hudson has elaborated his model to recreate three-dimensional asteroid shapes. With Toutatis, he was dealing with over 1,000 parameters. The application of Hudson's reconstruction technique to the Toutatis images was complicated by the asteroid's rotation. Unlike all other targets detected by radar, Toutatis was in a tumbling rotational state.

At the 1994 AAS Division for Planetary Science meeting in November 1994, Ostro and Hudson presented several movies of Castalia, including one in which the asteroid was portrayed as it might be viewed in space, complete with fictional optical illumination. The use of the older Castalia data was on purpose; it suggested the potential rewards of using higher resolution data. In addition to Castalia and Toutatis, Ostro and Hudson began working in 1994 on three-dimensional modeling of 1620 Geographos, an asteroid which the ill-fated Department of Defense's Clementine spacecraft was scheduled to observe during a flyby mission. Although a computer malfunction prevented the Clementine encounter, Ostro captured a detailed sequence of Geographos images at Goldstone only days before the scheduled flyby. Subsequent modeling of the data has yielded an impressive simulation of an asteroid flyby.[102]

---

102. Ostro 25 May 1994; Hudson, telephone conversation, 21 November 1994; Ostro, Jurgens, Rosema, R. Winkler, D. Howard, R. Rose, Slade, Yeomans, Campbell, Perillat, Chandler, Shapiro, Hudson, P. Palmer, and I. de Pater, "Radar Imaging of Asteroid 4179 Toutatis," *Bulletin of the American Astronomical Society* 25 (1993): 1126.

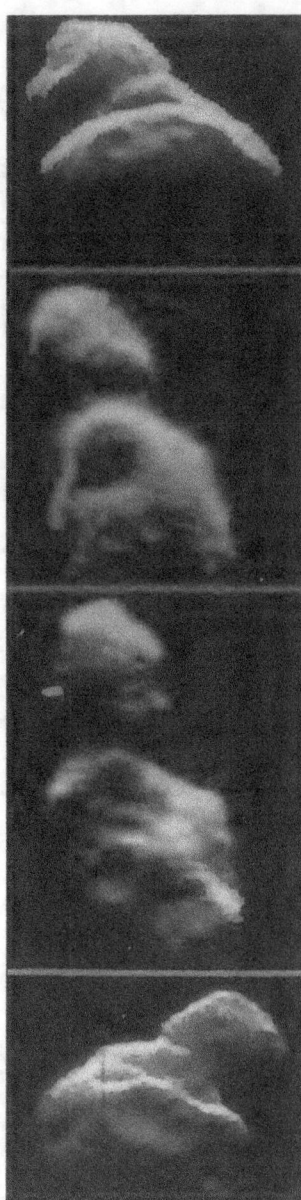

The state-of-the-art imaging and modeling of Castalia, Toutatis, and Geographos are feats that only a spacecraft flying by an asteroid could match. Although no probe deliberately set out to photograph an asteroid, Galileo, on its voyage to Jupiter, sent back the first spacecraft images of an asteroid, mainbelt object 243 Gaspra, on 29 October 1991 from a distance of about 16,200 km. Interestingly, Galileo also discovered that mainbelt asteroid 951 Ida had an orbiting satellite, recently named Dactyl.[103]

While the Toutatis images in themselves are spectacular witnesses to the ability of radar astronomy to solve problems left unsolved by other techniques, radar astronomy has achieved an equally great degree of success in another problem-solving area, asteroid orbits. Determining asteroid orbits with better degrees of accuracy and predictive reliability gained higher attention as scientists and the general public came to perceive asteroids as an ultimate threat to human civilization and to life itself on Earth. The perception grew out of the work of nuclear physicist and Nobel laureate Luis Alvarez, who first proposed that an asteroid was responsible for the extinction of the dinosaurs some 65 million years ago. Since then, evidence supporting the theory has accumulated, though not without arguments and evidence questioning the theory.

In a seminal paper published in *Nature*, Clark Chapman and David Morrison argued that the probability of a kilometer-size asteroid hitting Earth in the next century was 1 in 5,000. The collision would have a global effect, regardless of the impact site, because the dust blasted into the stratosphere would end agriculture for several years. Billions of people would starve to death.[104]

In order to detect a potentially civilization- and life-threatening asteroid, the scientific community proposed Spaceguard, a network of six optical telescopes dedicated to detecting asteroids. The name Spaceguard came from the book *Rendezvous with Rama*, in which its author Arthur C. Clarke envisioned an asteroid striking Earth in northern Italy in the year 2077. In response to the impact's devastation, the nations of Earth formed Project Spaceguard.

Creating a real Spaceguard has not been so straightforward. After asteroid 1989 FC came very close to the Earth in 1989, the American Institute of Aeronautics and Astronautics recommended to the House Committee on Science, Space, and Technology that it sponsor studies of asteroid detection and defense. Congress then commissioned NASA in 1990 to write reports on those subjects.

Figure 41
These radar images of Toutatis represent the highest resolution then achieved on an asteroid. A few impact craters, the first ever documented, are visible in the images. (Courtesy of Jet Propulsion Laboratory, photo no. P41525)

103. *The Spaceguard Survey: Report of the NASA International Near-Earth-Object Detection Workshop* (Pasadena: JPL, 25 January 1992), p. 19; NASA Press Release 94–158, 20 September 1994, Renzetti materials.
104. Chapman and Morrison, "Impacts on the Earth by Asteroids and Comets: Assessing the Hazard," *Nature* 367 (1994): 33–40.

NASA already had considered asteroid detection in a 1981 workshop held in Colorado, but 10 years later it acted in response to a Congressional mandate. The NASA 1991 workshop brought together 24 asteroid scientists from around the world, including radar astronomer Steve Ostro.[105]

Although Congress has not yet funded Spaceguard, a battle over how to defend the planet against a "killer asteroid" rages. The recent collision of Comet Shoemaker-Levy with Jupiter has driven home the point that the planets, Earth included, are susceptible to potentially life threatening impacts from comets and asteroids. The Spaceguard proposal came along just as the Department of Defense was seeking post-Cold War applications of its nuclear arsenal. The deflection of a menacing asteroid or comet with a series of nuclear explosions is, in the words of Carl Sagan and Steve Ostro, "a double-edged sword," which if wielded by the wrong hands could "introduce a new category of danger that dwarfs that posed by the objects themselves." They pointed out that a series of nuclear explosions capable of thwarting a dangerous asteroid is also capable of diverting a benign asteroid toward Earth.[106]

Regardless of the means used to defend Earth against asteroid hazards, radar is suited to play a vital role in identifying potentially hazardous objects. Radar is *the* essential tool for astrometry (position and movement); it can determine asteroid orbits with greater accuracy and reliability than any other method. After the detection of an asteroid and the determination of its orbit, astronomers extrapolate the orbit into the future. Without radar precision measurement, the uncertainty of that extrapolation increases strikingly. The role of radar in Spaceguard, consequently, is as the primary, post-discovery ground-based technique for refining asteroid orbits.

After Steve Ostro's experiences with the errors in the ephemerides provided for 1986 DA and 1986 JK, he and fellow JPL employees Don Yeomans and Paul Chodas, who were in charge of calculating ephemerides for space missions, including those for a potential future asteroid flyby mission, assessed the extent to which radar observations could improve the accuracy of near-Earth asteroid ephemerides. They wanted to know how useful radar ranging was for refining the orbits of Earth-crossing asteroids. Could radar improve the extrapolation of asteroid orbits into the future?

They studied four asteroids with different histories of optical and radar observations, 1627 Ivar, 1986 DA, 1986 JK, and 1982 DB. The radar data provided only a modest absolute improvement for Ivar, which had a long history of optical astrometric data, but rather dramatic reductions in the future ephemeris uncertainties of asteroids having only short optical-data histories. Those improvements were impressive ones, to three orders of magnitude.

Ray Jurgens, who had been observing asteroids at Goldstone since the 1970s, wrote a proposal to fund asteroid emphemeris work at JPL and persuaded Don Yeomans and Paul Chodas to help in the analysis of asteroid ephemerides. As Jurgens became overwhelmed by research and the rebuilding of the Goldstone radar, Steve Ostro took up the tasks of strengthening JPL's asteroid ephemeris program and advocating software tools and other measures for improving Goldstone's capability of detecting asteroids and improving the accuracy of asteroid orbit predictions.[107]

---

105. *The Spaceguard Survey*, pp. 1–3 and 49–52; Cunningham, pp. 113–116 and 141.

106. Sagan and Ostro, "Dangers of Asteroid Deflection," *Nature* 368 (1994): 501.

107. Ostro 25 May 1994; Yeomans, Ostro, and Paul W. Chodas, "Radar Astrometry of Near-Earth Asteroids," *The Astronomical Journal* 94 (1987): 189– 200; Ostro, "The Role of Ground-Based Radar in Near-Earth Object Hazard Identification and Mitigation," in *Hazards Due to Comets and Asteroids*, in press, p. 9. For a summary of asteroid radar astrometry, see Ostro, Campbell, Chandler, Shapiro, Hine, Vélez, Jurgens, Rosema, Winkler, and Yeomans, "Asteroid Radar Astrometry," *The Astronomical Journal* 102 (1991): 1490–1502; and Yeomans, Chodas, M. S. Keesey, Ostro, Chandler, and Shapiro, "Asteroid and Comet Orbits using Radar Data," *The Astronomical Journal* 103 (1992): 303–317.

The astrometric and imaging capabilities of radar soon will combine to reformulate the IAU circular that announces the discovery and orbit of a new asteroid. For newly spotted asteroids, Ostro has a vision of the kind of IAU circular that might be available before the end of the century. After astronomers discover and track an asteroid optically for a few nights and the orbit is at least crudely known, an IAU circular announces the object's existence. A few days later, the Arecibo or Goldstone radar observes the asteroid and takes range-Doppler data, refines the orbit, and images the object. The ephemeris is updated immediately. Streamlined software transforms the image data into a three-dimensional model of the asteroid, then produces a video simulation of the Sun-illuminated asteroid. This process yields a computer file that becomes the first post-discovery IAU circular: a finely-resolved video image of the object, as if made by a flyby spacecraft within a few days of discovery.[108] Here was the future of asteroid radar research and, to a dramatic degree, the future of planetary radar astronomy, as well.

---

108. Ostro 25 May 1994.

Conclusion

W(h)ither Planetary Radar Astronomy?

The dynamic interaction between epistemological (instruments and techniques) concerns and the kinds of problems radar astronomers seek to solve, which we have seen driving planetary radar astronomy to the present, also in all likelihood will continue to determine its future. Both new instruments and techniques will furnish the means for exploring new targets and for resolving problems, especially those left unresolved or unsatisfactorily resolved by optical means.

Planetary radar techniques developed recently perhaps hint at the sources of future techniques. Three examples are John Harmon's non-repeating code, which he adapted from Arecibo ionospheric research; Dewey Muhleman's use of radio astronomy imaging and arraying techniques at the VLA, as part of the bistatic Goldstone-VLA radar; and the planetary imaging technique developed by Scott Hudson and Steve Ostro. The Harmon and Muhleman techniques reflect the continuing, though diminished, influence of ionospheric research and radio astronomy on planetary radar astronomy.

A surprising number of new instruments may be available, too, many through the grouping of either the Goldstone or Arecibo antenna in tandem with a radio telescope to form a bistatic radar. The Goldstone-VLA radar appears to point the way to additional combinations with the soon-to-be-completed Green Bank radio telescope, or perhaps to the Goldstone X-band radar in tandem with a tracking station in the Soviet Union. Already, the Russian Yevpatoriya tracking station has made bistatic observations of the asteroid Toutatis in conjunction with the Effelsberg radio telescope, though without achieving the impressive results of the Arecibo and Goldstone antennas. Politics and funding will limit what, if any, future bistatic experiments take place outside the United States. Additional bistatic possibilities in the United States include the JPL Mars Station in combination with other Goldstone antennas, as well as an Arecibo-Goldstone link.

The bistatic possibilities are not limitless, however; not inconsequential institutional, political, and budgetary obstacles aside, the elementary technological need for compatible transmitting and receiving frequencies limits many bistatic options. Even more limited is the creation of new radars. Other countries continue to build antennas, such as the Arecibo-size dish planned in Brazil, but none anticipate a radar capability. No facility dedicated entirely to planetary radar astronomy ever has been built; nonetheless, Steve Ostro believes that it is time to build one in order to study asteroids. The cost of designing and building such a radar observatory would approach the modest level budgeted for NASA's Discovery space missions.[1] The role of this facility in the Spaceguard project aside, its potential scientific value in a short period of time would exceed that of any one Discovery flyby of an asteroid. Time will tell whether this worthwhile and economical project is realized.

---

1.   Ostro 25 May 1994.

These are all possible future planetary radar instruments. Nothing, especially not their scientific merit, either guarantees or favors their realization; budgets, not science, will determine their viability. With one exception, planetary radar astronomy always has subsisted on the budgetary margins, either by design (as at JPL) or by fate (as at Lincoln Laboratory). As budgets are trimmed, the freedom to fund bistatic experiments from discretionary funds diminishes, too. The only exception is Arecibo, where a five-year contract stabilizes the research budget, although within the shrinking NASA and NSF budgets. If one can say anything about the future of planetary radar astronomy with certainty, it is that the future is at the upgraded Arecibo telescope.

The Arecibo upgrade, as well as the potentially available novel instruments and techniques mentioned above, will bring new research targets within the reach of radar astronomers. Among the most striking new targets visible to the upgraded Arecibo observatory will be the satellites of Jupiter, Iapetus, Rhea, Amalthea, Dione, and Hyperion. The detections of those bodies very well may lead to radar solving new scientific problems. In addition, the upgraded Arecibo telescope will be able to map Jupiter's Galilean moons at much higher resolutions, perhaps down to 100 kilometers, and uncover fresh facts regarding Io and Saturn's moon Titan.[2]

Planetary radar may contribute as well to our understanding of the terrestrial planets through analysis of their polarization ratios; however, the greatest amount of research activity will be directed toward neither the terrestrial planets nor the systems of Jupiter and Saturn, but the asteroids. The space in which Earth turns abounds with thousands of asteroids. The largest, a kilometer or larger in diameter, number about 2,000, while those 100 meters or more in diameter number 150,000 or more, and those 10 meters and larger amount to some 300,000,000. These are estimates of the asteroid population; the number of asteroids actually observed increases continually. The likelihood that one of those asteroids might approach Earth perilously close has heightened interest in them.

In many ways, then, planetary radar astronomy has come full circle. It began with the study of large populations of meteors, and the observation and analysis of the large and varied asteroid population is carrying it into the future. Forty years ago, the forte of radar lay in its ability to determine accurately the radiants and speed of meteors and to ascertain unambiguously that they orbited around the Sun. Today, the value of radar is its ability to fix asteroid orbits with an accuracy and certainty that no other method can match.

Current planetary radar techniques, however, can do much more with asteroids than the earliest radar investigators at Jodrell Bank and the Canadian National Research Council were able to do with meteors using their pioneering techniques and far less sensitive radar equipment. Today's planetary radar astronomers can characterize asteroid composition, size, and shape and can provide a unique imaging ability. The ever-growing number of asteroid targets, combined with this wide range of epistemological tools and the present societal interest in a potential "killer asteroid," guarantees that the future of planetary radar astronomy will be asteroid research.

W(h)ither Planetary Radar Astronomy?

Asteroid literature and funding have grown markedly over the last fifteen years. In the past, similar rapid growth in ionospheric and radio astronomy research carried forward planetary radar astronomy. This growth has not yet reached radar studies of asteroids, however, and despite the expanding observational opportunities created by asteroid studies, the number of radar astronomers probably will not increase significantly. Steve

2. Ostro, "Benefits of an Upgraded Arecibo Observatory," pp. 238–239; Ostro 25 May 1994; Campbell 8 December 1993.

Ostro remains the sole full-time asteroid radar astronomer. Once the upgraded Arecibo radar becomes available, the number of radar investigators studying asteroids probably will increase, or rather, *must* increase, if an adequate number of observational opportunities are to be seized. Already, the Arecibo Observatory has hired a planetary astronomer with an interest in asteroidsy[3] who will also take part in radar observations of asteroids. Don Campbell will participate in those observations, as too may Dick Simpson of Stanford. The growth in asteroid science, then, may shift current radar researchers into the field, rather than provide a basis for expanding planetary radar astronomy.

As a scientific species, planetary radar astronomers have tended not to reproduce themselves. Hiring individuals from other fields yielded planetary radar astronomers in the 1960s and 1970s, but none in the last 15 years with the exceptions of Marty Slade at JPL and the recent hire at Arecibo. The number of planetary radar astronomers created through paid employment, therefore, may remain small and relatively stable. Being small yet may have its advantages in a future certain to be shaped by budget cuts in NASA and U.S. scientific research in general.

The other traditional career path into radar astronomy, university training, may furnish fresh practitioners, though. Gordon Pettengill at MIT and Don Campbell at Cornell directed many radar astronomy dissertations, although certainly not all of those students entered the field. The MIT-Cornell axis has supplied planetary radar astronomers since the 1960s, but the last Ph.D. to enter the field through that route (Steve Ostro) graduated in 1978. Moreover, with the retirement of Pettengill at MIT and the approaching retirement of Campbell at Cornell, who will train future planetary radar astronomers at Arecibo?

Outside of MIT and Cornell, only Caltech appears equipped or willing to train them. There, Dewey Muhleman has graduated one student, Bryan Butler, in 1994, who did doctoral research in radar astronomy. Although interested in pursuing radar research, Butler is at least equally excited by the prospect of planetary radio studies at the VLA, where he has taken a position. Muhleman is not interested in training additional radar astronomers.

In contrast, Steve Ostro at JPL teaches a course on radar astronomy at Caltech. That position gives him the ability to both recruit and train future radar astronomers. The key to training future radar astronomers in an academic setting like MIT or Cornell remains the master-disciple relationship. Replacing Pettengill and Campbell, then, is Ostro, who is in a unique position to carry the MIT-Cornell alliance one step further by linking Caltech, JPL, and Goldstone to it.

Ostro, a graduate of MIT who conducted his doctoral research on Cornell's Arecibo instrument, and a former member of the Cornell faculty, found Ray Jurgens and Marty Slade, graduates of the Cornell and MIT programs, respectively, when he began work at JPL. Ostro's arrival at JPL signalled a joining of the JPL and MIT-Cornell research groups. His opportunity to teach at Caltech and recruit radar astronomers, coming near the retirements of Pettengill and Campbell, assures the continuation of the master-disciple relationship as the source of future radar astronomers, but within a larger institutional (MIT, Cornell, Caltech-JPL) and instrumental complex that joins the Arecibo and Goldstone radars. The centering of Ostro within that complex also positions him to direct the future of radar astronomy.

Like the number of practitioners, the radar astronomy literature will remain at a low level as a result of both the small number of researchers and the nature of the science reported in those publications. The discoveries to be made on the terrestrial planets and the moons of Jupiter and Saturn will not generate a substantial number of articles, because those discoveries likely will not merit that level of scientific attention. The results

---

3.   Harmon 15 March 1994.

of asteroid research, moreover, will be described in articles that discuss the characteristics of a substantial population of asteroids and not the properties of just one or two asteroids. Consequently, the number of asteroid-related publications will remain limited.

Indeed, virtually the entire history of planetary radar astronomy has been one of limits. The number of practitioners has been limited, if not declining. As we saw, after the initial "explosion" of planetary radar activity in the early 1960s, as measured by the number of experimenters, publications, and instruments, the field of planetary radar astronomy assumed the manpower and publication dimensions of Little Science. After further shrinking during the 1970s, leaving a handful of researchers utilizing a single radar instrument in 1980, planetary radar astronomy stabilized at this lower level (the Arecibo radar being down for the duration of the upgrade). The circumscribed number of opportunities to train future radar practitioners in academia, as well as retirements (most current practitioners are at or near retirement age), will keep manpower levels low.

Little Science, Big Science

The practice of planetary radar astronomy as Little Science in the instrumental and institutional context of Big Science will likely continue into the future. A number of factors integral to the field have confined planetary radar astronomy to its existence as Little Science. To begin with, the field generally has operated at the limits of the technology (the instrument hardware). As soon as an instrument became available, radar astronomers sought to discover what new targets it could detect. Once the farthest target was reached, and the spatial limits of research defined, radar astronomers had insufficient sensitivity to achieve more than a detection. Imaging planetary surfaces always involved pushing the instrument's signal-to-noise ratio and resolution capability to the limit. These restrictions in turn prompted radar astronomers to continually press for hardware modifications that provided incremental increases in sensitivity. In the end, though, what could be done was limited by the capability of the instrument.

Another growth-limiting factor inherent in planetary radar astronomy is the availability of targets, a factor intimately linked to instrument capability. The planets and their moons cannot be detected unless they are within radar range. The sensitivity limits of planetary radars, such as the Arecibo telescope, prevent investigators from observing targets except when they approach Earth. At other points in their orbits around the Sun, they are too far away for radars to detect them. Thus, Venus is observed at inferior conjunction and Mars at opposition.

A related problem is that of declination. Although most planets rotate around the Sun more or less in the same plane, called the ecliptic, they are not visible in the sky at all times because of the Earth's motion about its own axis. A further complicating factor is the ability of the radar antenna to "see" a portion or all of the visible sky, that is, the so-called declination window of the antenna. The declination window of the Goldstone DSS-14 dish runs from 40° South to 80° North, while the Arecibo telescope is limited to solar system objects that pass within the far narrower band from 40° North to just below the equator.[4]

The combination of declination window and radar sensitivity restricts observational opportunities, so that planetary research demands only about five percent of total antenna time. The finite number of planetary targets and the narrow observational windows also tend to limit the number of radar researchers. Thus, the tendency at Arecibo was to establish a given target as the terrain or turf of a particular researcher. The number of radar investigators at JPL was always too small for such a division of targets, although

---

4.   Renzetti, Thompson, and Slade, "Relative Planetary Radar Sensitivities: Arecibo and Goldstone," *TDA Progress Report* no. 42–94 (Pasadena: JPL, April–June 1988): 292.

during the 1970s Ray Jurgens "specialized" in Venus and asteroids and George Downs in Mars, with Dick Goldstein continuing to do a little of everything.

The considerable and expanding number of known asteroids is too large for a single investigator. This will be especially true in the near future, once asteroid detection relies on CCD imaging and the Arecibo upgrade reaches completion. Then joining Ostro in radar observations of asteroids (many asteroid scientists hope) will be virtually all members of the small club of radar astronomy practitioners. Again, the expanded program of asteroid radar research that will take place throughout the remainder of this decade will not lead to a transformation of planetary radar astronomy into Big Science.

Planetary radar astronomy has been and likely will remain Little Science embedded in the matrix of Big Science. John Krige's study of British nuclear physics research in the period right after World War II provides a different case of Little Science being conducted with Big Science instruments. One can find another parallel example in the telegraph networks of the nineteenth-century United States.

The Western Union telegraph company, formed by the merger of several separate companies, absorbed both of its principal rivals in 1866 to become one of the nation's largest companies and thereby created the largest electrical communication network in the world. While not Big Science, this was Big Business and Big Technology. By the very nature of their position in the company, telegraph operators had access to the large-scale technological laboratory formed by the telegraph lines. Just as access to technology led to the emergence of planetary radar astronomy, so access to the telegraph network led these operators to perform electrical experiments on the lines. Out of those experiments came numerous inventions, many of which were patented.[5]

It is not going too far to draw this parallel between Little Science (planetary radar astronomy) and Little Technology (telegraph inventors), for several reasons. For one, most radar astronomers were trained as electrical engineers, not scientists. Also, radar astronomy was a science driven by technology, namely, the availability of radars capable of planetary exploration. It was through these instruments and their associated techniques of analysis, not through direct sensory observation, that radar astronomers conducted their experiments. They analyzed not sensory experience, but wave patterns of electromagnetic signals which analysis by computer software made "visible." Thus, not only were the instrumentation and techniques of radar astronomy dependent on technology, but so was the very content of the science.

Planetary radar astronomy historically has remained at the intersection of science and engineering. Attendance of radar astronomers at both IAU and URSI meetings during the 1960s reflected the dichotomous nature of radar astronomy, perched between radio engineering (URSI) and astronomical science (IAU). The dichotomy arose from the fact that radar astronomy is a set of techniques (engineering) used to generate data whose interpretation yields answers to scientific (planetary astronomy and geology) questions. Also as a result of this dichotomy, planetary radar astronomy concerns itself with two different but related sets of problems (in the Kuhnian sense discussed in Chapter Five). One set of problems is epistemological, that is, it deals with how radar astronomers know what they know and relates to the radar characteristics of the planets, such as surface scattering mechanisms, dielectric constants, and radar albedos; these problems arise out of

---

5. For a careful scholarly study of telegraph operators as inventors, see Paul Israel, *From Machine Shop to Industrial Laboratory: Telegraphy and the Changing Context of American Invention, 1830–1920* (Baltimore: Johns Hopkins University Press, 1992), which is based on the dissertation of the same title, Ph.D. diss., Rutgers University, 1989. For a discussion of the role of the entrepreneur in channeling the resources of large-scale organizations, specifically, the introduction of radio and radio research within the French military by Gustave Ferrié, see A. Butrica, "The Militarization of Technology in France: The Case of Electrotechnics, 1845–1914," paper read at the joint meeting of the American Historical Association and the History of Science Society, Cincinnati, December 1988.

the engineering side of radar astronomy. A second set of problems, such as planetary orbits and spin rates, arises out of the science side of the field.

The rooting of Little Science (of Little Technology) within large technological systems, such as the Western Union telegraph network or the Deep Space Network, suggests that it may be in the nature of large-scale "technosocial networks" or "systems" (to borrow the terminology of the social construction of technology mentioned in the Introduction) to sustain Little Science (or Little Technology). Large technological systems form a unified set of relations among individuals, objects, and ideas. As tightly "constructed" as these technosocial networks may be, the magnitude of the resources they encompass is of a sufficient extent to allow small-scale entrepreneurs (be they scientists, engineers, inventors) within the system to "socially construct" smaller technosocial networks within the larger.

Without the larger technosocial network, then, the smaller network is unthinkable. Planetary radar astronomy simply would not have existed without the enormous, powerful, highly sensitive radars on which the experiments were conducted and which were called into existence by the demands of the Cold War and Big Science. Another requisite, of course, was the radar experimenters themselves. The linking of research groups at MIT (Lincoln Laboratory), Cornell University, and (most recently) JPL (Caltech) has provided a means by which the Little Science planted in the interstices of large technological systems can perpetuate itself despite declining resources and limits to growth. For example, as planetary radar activity ceased at Haystack, it continued at the Arecibo Observatory. Given the symbiotic relationship between Big Science and the Little Science which depends on it, as well as the nature of that dependency, funding cutbacks intended to reduce Big Science also will diminish, or perhaps even eliminate, Little Science. Future research will have to determine how vast (and by what standard(s) that vastness is measured) a technosocial network must be in order to sustain Little Science.

The technological dependence of radar astronomy, and the availability of that technology within large technological systems, thus accounts for the emergence of radar astronomy within Big Science settings. The technological dependence of planetary radar astronomy, however, does not explain its utilitarian proclivity, namely, the tendency of radar astronomers to justify their research by its usefulness to space exploration. Nor does the training of most radar astronomers as electrical engineers, who must think in both theoretical and practical terms at the same time, illuminate that tendency. The rise of radar astronomy concurrently with the creation and rapid growth of NASA was perhaps not coincidental.

Although the space agency did not build research instruments outside NASA laboratories during the 1960s, its very existence from 1958 suggested the future availability of funds for instruments and research activity. The Endicott House Conference reflected those funding hopes. After 1970, when NASA funding became a reality, radar astronomy quickly began participating in NASA space missions, such as Viking, until radar astronomy *became* a space project, the Magellan radar mission to Venus. This close relationship to NASA space missions certainly amplified whatever utilitarian bent radar astronomy already had.

This utilitarian bent also arose from the very nature of conducting Little Science within the context of Big Science. Doing Little Science requires that scientists constantly defend the pragmatic value of their research. A good example is radar astronomy at JPL; it lived off the budgetary margins of NASA space missions until the 1980s. Because obtaining antenna time depended on securing the approval of a NASA mission, radar astronomers had to argue the value of their research on practical, mission-oriented terms.

In contrast, obtaining antenna time at the Arecibo Observatory depended on the scientific value of the radar experiment; its value to NASA was far less important, although research directly related to NASA space missions was carried out there. The primary difference between the JPL and Arecibo facilities was the official recognition granted radar astronomy at Arecibo from the start. The Arecibo telescope always had radar astronomy as one of its prime research objectives, while the JPL Goldstone antenna served mainly to track NASA launches, not conduct scientific experiments. NASA recognition for the scientific value of the Goldstone radar dish has yet to be realized fully or even established on a permanent foundation, though some preliminary steps have been taken.

We can conclude briefly the following about planetary radar astronomy. After a brief initial burst of activity, radar astronomy quickly developed the characteristics of Little Science in terms of manpower, instruments, and published literature. The field continued to shrink throughout the 1970s, reached a low plateau of activity around 1980, then rose slightly in the middle 1980s, as the Goldstone radar once again became available for research.

A number of factors kept planetary radar astronomy a Little Science. Radar sensitivity and target visibility within the declination window limited observational opportunities. The shortage of observational opportunities in turn restricted the number of investigators who could pursue radar astronomy on a full-time basis. Close ties to NASA space projects intensified radar astronomy's utilitarian tendency. The need to justify Little Science within a Big Science setting played at least an equal part in shaping that tendency. The case of the Arecibo Observatory, though, demonstrates the importance of securing institutional recognition for the conduct of Little Science from the outset. Finally, the subsistence of Little Science within Big Science niches and their symbiotic relationship may be a function of large-scale technological systems, whether they be the Western Union telegraph network of the nineteenth century or the big dishes of twentieth-century radio astronomy and space communications.

Planetary Radar Astronomy
Publications

At the beginning of this project, a bibliography of radar astronomy literature, consisting of 384 items arranged chronologically by year of publication and alphabetically by author within each year, was constructed from a search of the NASA STI Database (aeronautics and space) and a published bibliography, Jean E. Britton and Paul E. Green, Jr., *Radar Astronomy* (Cambridge: MIT Lincoln Laboratory Library, 1962), which Mr. Green generously made available. The NASA STI Database search alone resulted in a printout of 589 items published since 1963. To this initial bibliography were added additional publications uncovered in the researching and writing of this book.

The initial bibliography, with fewer than 400 entries, illustrated the diminutive character of planetary radar astronomy. In comparison, the radio astronomy literature of just the past two decades measures in the thousands. Because the extent, as well as the development, of the literature might help to characterize the progress of planetary radar astronomy over several decades, the bibliography was pruned and grafted in such a way as to reflect the published literature. Dissertations were missing from the bibliography, while publications by foreign researchers and abstracts abounded.

A number of rules were followed in including and excluding publications. Internal reports were omitted; these are not intended for consumption by the general public or the scientific community. Only works by American practitioners were included; British and Soviet titles were excluded. Planetary radar astronomy was defined more strictly than in the text; solar, lunar, meteor, auroral, and Earth radar studies; and synthetic radar aperture research were left out, because they are specializations unto themselves. Also excluded were items dealing with hardware, instruments, or techniques and those providing interpretations of radar results by individuals outside the field. For example, an article on the interpretation of radar topographic data, whose first author was a planetary geologist, was left out; however, if the first author was a radar astronomer, the article was added. Finally, abstracts were excluded, dissertations included.

The resulting planetary radar literature, spanning the period from 1958 to 1994 inclusively, amounted to 272 entries, or an annual average of about seven. Only twice did 15 or more items appear in a single year. A line chart (Table 9) showing the annual distribution of planetary radar publications indicates the explosion of radar astronomy activity during the 1960s. The remainder of the chart suggests the technological dependence of radar astronomy. A second spurt of growth appears following 1975, when the Arecibo Observatory S-band radar first became available, and a third spurt occurred around 1990, just after the Voyager upgrade of the Goldstone radar.

When the annual publication numbers are grouped by 5-year intervals, the sharp peaks and valleys of the annual chart are smoothed out and a new trend emerges (Table 10). The volatile growth of the 1960s remains, but what appeared to be seesaw-like growth around 1975 and 1990 disappears. Instead, a dip replaces the growth following 1975, and the literature reaches a plateau of activity. This plateau suggests that since 1980 the field has reached the limits to its growth.

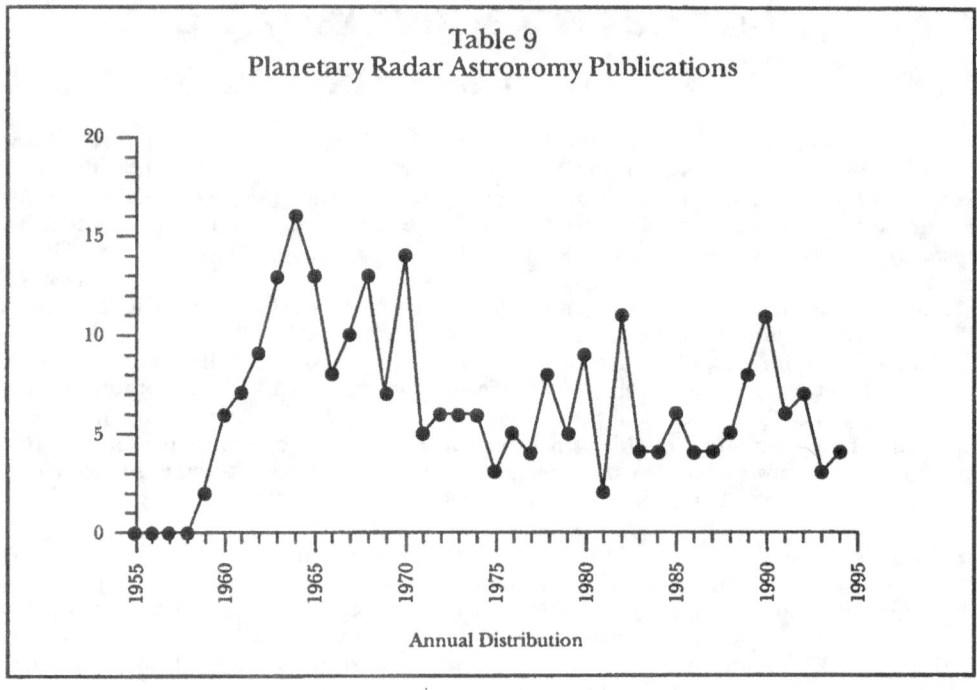

Table 9
Planetary Radar Astronomy Publications

Annual Distribution

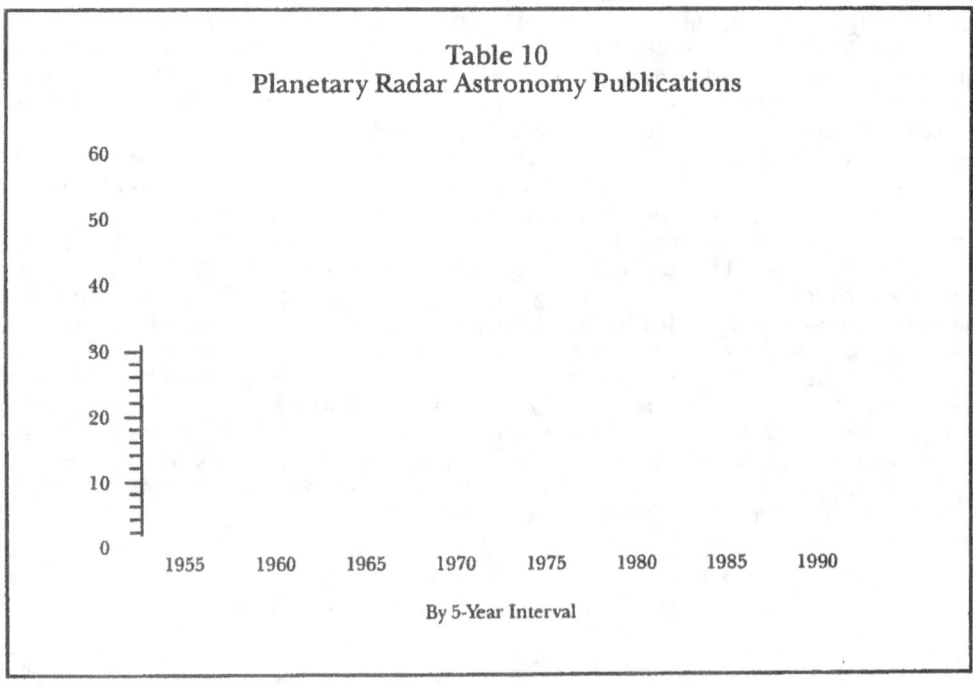

Table 10
Planetary Radar Astronomy Publications

By 5-Year Interval

A Note on Sources

For the early history of radar astronomy, a number of archival sources were consulted. The Historical Archives, U.S. Army Communications-Electronics Command, Ft. Monmouth, NJ, have several boxes of material on the pioneering lunar radar work of John DeWitt, but no such archival material was found on the radar work of Zoltán Bay, with the exception of the documents in the possession of his widow. The Naval Research Laboratory Historical Reference Collection, Office of the Historian, was not a ready source of information on the lunar radar work carried out there; much of the Laboratory's records remain classified. In contrast, the archives of Jodrell Bank, housed at the University of Manchester, contain a wealth of open information on radar astronomy, and a computerized index is available.

Radar research on meteors began at Stanford University as early as the 1950s. The university archives, however, hold no records relevant to either the early or later work done there. The only records available are those of the Stanford Center for Radar Astronomy, which for the most part consist of a large collection of offprints that document the Center's research results. Von Eshleman, the Center's director, was a far more important source of documentation.

Records relating to radar astronomy at the Arecibo Observatory are located for the most part in filing cabinets at the National Astronomy and Ionosphere Center (NAIC) offices on the Cornell University campus and are not normally open to researchers. Among the most useful of those records are the quarterly reports to the NSF and copies of Center for Radiophysics and Space Research (CRSR) research publications. The NAIC library retains copies of dissertations completed at the Arecibo Observatory. The CRSR, located in the same building, has the earlier ARPA reports. The library of the Arecibo Observatory contains additional reports, program plans, dissertations, and other materials. The minutes of the open sessions of the National Science Board were helpful, as were the archives of the AFCRL at Phillips Laboratory, Hanscom AFB, although the amount of documentation at each place was lean.

In contrast, an overwhelming abundance of documents relating to the history of radar astronomy were found at MIT and Lincoln Laboratory. The Lincoln Laboratory Library Archives contain both documents and photographs, while the MIT Institute Archives and Special Collections is a treasure trove of documentation, including NEROC materials. The Pusey Archives, Harvard University, hold additional NEROC documents. In general, the MIT, Lincoln Laboratory, and Harvard materials are available to researchers; examination of the Pusey papers requires written permission from the director of the Harvard College Observatory, though. A small building near the Haystack Observatory named for its first director, Paul Sebring, holds logbooks and other records relating to the Millstone and Haystack facilities, but those records normally are closed to researchers.

Documents relating to radar astronomy at Goldstone can be found in the Jet Propulsion Laboratory archives. Magellan materials, although somewhat organized, have not been fully integrated into that portion of the archives open to researchers. Also, a smaller batch of materials, initially removed from document storage for a history of the Deep Space Network and slated for integration into the JPL archives as the Peter Lyman Collection, was especially useful.

For further documentation of the NEROC saga, see the Archives of the Smithsonian Institution, in particular, the Office of the Secretary and the Under Secretary collections. The papers of William Brunk, at the NASA History Office, and the Historian's File, at the National Science Foundation, held valuable materials on the first upgrading of the Arecibo facility. The library of the National Science Foundation and the archives of the National Academy of Sciences also held useful secondary sources.

A significant number of documents relating to radar astronomy are in the possession of individuals who made the materials available exclusively for the writing of this history. Until the day arrives when (and if) those documents are entrusted to an archive, the above noted depositories will be the chief source of documentation for the history of radar astronomy. In addition, materials gathered or created in the process of writing this history, including photocopied documents, notes, and oral history transcripts, have been deposited with the NASA History Office for consultation by researchers.

Oral History Interviews

Because planetary radar astronomy is a relatively new field, virtually all of the founders, even those active during the 1940s, and practitioners are still with us. This project has been fortunate, too, in that with only one exception everyone approached agreed to be interviewed. Two-thirds of the interviews were taped and transcribed. The author alone conducted all interviews with the exceptions of Schaber, Soderblum, and Shoemaker, which were carried out jointly with Joseph Tatarewicz, as noted below. Copies of all transcripts are held by the NASA History Office and the JPL Archives; interviews of individuals formerly with MIT Lincoln Laboratory are also maintained at their archives.

Those interviews not transcribed, as well as the telephone interviews, consist of either notes or tapes on file at the NASA History Office and the JPL Archives. Additional interviews, carried out by José Alonzo and housed at the JPL Archives, were consulted; they are listed below, too.

Interviews

Transcribed Interviews

Interviewee	*Date of Interview*	*Place of Interview*
Donald B. Campbell	7 December 1993 8 December 1993 9 December 1993	Cornell University
Clark R. Chapman	28 June 1994	Flagstaff, Arizona
George Downs	4 October 1994	Lincoln Laboratory
Rolf B. Dyce	22 November 1994	Aguadilla, Puerto Rico
Von R. Eshleman	9 May 1994	Stanford University
John V. Evans	9 September 1993	NASA Headquarters
Thomas Gold	14 December 1993	Ithaca, New York
Richard M. Goldstein	14 September 1993	JPL
William E. Gordon	28 November 1994	Rice University
Paul E. Green, Jr.	20 September 1993	Hawthorne, New York
John K. Harmon	15 March 1994	Arecibo Observatory
Raymond F. Jurgens	23 May 1994	JPL
Sir Bernard Lovell	11 January 1994	Jodrell Bank
Duane O. Muhleman	19 May 1994 27 May 1994	California Institute of Technology
Steven J. Ostro	18 May 1994 25 May 1994	JPL
Gordon H. Pettengill	28 September 1993 29 September 1993 4 May 1994	MIT
Robert Price	27 September 1993	Lexington, Mass.
Alan E. E. Rogers	5 May 1994	Haystack Observatory
Irwin I. Shapiro	30 September 1993 1 October 1993 4 May 1994	Harvard-Smithsonian Center for Astrophysics
Richard A. Simpson	10 May 1994	Stanford University
Martin A. Slade	24 May 1994	JPL
William Boyd Smith	29 September 1993	Cambridge, Mass.

Thomas W. Thompson	29 November 1994	JPL
G. Leonard Tyler	10 May 1994	Stanford University
Herbert G. Weiss	29 September 1993	Cambridge, Mass.

Untranscribed Interviews
(Excluding Telephone Interviews)

Interviewee	Date of Interview	Place of Interview
Donald B. Campbell	10 March 1993	Arecibo Observatory
Robert Dickman	2 December 1992	NSF
Peter G. Ford	3 October 1994	MIT
Richard M. Goldstein	7 April 1993	JPL
Alice Hine	12 March 1993	Arecibo Observatory
Richard P. Ingalls	5 May 1994	Haystack Observatory
Raymond F. Jurgens	26 January 1993 28 April 1993	JPL
Duane O. Muhleman	8 April 1993	Caltech
Steven J. Ostro	1 April 1993	JPL
John E. B. Ponsonby	11 January 1994	Jodrell Bank
Martin A. Slade	26 January 1993	JPL
Robertson Stevens	14 September 1993	JPL

Telephone Interviews

Interviewee	Date of Interview
Roland L. Carpenter	14 September 1993
Robert Desourdis	22 September 1994
John H. DeWitt, Jr.	14 June 1993
Von R. Eshleman	26 January 1993
Daniel H. Herman	20 May 1994
R. Scott Hudson	21 November 1994
Benjamin Nichols	14 December 1993
Eberhardt Rechtin	13 September 1993

Paul Reichley	19 May 1994
Donald Spector	22 September 1994
E. Myles Standish	20 May 1994

Interviews with
Joseph N. Tatarewicz

Interviewee	*Date of Interview*	*Place of Interview*
Gerald Schaber	27 June 1994	Flagstaff, Arizona
Eugene M. Shoemaker	30 June 1994	Flagstaff, Arizona
Laurence A. Soderblom	26 June 1994	Flagstaff, Arizona

Interviews Conducted by José Alonzo

Interviewee	*Date of Interview*	*Place of Interview*
Richard M. Goldstein	19 September 1991 22 July 1992	JPL
Nicholas A. Renzetti	16 April 1992 17 April 1992 20 February 1992	JPL

Technical Essay
Planetary Radar Astronomy

The basic technology of planetary radar astronomy is, as the name implies, radar. Radar is an acronym for RAdio Detection And Ranging. U.S. Naval officers Lieutenant Commanders F. R. Furth and S. M. Tucker devised the acronym in 1940. By 1943, all allied forces had adopted the name, though it remained a classified term until after the second world war, when the acronym radar received general international acceptance, though more as a term than as an acronym.[1]

As the expression "radio detection and ranging" denotes, radar involves the use of radio for both detection (is it there?) and ranging (how far away is it?). Radar involves transmitting electromagnetic waves (commonly known as radio waves) toward a target and receiving the echoes from that target.

The wavelength of a radio or radar signal traveling through space is measured in meters and fractions of a meter (decimeter, centimeter, millimeter), and its frequency, that is, the number of waves per second, is expressed in hertz or multiples of hertz. One hertz is one wave per second. The high-frequency radar waves used in planetary research are expressed in megahertz (MHz, a million hertz) and gigahertz (GHz, a billion hertz).

There is no real difference between radio and radar waves. International treaties and regulatory agencies have set aside certain groups of radio frequencies, called bands, for specific radio uses, including radar applications (see Tables 11 and 12). Although the first radars to attempt detections of the Moon and Venus operated in the UHF band, planetary radar astronomy today uses only the S and X bands.

A radar system consists of a transmitter and a receiver, plus modulators, signal processors, and data processors. Generally, the transmitter and receiver share the same antenna. Such an arrangement is called a monostatic radar. When the transmitter and receiver do not share the same antenna, that is, when they are located in different places, it is called a bistatic radar.

All of the radars used since 1958 to study solar system objects have a parabolic or dish shape, with one exception. The exception is the Arecibo antenna, which is spherical.

Another difference among the planetary radars is their transmitters. The Millstone Hill radar, which Lincoln Laboratory investigators used to attempt a detection of Venus as early as 1958, was a pulse radar. Pulse radars transmit short bursts of energy and are best suited to tracking objects. Millstone, in fact, was an experimental prototype of a Ballistic Missile Early Warning System (BMEWS) radar used to detect and track potential incoming enemy missiles. In contrast are the continuous-wave radars. They transmit a continuous flow of energy and are better suited for communications applications. JPL's planetary radars at Goldstone are continuous-wave radars.

The output power of radars is expressed in watts. When comparing the power of pulse and continuous-wave radars, one must keep in mind that although pulse radars have relatively high peak power outputs (that is, the amount of power at the highest part of the pulse), their average power output, a measure more comparable to that of the continuous-wave radars, is much lower. Average power is what counts. Thus, while the Millstone pulse radar had a peak transmitting power of 265 kilowatts in 1958, the JPL transmitter in

---

1.   Louis A. Gebhard, *Evolution of Naval Radio-Electronics and Contributions of the Naval Research Laboratory,* Report 8300 (Washington: NRL, 1979), p. 170.

Table 11
Radar Frequency Bands and Usage

Letter Band	Frequency Range	Usage in Radar
HF	3–30 MHz	Over the horizon radar
VHF	30–300 MHz	Very-long-range surveillance
UHF	300–1000 MHz	Very-long-range surveillance
L	1–2 GHz	Long-range surveillance Enroute traffic control
S	2–4 GHz	Moderate-range surveillance Terminal air traffic control Long-range weather
C	4–8 GHz	Long-range tracking Airborne weather detection
X	8–12 GHz	Short-range tracking Missile guidance Mapping marine radar Airborne weather radar Airborne intercept
Ku	12–18 GHz	High-resolution mapping Satellite altimetry
K	18–27 GHz	Little used (water vapor)
Ka	27–40 GHz	Very-high-resolution mapping Short-range tracking Airport surveillance
V, W	40–110 GHz	Smart munitions Remote sensing
Millimeter	110 GHz	Experimental Remote sensing

Source

Fred E. Nathanson, *Radar Design Principles*, 2d ed. (New York: McGraw-Hill, 1991), p. 19.

Table 12
Standard Radar Frequency Bands

Letter Band	Frequency Range	Specific Frequencies Assigned to Radar
HF	3–30 MHz	None assigned (in practice, from just above the broadcast band, 1.605 MHz, to 40 MHz or higher
VHF	30–300 MHz	138–144 MHz 216–225 MHz
UHF	300–1000 MHz	420–450 MHz 890–942 MHz (at times included in the L band)
L	1000–2000 MHz	1215–1400 MHz
S	2000–4000 MHz	2300–2500 MHz 2700–3700 MHz
C	4000–8000 MHz	5250–5925 MHz
X	8–12 GHz	8.5–10.68 GHz
Ku	12–18 GHz	13.4–14.0 GHz 15.7–17.7 GHz
K	18–27 GHz	24.05–24.25 GHz
Ka	27–40 GHz	33.4–36.0 GHz
V	40–75 GHz	59–64 GHz
W	75–110 GHz	76–81 GHz 92–100 GHz
Millimeter	110-300 GHz	126–142 GHz 144–149 GHz 231–235 GHz 238–248 GHz

Source

Fred E. Nathanson, *Radar Design Principles*, 2d ed. (New York: McGraw-Hill, 1991), p. 19.

1961 was more powerful, though its average power output was only nine kilowatts. When the current Arecibo upgrade is completed, it will have the highest continuous-wave transmitter output available, one megawatt (1,000 kilowatts).

Radar sensitivity relates to the ability to receive signals. One of the limits to radar sensitivity is the noise created by the antenna and receiver systems, not to mention cosmic background and extraneous terrestrial radiation, all of which is expressed as noise "temperature" in Kelvins (abbreviated K), analogous to the temperature scale of the same name. The higher the system temperature in Kelvins, the noisier the radar and the lower its sensitivity. In 1958, the Millstone radar had an overall system temperature of 170 Kelvins, while the more sensitive JPL radar receiver had an overall system temperature of 64 K in 1961. Although impressively low in their day, these temperatures today are judged intolerably high.

Planetary radar astronomy borrows much of its terminology from optical astronomy, although not always retaining the original meaning. Facilities for conducting radar astronomy research are called observatories and the instruments telescopes. Radar telescopes "illuminate" the surface of targets. The reflecting geometries of radar telescopes, called their "optics," take their names from optical instruments (Cassegrainian and Gregorian subreflectors, for example).

Detection and ranging are two of the elemental observations made by planetary radar astronomers. A detection occurs when a radar antenna transmits waves toward a suspected target, the target reflects those waves, and a radar antenna receives the reflected waves (echoes) from the target. If we were crossing the Atlantic Ocean aboard a fictional ship, say the U.S.S. Marconi, we could use a radio transmitter and receiver, acting as a simple radar system, to detect the presence of icebergs or other ships. The radio pioneer Guglielmo Marconi suggested doing precisely that in a speech delivered in 1922.

If our U.S.S. Marconi radar were to detect the presence of another ship, we could determine the distance from the Marconi to the other ship with our radar. Measurements of the distance to the target are called range, time-delay, or delay measurements. The ability to use radar to measure range is based on the knowledge that radio waves travel at a constant speed, namely, the same speed as light.

In order to determine how far away a target is, we simply measure how long it takes the echoes to arrive at the receiver antenna. The greater the distance to the target, the longer the echoes take to appear in the receiver. Conversely, the shorter the distance to the target, the less time the echo takes to appear in the receiver. The time between the moment of transmission and the moment the echo is detected can vary considerably. For the farthest bodies detected by radar astronomers, such as Saturn's rings, the signal round-trip travel time is about two and a half hours, while the round-trip travel time to some asteroids detected close to Earth is about two and a half seconds.

Another basic planetary radar measurement is Doppler frequency shift. Whereas range measurements indicate the distance between the radar observer and the target, Doppler shift indicates the motion of the target relative to the observer. With our fictional U.S.S. Marconi radar, we can determine not only the presence and distance of another ship, but its speed toward or away from us as well.

Radar transmitters send waves at a specific frequency. A perfect reflection from a motionless target appears at the radar receiver (after Fourier transformation, see below) as an almost line-like peak. The echo from an actual solar system target, however, is spread over a range of frequencies. This frequency spread is called a spectrum (plural spectra). In radar astronomy experiments, the motions of the Earth, on which the planetary radar sits, and the motions of the target are far more complex. The Earth spins on its axis and rotates around the Sun, while the target planet similarly spins and rotates. The relative motions of the Earth and target planet cause what is known as the Doppler effect or Doppler shift (or even Doppler offset).

Simply stated, the Doppler effect causes the frequency of radar echoes to differ from the transmitted frequency. The Doppler effect on sound waves is a rather common experience around high-speed transport. If we stand alongside railroad tracks, or a freeway, we can detect the Doppler effect with our ears. As a train rapidly approaches, the sound of the train seems to rise in frequency, that is, in pitch; as the train travels away from us, its sound seems to fall in frequency. The same Doppler effect occurs in radar. Depending on the line-of-sight motion of a planet (or whether the object is approaching or moving away from the observing radar), the frequency of the planet's echo will be higher or lower than the transmitted frequency.

Planetary radar astronomers want to remove the average Doppler effect in order to analyze the information contained in the Doppler spectrum spread, so they use a radar ephemeris program. Although the average Doppler effect sometimes is removed in the transmitter, such as when several antennas receive, in general it is removed in the receiver. An ephemeris (plural ephemerides) is an astronomical term that refers to a set of tables that indicate the position of a planet or other body in the sky. A radar ephemeris program is computer software linked to the radar receiver that automatically adjusts the incoming signal for the expected Doppler shift. The amount of Doppler shift predicted by the ephemeris program must be accurate enough to avoid smearing the echo in frequency, and this requirement places stringent demands on the quality of the radar ephemeris.

Once we know the range and Doppler shift values of a solar system target, we can construct a two-dimensional radar image of the target called a range-Doppler or delay-Doppler image. Maps made from these images are vital to the exploration of the solar system, especially the planet Venus, whose surface is obscured by clouds.

In range-Doppler imaging, we assume that the target is a perfect sphere. An exception is the case of asteroids, whose nonspherical shapes require special modeling techniques. The transmitted radar waves arrive first at the area on the planet's surface that is closest to Earth. This area of initial impact is circular, because we have assumed that the target is spherical. The point on the planet's surface that is closest to the observer is called the subradar point. If we could look at the target with radar sensitive eyes, we would see a relatively dark sphere with a small bright spot in the middle, rather like a shiny ballbearing being held up to the light.

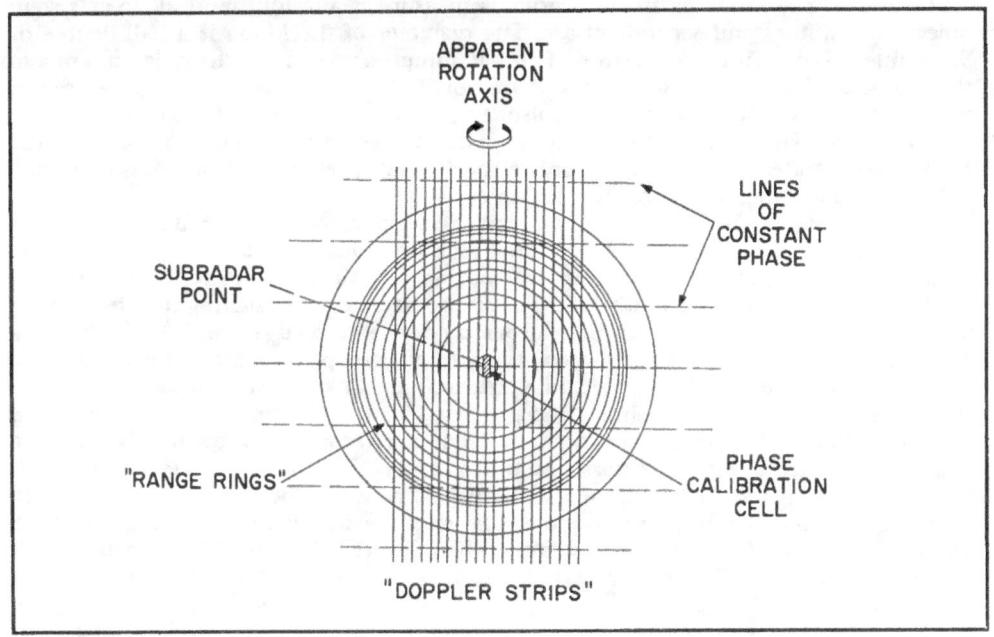

Figure 42

Diagram showing intersection of range rings and Doppler strips to form a planetary range-Doppler image. The lines of constant phase permit resolution of north-south ambiguity. (Courtesy of Alan E. E. Rogers.)

Echoes from the area beyond the subradar point are fewer than those that produce the central bright spot. Moreover, they reach those areas later than the waves striking the subradar point, because they have a greater distance to travel. The range values for those areas toward the limbs, then, are greater than those for the subradar region. Looking again at the target with our radar sensitive eyes, we see that the areas at a constant distance (range) from the radar transmitter form rings around the subradar point. These are called range rings.

As the planet spins on its axis toward or away from the oncoming radar waves, the spinning motion creates a Doppler effect. The Doppler frequency shift is the same along a strip or slice running across the planet's surface, because within each Doppler strip of the planet's surface, the motion relative to the observer is the same. When range and Doppler measurements made at the same time are combined, the strips of equal Doppler shift intersect the range rings to form "cells." Each range-frequency cell (or resolution cell) corresponds to a particular area on the planet's surface. The amount of area in a particular cell represents the amount of resolution of the radar image.

In range-Doppler imaging, any given range ring passes through the same Doppler strip at two points. One point is in the northern hemisphere, the other in the southern hemisphere of the planet. The two points have the same range and Doppler values, because they are in the same range ring and the same Doppler strip. As a result, these two points are indistinguishable in the radar image although they are in different hemispheres. Radar astronomers call this problem north-south ambiguity.

We can resolve the north-south ambiguity on the Moon by using a radar whose beamwidth is narrower than the diameter of the Moon. The beamwidth is the area of sky subtended by the radar beam. Astronomers measure beamwidth, and all solar system objects, in minutes and seconds of arc. The diameter of the Moon is a half degree or 30 minutes of arc. With a beamwidth of only 10 minutes, we can aim the radar antenna so that the subradar point is 10 minutes of arc north of the lunar equator. Echoes are not received from most of the southern hemisphere, so that echoes from the two hemispheres do not overlap. However, the technique is applicable to only the Moon. Compared to the Moon's 30 minutes of arc, Venus is only a speck; its diameter is but one minute of arc. Asteroids are less then one second of arc across.

In order to resolve north-south ambiguity on planetary targets, radar astronomers sometimes use a technique called interferometry. An optical interferometer is an instrument for analyzing the light spectrum by studying patterns of interference, that is, how light waves interact with each other. Radio astronomers began designing interferometers in the late 1950s. Planetary radar interferometry derived directly from those interferometers. Radio interferometers use two or more radio telescope antennas arranged along a line (called the base line). The separate antennas are linked electronically, so that the signals received at different points along the base line can be combined, compared, and studied with elaborate computer programs. Radar interferometers are somewhat simpler.

A radar interferometer consists of two antennas. The primary antenna transmits signals to the target and receives them. A secondary antenna, located not far (say, 1 to 10 km distant) from the primary antenna, also receives the echoes. Although a three-antenna radar interferometer was attempted between 1977 and 1988,[2] in practice radar interferometers use only two antennas.

---

2.   Jurgens, Goldstein, Rumsey, and R. Green, "Images of Venus by Three-Station Radar Interferometry: 1977 Results," *Journal of Geophysical Research* 85 (1980): 8282–8294.

The echoes received by both antennas are fed into a complex computer program that combines the echoes and obtains the fringe size (amplitude) and phase for each range-Doppler resolution cell. The computer program rotates the fringe pattern so that the lines of constant phase are perpendicular to the strips of equal Doppler value. The north-south ambiguity now is resolved, because the phase at points A and B have distinct phase values.

During the 1950s, researchers underwritten by the military developed a similar radar imaging process that used both range and Doppler data. However, that process involved imaging the Earth from aircraft and relied on developing a radar "history" of the target to create an image, while planetary range-Doppler mapping created a "snapshot" of a planetary surface from a ground-based radar. The airborne imaging process, called synthetic aperture radar, has since played a key role in the mapping of Venus by the Magellan spacecraft.

Radar astronomers do not depend entirely on range and Doppler data, however. The echo from a solar system target exhibits a number of attributes. From their analysis of those attributes, radar astronomers draw conclusions about the characteristics of the target. For example, the shape of power spectra can provide information about a target. If we aim at an asteroid and get an echo with two major peaks, called a bimodal echo, we can interpret the echo as possibly indicating a bifurcated shape, perhaps two asteroids joined together. Radar observations of asteroid 4769 Castalia (1989 PB), for instance, revealed it to be a contact binary asteroid.[3]

Small detail features on power spectra also can reveal vital information about a target's motion. For example, radar observations of Venus made in 1964 indicated that planet's rotational rate and direction. The radar instrument was both sufficiently powerful and sensitive that a large feature on the planet's surface showed up in the power spectra as an irregularity or "detail." The detail resulted from the fact that the surface feature scattered back to the radar antenna more energy than the surrounding area.

On close examination, one irregularity in the power spectra persisted day after day and appeared to change its position slowly. A study of the irregularity's movement led to a calculated rotational rate for the planet, but not immediately its prograde (forward) or retrograde motion. That information came from measurements of the width of the lower portion of the power spectra. Those widths were compatible with only a retrograde motion.

In the normal, round-trip journey of a radar wave from transmitter to target to receiver, a certain amount of power is lost. The amount of that loss is given by the so-called radar equation. The amount of power that reaches a target is inversely proportional to the square of the distance to the target, but the amount of power returned from the target to the receiving antenna also varies inversely proportional to the square of the distance to the target. After the complete round-trip from transmitter to receiver, the amount of power that arrives at the receiving antenna varies inversely with the distance to the target raised to the fourth power, that is, the square of the square of the distance. The radar equation shows that large amounts of power (hundreds of kilowatts) must be radiated into space in a very narrow beam in order to detect a target.

The amount of power returned from a target can reveal much about its surface characteristics. The total power returned from a target is a function of its radar cross section

---

3.   Hudson and Ostro, "Shape of Asteroid 4769 Castalia (1989 PB) from Inversion of Radar Images," *Science* 263 (1994): 940–943.

or backscattering coefficient, that is, the target's ability to reflect energy to the radar receiving antenna. Radar astronomers express the radar cross section of a target in terms of an equivalent, perfectly reflecting surface. If a target scatters power equally in all directions, its cross section is equal to the geometric area of the target. That is the case of our ideal ballbearing target. For a perfectly reflecting spherical target, the radar cross section is one fourth the total surface. Surface irregularities affect the amount of power returned (or scattered back) from a target. Radar echoes have two scattering components, called quasispecular and diffuse. The quasispecular component arises from mirror-like reflections from parts of a flat or gently undulating surface. Those surface facets are perpendicular to the line of propagation, so they direct a large amount of energy back toward the observer. Such echoes concentrate at the center of the planet's visible disk, that is, around the subradar point, because the likelihood of finding favorably oriented facets is highest where the surface is perpendicular to the incoming radar beam. The diffuse scattering component comes from objects and structures with irregular shapes and therefore facets that redirect much of the radar beam away from the observer. The signal returned from the areas toward the limbs is called diffuse.

The amount of power returned from a target is, therefore, a consequence of the scattering component, quasispecular or diffuse, and the angle of a surface facet relative to the line of propagation of the radar wave. Flat surfaces perpendicular to the line of propagation return power directly back to the radar. The greatest amount of power returned from a target, then, comes from flat surfaces that are perpendicular to the line of propagation. If the reflecting surface is not perpendicular to the line of propagation, then power will be reflected away from the radar, and the amount of power returned to the radar antenna will diminish. The reduction in power will increase as the surface is tilted away from the radar.

For example, if a planetary target has mountains or craters, a portion of the radar power will be reflected away from the return path, depending on the angle, that is, the amount of slope, of the mountain or crater. The more power returned, the gentler is the slope or angle of the surface. Factors other than surface slope can affect the amount of power returned from a target, too.

If a planetary surface is covered by boulders or other material with multiple sides, a complex scattering process takes place. Some power is returned to the radar, some power is deflected away from the radar return path, while some power scatters among the boulders. If the surface is covered by material significantly smaller than boulders, say volcanic ash, the loss of power from scattering in directions other than the return path can be considerable.

Although range-Doppler mapping techniques provide one means for correlating echo power spectra and surface features, they are not always practical. Other methods must be used. For example, Mars rotates much faster than Venus, whose slow retrograde motion makes it an ideal radar target. Mars is what radar astronomers call an overspread target. The rapid rotation of that planet means that the signal from one range ring spreads over into the next ring, or the signal from one Doppler strip spreads over into the next strip. Also, the echo from Mars is much weaker, because the distance to Mars is greater than to Venus.

In helping to select a landing site for the Viking lander, for example, radar astronomers relied on a different approach to interpret the amount of power returned from an area of the surface. In this approach, a geometric model for the entire visible surface of the planet was assumed. These models, or scattering laws as they are called, also can be derived empirically from actual radar observations of the target, if the target surface is sufficiently well known. The most commonly used model is the Hagfors scattering law, named for its originator, Tor Hagfors, an ionosphericist and radar astronomer who is currently Director of the Max Planck Institut für Aeronomie.

In studying candidate sites for the Viking lander, radar astronomers at Haystack, Arecibo, and Goldstone applied the Hagfors scattering law and determined the slope, that is, the degree of tilt from the line of propagation, of various areas of the Martian surface. Site candidates had to be inclined no more than 19 degrees; otherwise, the lander would topple over. They also had to be free of rocks and other objects larger than 22 cm, the height of the lander vehicle.

Radar data were capable of indicating the roughness of the Martian surface down to a few centimeters, while photographs had a resolution of roughly 100 meters, larger than a football field. The radar data, however, was not expressed visually, like the photographs, but mathematically as the root-mean-square (rms) slope. The rms is a special mathematical method for averaging. The rms slope gives an indication of the average slope or inclination of a given area of the Martian surface. When applying the Hagfors scattering law, the value for the rms slope varies in theory up to three degrees, the upper limit for the validity of the assumptions underlying the model. In practice, however, the Hagfors scattering law yields much higher values of rms slope.

One of the most important signal parameters used in planetary radar astronomy today is polarization. The earliest lunar radar experiments carried out at Jodrell Bank used a linear antenna feed. Antenna feeds are either linear or circular, and the feed shape determines the polarization of the transmitted wave. When Jodrell Bank investigators sent radar signals to the Moon, they discovered two patterns of signal fading. Normal lunar libration caused the slowly fading echoes, while the rapidly fading signals, they concluded, resulted from the radar waves passing through the Earth's ionosphere.

In the 19th century, the British scientist Michael Faraday discovered that a magnetic field could alter the plane of polarization. The effect since has come to be known as Faraday rotation. In the case of the lunar radar signals, the Earth's magnetic field rotated the signals' plane of polarization as they passed through the ionosphere.

A radar target also can change the handedness, or rotational sense, of circularly polarized waves. If we transmit circular waves with right-handed polarization to a perfectly reflecting target, the power returns with a left-handed polarization. If we transmit right-handed polarization and adjust the antenna feed to accept right-handed polarization, the antenna will detect little or no returned power. In practice, when dealing with the inner planets (Mercury, Venus, and Mars), most power returns from a planetary target in the opposite sense of polarization in which it is transmitted.

Exceptions to this rule arose when radar astronomers began exploring the Galilean moons of Jupiter and other icy targets. When radar signals return from the surfaces of those targets, somewhat more of the power is received in the same sense of polarization. In other words, if we transmit right-hand polarization, more of the power will return with right-handed polarization. The peculiar nature of these radar targets has elevated the importance of the ratio of same sense to opposite sense polarization as a radar measurement. Radar astronomers now transmit one sense of polarization and receive both right-hand and left-hand polarization, then they compare the right-hand and left-hand values. While fractured ice is the apparent cause of this polarization phenomenon, the mechanism that gives rise to it is only now beginning to be understood.

Planetary range-Doppler radar experiments take place in several stages: data taking, decoding, rotating the matrix, the Fourier transform, conversion into latitudes and longitudes to create maps. The first stages, especially data taking and decoding, are rather routine and standardized; software specialization tends to take place in the last stages of the process, such as converting the data into latitudes and longitudes.

While data taking involves the combined use of the radar system and an associated computer, the remaining stages take place entirely on a computer. Computer time accounts for most of the processing time spent on a range-Doppler experiment. With modern computer technologies that accelerate processing time, data reduction takes far less time than before. At the Arecibo telescope, a typical run of observations on Mercury takes about 10 minutes. Data processing of those 10 minutes of radar activity consumes another hour and a quarter to an hour and a half of computer time. Roughly, then, every minute spent making radar observations translates into eight minutes of processing time.

Without those special accelerating technologies, a mainframe computer takes far more time; it is almost 80 times slower. One run, then, might take an entire day to process. Older mainframe computers were even slower. The competition for computer time was often as intense as it was for antenna time. Moreover, these computer times apply only to data reduction, the initial preparation of the data. Analysis, modeling, and interpretation can be far more time consuming.

The first stage of a planetary radar experiment is the recording of the raw echoes as they come from the antenna through the receiver. In the earliest lunar radar experiments conducted at Jodrell Bank in the 1950s, the echoes were observed on an oscilloscope and recorded with a cinema camera. The films are extant and form part of the archives deposited with the University of Manchester. Beginning with the first attempts on Venus in 1958, the raw signals were recorded on magnetic tape. In addition, they were routinely converted from analog into digital signals for processing. Today, all planetary radar astronomy is carried out digitally.

The unprocessed echoes are usually too weak and too noisy to process, so the echoes are accumulated together. The next stage is to decode the signals. Before transmission, the signal is encoded with a repeating binary code. Pulse radars achieve binary coding by turning the signal off and on. With continuous-wave radars, binary coding is accomplished by changing the phase of the signal. These off-on states and phase changes in the coded transmission tell the radar astronomer which part of the wave is being examined. An accompanying time code identifies the location on the planet where the particular echoes originate.

The next step in forming a planetary radar map is to rotate the matrix. Once the codes have been removed from the echoes, the signals are arranged in a matrix that corresponds to the various range rings on the planet. The computer software looks at each code cycle and considers each range ring separately. The values for a given range ring are a function of time. The software now must decide which frequencies are present, in order to find the Doppler delay values.

A Fourier transform sorts echoes from a given range ring into frequency bins. A Fourier transform is a specific type of transform, a powerful mathematical expression that transforms (hence the name) one geometrical figure or analytical expression into another. During the 1960s, the Fast Fourier transform was devised by engineers in the field of signal processing. As a result, a mathematical operation that previously took 30 minutes on an IBM 8094 mainframe computer now took only about five seconds.[4]

The result of the Fourier transform is a range-Doppler, two-dimensional picture. Additional analysis with various computer algorithms written in the software can yield a three-dimensional picture. However, the three-dimensional picture requires adding further information to the range-Doppler map.

This succinct cursory overview of a planetary radar experiment is limited to only range-Doppler mapping. Radar astronomers carry out several other types of experiments. Most experiments are routine and rely on cookbook software and processing. Radar

---

    4.   Gwilym M. Jenkins and Donald G. Watts, *Spectral Analysis and its Applications* (San Francisco: Holden-day, 1968), pp. 313–314.

interferometry and the random code technique, adapted from ionospheric radar research, for example, employ specialized software and processing techniques.

A number of articles and book chapters on radar astronomy published since 1960 discuss the field, its accomplishments, and its techniques. Their intended audience runs the gamut from general to specialized. They are recommended to those wishing information on radar astronomy beyond that provided here.

For Further Reading

Eshleman, Von R. "Radar Astronomy: Exploration of the Solar System Using Man-made Radio Waves." In *Aeronautics and Astronautics: Proceedings of the Durand Centennial Conference, Stanford, August 5, 1959*, edited by N. J. Hoff and W. G. Vincenti, 207–226. London: Pergamon Press, 1960.

Eshleman, Von R. and Alan M. Paterson. "Radar Astronomy." *Scientific American* 203 (1960): 50–54.

Evans, John V. "Radar Astronomy." *Contemporary Physics* 2 (1960): 116–142.

Evans, John V. "Radar Astronomy." *Science* 158 (1967): 585–597.

Green, Paul E., Jr., and Gordon H. Pettengill. "Exploring the Solar System by Radar". *Sky and Telescope* 20 (1960): 9–14.

Hagfors, Tor, and Donald B. Campbell. "Mapping of Planetary Surfaces by Radar". *Proceedings of the IEEE* 61 (1973): 1219–1225.

Jurgens, Raymond F. "Earth-based Radar Studies of Planetary Surfaces and Atmospheres." *IEEE Transactions on Geoscience and Remote Sensing* GE-20 (1982): 293–305.

Kippenhahn, Rudolf. *Bound to the Sun: The Story of Planets, Moons, and Comets*, trans. Storm Dunlop. New York: W. H. Freeman and Company, 1990, pp. 259–272.

Muhleman, Duane O., Richard M. Goldstein, and Roland Carpenter. "A Review of Radar Astronomy, Parts 1 and 2." *IEEE Spectrum* 2 (1965): 44–55 and 78–89.

Ostro, Steven J. "Planetary Radar Astronomy." *Reviews of Geophysics and Space Physics* 21 (1983): 186–196.

Ostro, Steven J. "Planetary Radar Astronomy." In *Encyclopedia of Physical Science and Technology*, edited by Robert A. Meyers, 10:611–634. Orlando: Academic Press, 1987.

Ostro, Steven J. "Planetary Radar Astronomy." *Reviews of Modern Physics* 65 (1993): 1235–1279.

Ostro, Steven J. "Radar Astronomy." In *McGraw-Hill Encyclopedia of Astronomy*, edited by Sybil P. Parker and Jay M. Pasachoff, 347–348. New York: McGraw-Hill, 1993.

Pettengill, Gordon H. "Radar Astronomy." *Transactions of the American Geophysical Union* 44 (1963): 453–455.

Pettengill, Gordon H. "Planetary Radar Astronomy." In *Solar System Radio Astronomy*, edited by Jules Aarons, 401–411. New York: Plenum Press, 1965.

Pettengill, Gordon H. and Irwin I. Shapiro. "Radar Astronomy." *Annual Review of Astronomy and Astrophysics* 3 (1965): 377–410.

Pettengill, Gordon H. "Radar Astronomy." *International Science and Technology* 58 (1966): 72–74, 76, 78, 80–82.

Shapiro, Irwin I. "Planetary Radar Astronomy." *IEEE Spectrum* 5 (1968): 70–79.

Thomson, John H. "Planetary Radar." *Quarterly Journal of the Royal Astronomical Society* 4 (1963): 347–375.

Thomson, John H. "Planetary Radar." *Science Progress* 53 (1965): 183–190.

Abbreviations

For the most part, when an abbreviation is first introduced, it is preceded by the full spelling. Most abbreviations found in this text are commonly used, such as AFB for Air Force Base, while others are more apt to be seen in technical literature, such as K for Kelvin. In the case of organizations, they are always introduced by their full name; abbreviations appear only in subsequent references, though not all subsequent references are abbreviated. Most readers will recognize many organizational abbreviations immediately, such as NASA.

Interviews are referenced in the notes in an abridged form. Only the interviewee's surname and the interview date (in the form date/month/year) are noted; complete interview information is located in Appendix Two, Oral History Interviews.

A few reports that appear serially and are available from only a limited source are routinely cited in the notes in abbreviated form. The annual reports published by JPL and found in the JPL Archives are cited as JPL Annual Reports, JPLA. The quarterly reports of the NAIC were found only at the NAIC. These reports consist of a small number of typed pages submitted to the NSF, the oversight agency for the Arecibo Observatory. In order to cite them simply, the following form was used: NAIC QR Q2/1987, 4–5, in which pages 4–5 of the second QR (quarterly report) of 1987 are referenced. Also, the minutes of the meetings of the Goldstone Solar System Radar, furnished to the author by Steven Ostro, are referred to as GSSR Min., followed by the date of the meeting in the form date/month/year.

The GSSR minutes were only one of many items provided by several individuals from their personal materials. In the notes, these are referred to as "Smith materials," after the surname of the individual providing them. The Acknowledgments (p. iv) section above indicates the full names of those furnishing such materials.

Documents in open archives are cited in one of two ways, depending on whether the archives had assigned numbers to files. When a file number is available, the citation provides a description of the item and its date, followed by the file number, the box number, and the accession number (or collection name), and lastly the name of the archival repository. Otherwise, the description of the item and its date are followed by the folder title in quotes, the box number and accession number (or collection name), and the name of the archival repository.

References to unpublished materials use the following abbreviations to indicate archival repositories and specific collections within those archives.

AIO	Arecibo Ionospheric Observatory
AIOL	Library, Arecibo Observatory
CRSR	Center for Radiophysics and Space Research, Cornell University
HAUSACEC	Historical Archives, U.S. Army Communications-Electronics Command, Ft. Monmouth, NJ
JBA	Jodrell Bank Archives, University of Manchester
JPLA	Jet Propulsion Laboratory Archives

JPLMM	Jet Propulsion Laboratory Archives, Magellan materials
JPLPLC	Jet Propulsion Laboratory Archives, Peter Lyman Collection
LLLA	Lincoln Laboratory Library Archives
MITA	MIT Institute Archives and Special Collections
NAIC	National Astronomy and Ionosphere Center
NAS	Archives of the National Academy of Sciences
NHO	NASA History Office
NHOB	NASA History Office, Brunk Papers
NRLHRC	Naval Research Laboratory Historical Reference Collection, Office of the Historian, NRL, Washington
NSFHF	NSF Historian's File, National Science Foundation
NSFL	Library, National Science Foundation
PAHU	Pusey Archives, Harvard University
RLSEL	Radioscience Laboratory, Stanford Electronics Laboratories
SCRA	Stanford Center for Radar Astronomy
SEBRING	NEROC, Haystack Observatory, materials held in the Sebring building
SIA	Archives of the Smithsonian Institution
SIAOS	Archives of the Smithsonian Institution, Office of the Secretary Collection
SIAUSC	Archives of the Smithsonian Institution, Under Secretary Collection
SUA	Stanford University Archives

Index

About the Author

Andrew J. Butrica, a graduate of the doctoral program in the history of science and technology at Iowa State University, is a research historian and author of numerous articles and papers on the history of electricity and electrical engineering in the United States and France and the history of science and technology in nineteenth-century France. He is the author of a corporate history, *Out of Thin Air: A History of Air Products and Chemicals, Inc., 1940–1990*, published by Praeger in 1990, and a co-editor of *The Papers of Thomas Edison: Vol. I: The Making of an Inventor, 1847–1873*, published by Johns Hopkins University Press in 1989.

Prior to writing this history of planetary radar astronomy, Dr. Butrica was a research fellow with the Center for Research in the History of Science and Technology, Cité des Sciences et de l'Industrie (La Villette), Paris, thanks to a grant from the International Division of the National Science Foundation (1991–1992) and an earlier fellowship from the Centre National de la Recherche Scientifique (1987–1988). Butrica also has undertaken public history work, including the researching, conducting, and editing of oral history interviews for chemical company and hospital histories.

Dr. Butrica has been an invited lecturer at the Ecole des Hautes Etudes en Sciences Sociales (Paris), the University of Paris (Sorbonne), and Nottingham (England) University, as well as at Rutgers University, and has been a visiting scholar at the Deutsches Museum (Munich), the University of Pennsylvania, and Lehigh University. He is a member of several professional bodies, including the American Historical Association, the History of Science Society, the Society for the History of Technology (Robinson Prize Committee), the Society for French Historical Studies, the Institute of Electrical and Electronic Engineers, and the Association pour l'Histoire de l'Electricité en France.

The NASA History Series

Reference Works, NASA SP-4000:

Grimwood, James M. *Project Mercury: A Chronology.* (NASA SP-4001, 1963).

Grimwood, James M., and Hacker, Barton C., with Vorzimmer, Peter J. *Project Gemini Technology and Operations: A Chronology.* (NASA SP-4002, 1969).

Link, Mae Mills. *Space Medicine in Project Mercury.* (NASA SP-4003, 1965).

Astronautics and Aeronautics, 1963: Chronology of Science, Technology, and Policy. (NASA SP-4004, 1964).

Astronautics and Aeronautics, 1964: Chronology of Science, Technology, and Policy. (NASA SP-4005, 1965).

Astronautics and Aeronautics, 1965: Chronology of Science, Technology, and Policy. (NASA SP-4006, 1966).

Astronautics and Aeronautics, 1966: Chronology of Science, Technology, and Policy. (NASA SP-4007, 1967).

Astronautics and Aeronautics, 1967: Chronology of Science, Technology, and Policy. (NASA SP-4008, 1968).

Ertel, Ivan D., and Morse, Mary Louise. *The Apollo Spacecraft: A Chronology, Volume I, Through November 7, 1962.* (NASA SP-4009, 1969).

Morse, Mary Louise, and Bays, Jean Kernahan. *The Apollo Spacecraft: A Chronology, Volume II, November 8, 1962–September 30, 1964.* (NASA SP-4009, 1973).

Brooks, Courtney G., and Ertel, Ivan D. *The Apollo Spacecraft: A Chronology, Volume III, October 1, 1964–January 20, 1966.* (NASA SP-4009, 1973).

Ertel, Ivan D., and Newkirk, Roland W., with Brooks, Courtney G. *The Apollo Spacecraft: A Chronology, Volume IV, January 21, 1966–July 13, 1974.* (NASA SP-4009, 1978).

Astronautics and Aeronautics, 1968: Chronology of Science, Technology, and Policy. (NASA SP-4010, 1969).

Newkirk, Roland W., and Ertel, Ivan D., with Brooks, Courtney G. *Skylab: A Chronology.* (NASA SP-4011, 1977).

Van Nimmen, Jane, and Bruno, Leonard C., with Rosholt, Robert L. *NASA Historical Data Book, Vol. I: NASA Resources, 1958–1968.* (NASA SP-4012, 1976, rep. ed. 1988).

Ezell, Linda Neuman. *NASA Historical Data Book, Vol II: Programs and Projects, 1958–1968.* (NASA SP-4012, 1988).

Ezell, Linda Neuman. *NASA Historical Data Book, Vol. III: Programs and Projects, 1969–1978.* (NASA SP-4012, 1988).

Astronautics and Aeronautics, 1969: Chronology of Science, Technology, and Policy. (NASA SP-4014, 1970).

Astronautics and Aeronautics, 1970: Chronology of Science, Technology, and Policy. (NASA SP-4015, 1972).

Astronautics and Aeronautics, 1971: Chronology of Science, Technology, and Policy. (NASA SP-4016, 1972).

Astronautics and Aeronautics, 1972: Chronology of Science, Technology, and Policy. (NASA SP-4017, 1974).

Astronautics and Aeronautics, 1973: Chronology of Science, Technology, and Policy. (NASA SP-4018, 1975).

Astronautics and Aeronautics, 1974: Chronology of Science, Technology, and Policy. (NASA SP-4019, 1977).

Astronautics and Aeronautics, 1975: Chronology of Science, Technology, and Policy. (NASA SP-4020, 1979).
Astronautics and Aeronautics, 1976: Chronology of Science, Technology, and Policy. (NASA SP-4021, 1984).

Astronautics and Aeronautics, 1977: Chronology of Science, Technology, and Policy. (NASA SP-4022, 1986).

Astronautics and Aeronautics, 1978: Chronology of Science, Technology, and Policy. (NASA SP-4023, 1986).

Astronautics and Aeronautics, 1979–1984: Chronology of Science, Technology, and Policy. (NASA SP-4024, 1988).

Astronautics and Aeronautics, 1985: Chronology of Science, Technology, and Policy. (NASA SP-4025, 1990).

Gawdiak, Ihor Y. Compiler. *NASA Historical Data Book, Vol. IV: NASA Resources, 1969–1978.* (NASA SP-4012, 1994).

Noordung, Hermann. *The Problem of Space travel: The Rocket Motor.* Ernst Stuhlinger, and J.D. Hunley, with Jennifer Garland. Editors. (NASA SP-4026, 1995).

Management Histories, NASA SP-4100:

Rosholt, Robert L. *An Administrative History of NASA, 1958–1963.* (NASA SP-4101, 1966).

Levine, Arnold S. *Managing NASA in the Apollo Era.* (NASA SP-4102, 1982).

Roland, Alex. *Model Research: The National Advisory Committee for Aeronautics, 1915–1958.* (NASA SP-4103, 1985).

Fries, Sylvia D. *NASA Engineers and the Age of Apollo* (NASA SP-4104, 1992).

Glennan, T. Keith. *The Birth of NASA: The Diary of T. Keith Glennan,* edited by J.D. Hunley. (NASA SP-4105, 1993).

Project Histories, NASA SP-4200:

Swenson, Loyd S., Jr., Grimwood, James M., and Alexander, Charles C. *This New Ocean: A History of Project Mercury.* (NASA SP-4201, 1966).

Green, Constance McL., and Lomask, Milton. *Vanguard: A History.* (NASA SP-4202, 1970; rep. ed. Smithsonian Institution Press, 1971).

Hacker, Barton C., and Grimwood, James M. *On Shoulders of Titans: A History of Project Gemini.* (NASA SP-4203, 1977).

Benson, Charles D. and Faherty, William Barnaby. *Moonport: A History of Apollo Launch Facilities and Operations.* (NASA SP-4204, 1978).

Brooks, Courtney G., Grimwood, James M., and Swenson, Loyd S., Jr. *Chariots for Apollo: A History of Manned Lunar Spacecraft.* (NASA SP-4205, 1979).

Bilstein, Roger E. *Stages to Saturn: A Technological History of the Apollo/Saturn Launch Vehicles.* (NASA SP-4206, 1980).

Compton, W. David, and Benson, Charles D. *Living and Working in Space: A History of Skylab.* (NASA SP-4208, 1983).

Ezell, Edward Clinton, and Ezell, Linda Neuman. *The Partnership: A History of the Apollo-Soyuz Test Project.* (NASA SP-4209, 1978).

Hall, R. Cargill. *Lunar Impact: A History of Project Ranger.* (NASA SP-4210, 1977).

Newell, Homer E. *Beyond the Atmosphere: Early Years of Space Science.* (NASA SP-4211, 1980).

Ezell, Edward Clinton, and Ezell, Linda Neuman. *On Mars: Exploration of the Red Planet, 1958–1978.* (NASA SP-4212, 1984).

Pitts, John A. *The Human Factor: Biomedicine in the Manned Space Program to 1980.* (NASA SP-4213, 1985).

Compton, W. David. *Where No Man Has Gone Before: A History of Apollo Lunar Exploration Missions.* (NASA SP-4214, 1989).

Naugle, John E. *First Among Equals: The Selection of NASA Space Science Experiments* (NASA SP-4215, 1991).

Wallace, Lane E. *Airborne Trailblazer: Two Decades with NASA Langley's Boeing 737 Flying Laboratory.* (NASA SP-4216, 1994).

Center Histories, NASA SP-4300:

Rosenthal, Alfred. *Venture into Space: Early Years of Goddard Space Flight Center.* (NASA SP-4301, 1985).

Hartman, Edwin, P. *Adventures in Research: A History of Ames Research Center, 1940–1965.* (NASA SP-4302, 1970).

Hallion, Richard P. *On the Frontier: Flight Research at Dryden, 1946–1981.* (NASA SP-4303, 1984).

Muenger, Elizabeth A. *Searching the Horizon: A History of Ames Research Center, 1940–1976.* (NASA SP-4304, 1985).

Hansen, James R. *Engineer in Charge: A History of the Langley Aeronautical Laboratory, 1917–1958.* (NASA SP-4305, 1987).

Dawson, Virginia P. *Engines and Innovation: Lewis Laboratory and American Propulsion Technology.* (NASA SP-4306, 1991).

Dethloff, Henry C. *"Suddenly Tomorrow Came . . .": A History of the Johnson Space Center, 1957–1990.* (NASA SP-4307, 1993).

Hansen, James R. *Spaceflight Revolution: NASA Langley Research Center from Sputnik to Apollo* (NASA SP-4308, 1995).

General Histories, NASA SP-4400:

Corliss, William R. *NASA Sounding Rockets, 1958-1968: A Historical Summary.* (NASA SP-4401, 1971).

Wells, Helen T., Whiteley, Susan H., and Karegeannes, Carrie. *Origins of NASA Names.* (NASA SP-4402, 1976).

Anderson, Frank W., Jr., *Orders of Magnitude: A History of NACA and NASA, 1915–1980.* (NASA SP-4403, 1981).

Sloop, John L. *Liquid Hydrogen as a Propulsion Fuel, 1945–1959.* (NASA SP-4404, 1978).

Roland, Alex. *A Spacefaring People: Perspectives on Early Spaceflight.* (NASA SP-4405, 1985).

Bilstein, Roger E. *Orders of Magnitude: A History of the NACA and NASA, 1915–1990.* (NASA SP-4406, 1989).

Logsdon, John M. Logsdon, with Lear, Linda J., Warren-Findley, Jannelle, Williamson, Ray A., and Day, Dwayne A. *Exploring the Unknown: Selected Documents in the History of the U.S. Civil Space Program, Volume I: Organizing for Exploration.* (NASA SP-4407, 1995).

"New Series in NASA History," published by The Johns Hopkins University Press:

Cooper, Henry S. F., Jr. *Before Lift-Off: The Making of a Space Shuttle Crew.* (1987).

McCurdy, Howard E. *The Space Station Decision: Incremental Politics and Technological Choice.* (1990).

Hufbauer, Karl. *Exploring the Sun: Solar Science Since Galileo.* (1991).

McCurdy, Howard E. *Inside NASA: High Technology and Organizational Change in the U.S. Space Program.* (1993).

Lambright, W. Henry. *Powering Apollo: James E. Webb of NASA.* (1995).

New in the NASA History Series

Exploring the Unknown: Selected Documents in the History of the U.S. Civil Space Program, Volume I: Organizing for Exploration
Edited by John M. Logsdon, with Linda J. Lear, Jannelle Warren-Findley, Ray A. Williamson, and Dwayne A. Day, NASA SP-4407, 1995

The first of a projected three volume selection of key documents in the history of the U.S. civil space program, this volume prints more than 200 key documents relative to the theme of organizing for exploration, each with a headnote providing background and bibliographical information. These are organized into four sections introduced by an essay that gives context for the larger history of the space program. This is a major new resource for those who seek to understand the development of the entry of the United States into space exploration.

Spaceflight Revolution: NASA Langley Research Center from Sputnik to Apollo
James R. Hansen, NASA SP-4308, 1995

Focusing on Langley Research Center in Hampton, Virginia, during the late 1950s and 1960s, this book assesses the rapid transformation of a government research laboratory during a pivotal era. Langley, established as the original laboratory of the National Advisory Committee for Aeronautics in 1917, had long been involved in cutting-edge aeronautical research and development. The flight of Sputnik I in 1957, however, prompted important changes in the center's focus and method of operation. It became part of NASA in 1958, and its leaders shifted the workload from almost exclusively center-unique aeronautical research to efforts that involved other research facilities and dealt much more fully with the challenges of spaceflight.

The Problem of Space Travel: The Rocket Motor
Edited by Ernst Stuhlinger and J.D. Hunley with Jennifer Garland, NASA SP-4026, 1995

This is the first fully edited, complete English translation of Hermann Noordung's 1929 classic book, *Das Problem der Befahrung des Weltraums*, treating the engineering details of an orbital space station. "For those who do not read German (or who do not have access to the original), and who are interested in the early history of space travel concepts, this is an invaluable addition to one's reference library. It's been a long time coming, but well worth the wait."—*Quest: The Magazine of Spaceflight History*, Spring 1995.

www.ingramcontent.com/pod-product-compliance
Lightning Source LLC
Chambersburg PA
CBHW081613200526
45167CB00020B/3624